Using Technology to Support Evidence-Based Behavioral Health Practices

To my Dad,
 Thanks for all of your love, support, mentorship and guidance over the years. I think this technology stuff might really catch on!

Much Love

Jen
xoxo

Using Technology to Support Evidence-Based Behavioral Health Practices
A Clinician's Guide

Edited by
Michael A. Cucciare and Kenneth R. Weingardt

Routledge
Taylor & Francis Group
New York London

Routledge
Taylor & Francis Group
270 Madison Avenue
New York, NY 10016

Routledge
Taylor & Francis Group
27 Church Road
Hove, East Sussex BN3 2FA

© 2010 by Taylor and Francis Group, LLC
Routledge is an imprint of Taylor & Francis Group, an Informa business

Printed in the United States of America on acid-free paper
10 9 8 7 6 5 4 3 2 1

International Standard Book Number: 978-0-415-99474-3 (Hardback)

For permission to photocopy or use material electronically from this work, please access www.copyright.com (http://www.copyright.com/) or contact the Copyright Clearance Center, Inc. (CCC), 222 Rosewood Drive, Danvers, MA 01923, 978-750-8400. CCC is a not-for-profit organization that provides licenses and registration for a variety of users. For organizations that have been granted a photocopy license by the CCC, a separate system of payment has been arranged.

Trademark Notice: Product or corporate names may be trademarks or registered trademarks, and are used only for identification and explanation without intent to infringe.

Library of Congress Cataloging-in-Publication Data

Using technology to support evidence-based behavioral health practices : a clinician's guide / edited by Michael A. Cucciare, Kenneth R. Weingardt.
        p. cm.
    Includes bibliographical references and index.
    ISBN 978-0-415-99474-3 (hbk. : alk. paper)
    1. Evidence-based psychotherapy--Technological innovations. 2. Mental health services--Technological innovations. 3. Mental illness--Treatment. I. Cucciare, Michael A., 1976- II. Weingardt, Kenneth R., 1968- III. Title.

RC455.2.E94U85 2009
616.89--dc22   2009023507

Visit the Taylor & Francis Web site at
http://www.taylorandfrancis.com

and the Routledge Web site at
http://www.routledgementalhealth.com

Contents

Editors' introduction vii
Contributors xi

Section I: Technology and behavioral health

Chapter 1   Mood disorders 3
            Judith A. Callan and Jesse H. Wright

Chapter 2   Anxiety disorders 27
            Michelle G. Newman, Darryl F. Koif, Amy Przeworski, and
            Sandra J. Llera

Chapter 3   Posttraumatic stress disorder 45
            Carmen P. McLean, Maria M. Steenkamp, Hannah C. Levy,
            and Brett T. Litz

Chapter 4   Schizophrenia 69
            Armando J. Rotondi

Chapter 5   Substance use disorders 91
            Daniel D. Squires and Monte D. Bryant

Chapter 6   Smoking cessation via the Internet 109
            Yan Leykin, Alinne Z. Barrera, and Ricardo F. Muñoz

Chapter 7   Pain management 133
            Jeffrey J. Borckardt, Alok Madan, Arthur R. Smith, and
            Stephen Gibert

Chapter 8   Body image and eating disorders 151
            Susan J. Paxton and Debra L. Franko

Chapter 9 Obesity 169
Rebecca A. Krukowski, Jean Harvey-Berino, and
Delia Smith West

Chapter 10 Diabetes management 199
Jun Ma, Sarah B. Knowles, and Sandra R. Wilson

Section II: Issues concerning implementation and evaluation

Chapter 11 Implementation 225
Michael A. Cucciare

Chapter 12 Evaluation 247
Kenneth R. Weingardt

Chapter 13 Ethics in technology and mental health 267
Elizabeth Reynolds Welfel and Kathleen (Ky) T. Heinlen

Index 291

Editors' introduction

Michael A. Cucciare
Kenneth R. Weingardt

Over the past two decades, computer technology has transformed virtually every aspect of daily life. From the way we bank, shop, and learn, to the way that we socialize, communicate, and entertain ourselves, the widespread availability of affordable computer technology has dramatically changed the way we live. The way we deliver behavioral health care is no exception. Personal computers, smart cell phones, wireless technology, handheld devices, virtual reality, Internet technologies, and audio and visual media such as CD-ROMs and DVDs are increasingly enhancing the delivery of behavioral health services by supporting providers in the identification, assessment, and treatment of various mental and physical health problems.

Fortunately, many applications of technology in mental and behavioral health have been developed and evaluated by academic researchers. This has predictably resulted in a dramatic increase in the number of journal articles, literature reviews, special sections in scientific journals, and entirely new scientific journals devoted to this topic. The primary aim of this book is to synthesize and organize the current research literature in this exciting and rapidly evolving area, while providing practical recommendations for clinicians interested in integrating various forms of technology into their clinical practice.

More specifically, this volume focuses on the many ways in which technology can be used as an adjunct to traditional face-to-face psychotherapy or counseling. Although self-help applications are presented and discussed in some of the chapters, we have asked our authors to discuss these applications in the context of how providers might use them to complement traditional face-to-face psychotherapy or counseling. We hope that this perspective will help practicing clinicians come to view emerging technologies as tools that may be incorporated into their work rather than as potential competition.

As in so many other areas of our lives, technology is likely to play an increasingly important role in the future of behavioral health. From our perspective, this evolving role provides us with a variety of excellent opportunities to improve the quality of behavioral health services. First, as many of the chapters in this volume attest, evidence-based treatments, such as cognitive-behavior therapy (CBT), are often highly structured and lend themselves particularly well to delivery via computer. By designing software that supports both the patients who are receiving evidence-based treatment (e.g., Web-based training modules, structured self-monitoring, tracking, recording) as well as the clinicians who are delivering it (e.g., clinical training materials, reports of patient progress), technology has enormous potential to support competent clinicians in the delivery of evidence-based treatment practices.

The diffusion of technological innovation into behavioral health care also presents us with an opportunity to better support integrated models of care. In recent years, major health care systems have begun to shift the delivery of mental and behavioral health services from the traditional specialty care setting (e.g., outpatient mental health clinic or inpatient psychiatry ward) to general medical settings (e.g., outpatient primary care clinic or inpatient medical and surgical ward). This integration and coordination of mental, behavioral, and physical health care services in medical care settings is often referred to as *integrated behavioral health care*. Such integrated care presents unique challenges as it puts behavioral health care providers in the position of caring for patients who not only present with mental health problems (e.g., depression, anxiety) but may also present with chronic physical health conditions that have a substantial behavioral component (e.g., obesity, diabetes, and chronic pain). Thus, integrated behavioral health care providers not only must be familiar with identifying and treating common mental health problems, but must also be ready to apply their clinical skills to help patients more effectively manage a wide variety of common chronic physical health conditions. Technologies of the type described in this volume give behavioral health providers the tools necessary to help their patients manage complex comorbid physical and mental health concerns.

The development of new technologies to support evidence-based behavioral health care also presents an opportunity to increase the efficiency and cost effectiveness of treatment delivery. Computers have been used to deliver interactive and engaging psychoeducational content for both provider and patient. They have also been used to support ongoing self-monitoring of symptoms and to provide a convenient and confidential way to complete therapeutic homework assignments. Alone or in various combinations, these technology applications may give behavioral health providers the opportunity to use face-to-face sessions more efficiently, and potentially to even better effect. In fact, several recent studies suggest that technology can be used to enhance the cost-effectiveness of behavioral health services without compromising treatment efficacy. Although more research is clearly needed, the cost savings that may be achieved by integrating various types of technology into the delivery of mental health services may ultimately prove to be quite large.

Finally, the adoption of new technologies in behavioral health care provides us with the opportunity to meet our clients' evolving expectations. As we have often experienced in our own work at the U.S. Department of Veterans Affairs in designing Web-based resources to support veterans of the wars in Iraq and Afghanistan, the younger generation is virtually immersed in technology. Texting, instant messaging, Facebook, Xbox—for some, the Internet and other technologies have become as essential to life as food, water, and air. Whether we as clinicians are ready to adopt these new technologies or not, many of our clients already have.

To help guide the reader through this literature, we have divided this book into two sections. The first section is divided into specific disorder or problem areas, while the second section presents issues concerning implementing and evaluating such tools in clinical practice and important ethical issues to consider when doing so. Although we have given the chapter authors flexibility in terms of the structure of each chapter, each chapter generally provides information on the following topics: a brief discussion of the specific diagnostic focus; a thoughtful summary of the research evidence demonstrating the efficacy of using computer-based technology to treat specific diagnoses; various aspects of the treatment process that can be enhanced by using technology, and how providers might incorporate these into their practice; key practical clinical or technical skills required to successfully integrate computer-based applications into one's practice; and when available, specific resources that can be accessed through the Internet or links to Web sites that provide various computer-based applications.

Contributors

Alinne Z. Barrera
University of California, San Francisco
San Francisco, California

Jeffrey J. Borckardt
Medical University of South Carolina
Charleston, South Carolina

Monte D. Bryant
Brown University
Providence, Rhode Island

Judith A. Callan
University of Pittsburgh
Pittsburgh, Pennsylvania

Michael A. Cucciare
Stanford University School of Medicine
Palo Alto, California

Debra L. Franko
Northeastern University
and
Massachusetts General Hospital
Boston, Massachusetts

Stephen Gibert
Medical University of South Carolina
Charleston, South Carolina

Jean Harvey-Berino
University of Vermont
Burlington, Vermont

Kathleen (Ky) T. Heinlen
Heinlen & Hanson, LLC
Rocky River, Ohio

Sarah B. Knowles
Palo Alto Medical Foundation Research Institute
Palo Alto, California

Darryl F. Koif
The Pennsylvania State University
University Park, Pennsylvania

Rebecca A. Krukowski
University of Arkansas for Medical Sciences
Little Rock, Arkansas

Hannah C. Levy
Boston VA Health Care System
Boston, Massachusetts

Yan Leykin
University of California, San Francisco
San Francisco, California

Brett T. Litz
Boston VA Health Care System
Boston, Massachusetts

Sandra J. Llera
The Pennsylvania State University
University Park, Pennsylvania

Jun Ma
Palo Alto Medical Foundation Research Institute
Palo Alto, California

Alok Madan
Medical University of South Carolina
Charleston, South Carolina

Carmen P. McLean
Boston VA Health Care System
Boston, Massachusetts

Ricardo F. Muñoz
University of California, San Francisco
San Francisco, California

Michelle G. Newman
The Pennsylvania State University
University Park, Pennsylvania

Susan J. Paxton
La Trobe University
Melbourne, Victoria, Australia

Amy Przeworski
Case Western Reserve University
Cleveland, Ohio

Armando J. Rotondi
University of Pittsburgh
Pittsburgh, Pennsylvania

Arthur R. Smith
Medical University of South Carolina
Charleston, South Carolina

Daniel D. Squires
Brown University
Providence, Rhode Island

Maria M. Steenkamp
Boston University
Boston, Massachusetts

Kenneth R. Weingardt
Center for Health Care Evaluation
Palo Alto, California

Elizabeth Reynolds Welfel
Cleveland State University
Cleveland, Ohio

Delia Smith West
University of Arkansas for Medical Sciences
Little Rock, Arkansas

Sandra R. Wilson
Palo Alto Medical Foundation Research Institute
Palo Alto, California

Jesse H. Wright
University of Louisville
Louisville, Kentucky

section one

Technology and behavioral health

chapter one

Mood disorders

Judith A. Callan and Jesse H. Wright

Introduction

Mood disorders are disabling and common conditions. The lifetime prevalence of major depression (major depressive disorder, MDD) and bipolar disorder together is approximately 21% (Kessler, Berglund, Demler, Jin, & Walters, 2005). The negative impact of mood disorders on functional, emotional, occupational, and social functioning is even greater than many debilitating physical disorders (Kessler & Frank, 1997; Mintz, Mintz, Arruda, & Hwang, 1992; Wells et al., 1989), and the economic burden of depression is also excessive with estimates as high as $83.1 billion per year (Greenberg et al., 2003). Depression is the highest predictor of suicide in adults aged 30 to 65 years (Maris, 1992). Because information technology efforts have been directed to date almost completely on MDD, the focus of this chapter is primarily on this form of mood disorder.

Traditional treatment for MDD has primarily included medication and psychotherapy. However, there are many problems in the provision of these treatments. Over 60% of individuals with MDD in the United States receive no treatment in the year after they first develop this disorder (Wang et al., 2005). There are multiple and complex reasons for delays in treatment, including lack of access to different types of therapy, or failure to receive any treatment (Regier et al., 1993; Wang et al. 2005). One of the possible barriers that has been suggested is lack of sufficient numbers of therapists who are trained in evidence-based psychotherapies such as cognitive-behavior therapy (CBT) or interpersonal psychotherapy (IPT; Wilson, 1996; Wright et al., 2005). Information technology would appear to have the potential to improve access to effective psychotherapy for MDD by assisting clinicians, improving the efficiency of treatment, and/or reducing the cost of treatment (Wright, 2004). However, large-scale adoption of computer-assisted therapy, telepsychiatry, and other technological advances has not yet occurred.

In this chapter we examine the use of technology in five principal applications: diagnostic and symptom assessment, computer-assisted psychotherapy, self-help via the Internet, telepsychiatry, and symptom monitoring. The topic of online psychotherapy is covered within the computer-assisted psychotherapy and telepsychiatry sections. Because the emphasis of this book is on technology that supports clinical practice, we only briefly discuss the various self-help materials and programs available on the Internet. Some of these resources are listed so that clinicians can tap them for self-help assignments for their patients.

Computer-aided diagnostic and symptom assessment

The use of technology to assist in diagnostic evaluation, depression screening, and severity assessment began in the late 1960s and 1970s. However, these early programs were typically not made available to practicing clinicians and were found to be of limited value (Das, 2002; Sletten, Altman, & Ulett, 1971). Farvolden and coworkers (Farvolden, Denisoff, Selby, Babgy, & Rudy, 2005), more recently used a Web-based anxiety and depression test (Structured Clinical Interview for DSM-IV Axis 1 Disorders [SCID-I]), which was able to provide diagnoses equivalent to most anxiety and depressive diagnoses with good sensitivity and specificity. At present, there is no widespread use of computer-assisted psychiatric diagnostic software. Nevertheless, Emmelkamp (2005) concluded that online assessment can have psychometric properties comparable to in-person evaluation and may have the advantage of promoting self-disclosure.

Attempts to improve evaluation of depression symptom severity have led to the use of interactive voice response (IVR) technology. Kobak et al. (2000) reported on the use of IVR to evaluate the use of the Hamilton Depression Rating Scale (HDRS) in 10 studies with a total of 1,761 subjects. Patients access the system by calling a toll-free number and then entering a password. Each item of the HDRS is evaluated by multiple queries. Cronbach's alpha, as the measure of internal consistency, yielded ranges of .60 to .91 compared to clinician-rated values of −.41 to .91. Test–retest reliability was examined after 24 hours with the IVR and clinician-rated HDRS. The IVR method for the HDRS had a test–retest reliability of .74 compared to .98 for the clinician-rated HDRS. The overall correlation between the clinician-rated assessments and the IVR system was .81 ($p < .001$). There was a strong level of satisfaction with the use of these technologies to evaluate depression severity. However, 75% of subjects still preferred in-person evaluation.

A method of screening the public for depression on the Internet was accomplished by using the CES-D adapted for online screening and placed on the Intelihealth Web site (Houston et al. 2001). Over an 8-month period,

Chapter one: Mood disorders

24,479 people completed the Center for Epidemiologic Studies Depression Scale (CES-D). A total of 58% of those completing the online evaluation screened positive for depression. Less than half of those screening positive had never been treated for depression. A follow-up evaluation found that 56% of those who screened positive for depression did seek a subsequent evaluation for treatment. The authors concluded that this platform for anonymous screening for depression was an effective method, especially for younger age groups who have not been typically reached by traditional methods. The Web site is available at http://www.intelihealth.com.

Computer-assisted psychotherapy for depression

Computer-assisted therapy (CAT) can be defined as "psychotherapy that utilizes a computer program to deliver a significant part of the therapy content or uses a computer program to assist the work of a therapist" (Wright, 2008, p. 14). In CAT, a clinician evaluates the patient, prescribes the therapy program, and provides at least minimal supervision and guidance for use of the computer-delivered portion of treatment. In some applications, there is a highly integrated human–computer *team* that maximizes the use of technology to improve the efficiency of treatment delivery. In addition to potential advantages of efficiency and cost reduction, computer-assisted therapy may enhance learning, reduce the burden on therapists to perform repetitive tasks, provide systematic delivery of core therapy concepts and skills, deliver effective feedback to users, store and analyze patient responses, and promote use of homework (Wright, 2004).

Initially, CAT was only delivered in clinical settings. But more recently, some programs have been available on the Internet or in DVD-ROM editions that patients can use at home or wherever they prefer. As Internet delivery has become more common, the definition of computer-assisted therapy has become blurred. How much clinician involvement is required if the program is to be considered true psychotherapy? If a clinician simply recommends that a patient use an educational Web site for self-help, is this *psychotherapy*? If a person with depression finds a Web site by browsing and spends a few minutes looking at the program, is this psychotherapy? Although some developers of Web sites that are used without a clinician's supervision may consider the learning experience to be psychotherapy, we believe that a more traditional definition should be retained.

Although self-help programs have some definite advantages (e.g., available 24 hours a day, low cost or no cost, able to reach people who have no access to trained clinicians), research has shown that they are less effective than CAT in reducing symptoms of depression. A meta-analysis of 12 studies of Internet-delivered programs that used cognitive-behavior therapy principles found that when therapist support was provided, the effect size was d = 1.00 as compared with d = .27 when therapist support was not involved (Spek et al., 2007). Another problem with Internet-delivered self-help programs is that relatively few people complete the full program. The completion rate can be lower than 1% when there is open access to therapeutic Web sites, a research study is not being conducted, and there is no clinician contact (Christiansen, Griffiths, Groves, & Korten, 2006; Eysenbach, Powell, Rizo, & Stern, 2004). In contrast, completion rates for CAT with significant clinician involvement, either in clinical offices or over the Internet, is typically 70% or higher (Litz, Engel, Bryant, & Papa, 2007; Wright et al., 2002).

In reviewing studies of CAT and self-help computer programs, we conclude that the mode of delivery (i.e., CD-ROM, DVD-ROM, or Internet) is not the critical determinant in success—the key feature is clinician involvement. Thus, we focus most of our effort on describing computer-assisted therapy for depression as compared to programs that have been designed to be incorporated into clinical practice.

Early programs for CAT for depression

Efforts to develop and test computer programs for treatment of depression began in the 1980s with the work of Selmi and coworkers (Selmi, Klein, Greist, & Harris, 1982; Selmi, Klein, Greist, Sorrell, & Endman, 1990), who introduced software that instructed patients on the basic concepts of cognitive-behavior therapy (CBT). The Selmi group's software used text on computer screens to educate patients, perform depression ratings, and participate in simple interactive exercises. Although this software did not have a sophisticated interface, advanced graphics, or multimedia elements, which are standard components of contemporary programs for computer-assisted CBT (CCBT) of depression, the program fared well in a randomized, controlled trial. Patients with MDD treated with the Selmi and coworkers' program for both CCBT or standard CBT both had significantly greater improvement in depression ratings than a wait list control, and there were no significant differences found between the two active therapies (Selmi et al., 1990). This program is now obsolete and is no longer available for clinical use.

Another early program for computer-assisted therapy of depression was developed by Colby and Colby (1990). As with the Selmi et al. program (1982, 1990), the software is text based and contains no multimedia components. The Colby and Colby program ("Overcoming Depression") has two main elements: a psychoeducational module loosely based on CBT and a "natural language" section that attempts to conduct a nondirective therapeutic interview. The latter feature of the Colby and Colby software has roots in much earlier efforts to program a computer to perform like a human therapist—asking and responding to questions, understanding the meaning of the patient's responses, and giving empathic and insightful comments (O'Dell & Dickson, 1984; Weizenbaum, 1996). Such programs have not yet been able to reliably replicate therapeutic communication between therapists and patients (Wright, 2004).

Problems with the natural language module may have contributed to the negative findings of a study of the Colby and Colby software for depression (Bowers, Stuart, MacFarlane, & Gorman, 1993; Stuart & LaRue, 1996). In a small investigation with 22 depressed inpatients, subjects assigned to therapy with the Colby and Colby program (1990) had less symptomatic improvement than those treated with standard CBT (Bowers et al., 1993). This is the only study of computer-assisted CBT for depression in clinical settings that did not show a positive effect for the computerized treatment intervention.

Currently available programs for CAT for depression

More contemporary programs for computer-assisted therapy of depression have moved away from natural language applications and have focused instead on using multimedia to enhance learning and skill acquisition. All of the newer computer programs for psychotherapy of depression utilize a CBT approach. CBT is well suited for computer-assisted therapy because of its information processing theories, psychoeducational emphasis, practical methods, skill-based approach, and use of homework (Wright, 2004).

Several uncontrolled trials of multimedia forms of CCBT have shown excellent patient acceptance (Cavanaugh et al., 2006; Whitfield, Winshelwood, Pashely, Williams, & Campsie, 2006; Wright et al., 2002). For example, Wright and coworkers (2002) reported that 78.1% of a series of 96 outpatients and inpatients treated with a multimedia program for CBT (later named Good Days Ahead: The Multimedia Program for Cognitive Therapy) completed the entire program, while 93.4% reached at least the midpoint.

The affinity of patients for using the Good Days Ahead software was assessed with a scale including responses such as "I liked the program," "The program helped me," and "I would recommend the program to others." The range for mean scores for individual questions was 4.3 +/– 0.6 to 4.5 +/– 0.6 (five = highest possible rating). Mean scores on a measure of knowledge of cognitive therapy (Cognitive Therapy Awareness Scale [CTAS]) increased significantly from 24.2 +/– 4.2 to 32.5 +/– 3.7 in study completers (Wright et al., 2002). Although, this uncontrolled study was not designed to measure the effectiveness of using the Good Days Ahead program, symptoms of depression and anxiety were measured to obtain information on possible effects of the software when used along with other treatments. Beck Depression Inventory (BDI), Beck Anxiety Inventory (BAI), and Automatic Thoughts Questionnaire (ATQ) scores all dropped significantly (Wright et al., 2002).

Two research groups (Proudfoot et al., 2003; Wright et al., 2005) have performed randomized, controlled trials to investigate the efficacy of modern forms of computer-assisted CBT that employ multimedia technology and have been used in clinical settings. Proudfoot and coworkers (Proudfoot et al., 2003, 2004; McCrone et al., 2004) have examined the usefulness of computer software (called Beating the Blues) developed in the United Kingdom in a large sample (n = 274) of primary care patients with mixed depression and anxiety. Subjects who scored 4 or more on the General Health Questionnaire and 12 or more on the Clinical Interview Schedule-Revised were randomly assigned to CCBT plus treatment as usual (TAU) or TAU alone. Treatment with antidepressants was not controlled. Baseline BDI-II scores were similar in both groups (CCBT = 24.9; TAU = 24.9), but CCBT patients had lower BDI-II scores 3 months after treatment (CCBT = 12.1; TAU = 16.4).

The Beating the Blues software uses video simulations to educate patients on symptoms of depression and anxiety. It also includes a variety of animations, audio instructions, and interactive exercises to help patients learn to use CBT to reduce symptoms. A limited amount of clinician assistance was provided for the computer-assisted component of the Proudfoot et al. study (2003, 2004). Nurses could spend up to 10 minutes in each of 8 sessions assisting primary care patients with the computerized therapy.

Wright and coworkers (2005) have conducted a randomized, controlled trial of CCBT using the Good Days Ahead software in patients with major depressive disorder (MDD). In this study, 45 drug-free patients with MDD were randomly assigned to CCBT, standard cognitive behavior therapy (CBT), or a wait list. CCBT was delivered in 9 sessions over 8 weeks. Therapist contact was reduced in CCBT by utilizing 25-minute sessions instead of 50-minute sessions after the first visit. The total amount of therapist time in CCBT was about 4 hours or less. There were no significant differences found in primary outcome measures between CCBT and CBT, and both forms of CBT were significantly better than the wait list in reducing symptoms of depression. BDI-II change scores from baseline to 8 weeks of treatment were 17.5 for CCBT, 14.7 for standard CBT, and 5.8 for the wait list. Large effect sizes were found for both CCBT (1.14) and CBT (1.04). Improvement was sustained in both active treatments at the 3- and 6-month follow-up evaluations (Wright et al., 2005).

Interestingly, there appeared to be possible advantages for CCBT on measures of learning CBT and in changing core beliefs. Scores on the Cognitive Therapy Awareness Scale (CTAS) increased significantly more

Chapter one: Mood disorders

in patients treated with CCBT than standard CBT. Also, patients treated with CCBT had significantly greater improvement in Dysfunctional Attitude Scale (DAS) scores than did the wait list subjects, while patients who received standard CBT did not (Wright et al., 2005). Although these findings require further investigation, it is possible that computer-assisted therapy may provide an engaging and effective learning environment that may boost acquisition of knowledge or skills beyond the levels reached in standard treatment.

The software for CCBT developed by Wright and coworkers (Wright, Wright, & Beck, 2003; Wright 2005, 2002) uses multimedia (video, voice-overs, graphs, and illustrations), self-help exercises, checklists, multiple-choice questions, and extensive feedback to engage users and convey basic concepts of CBT. Users see scenes from the life of a main character who is struggling with depression and anxiety. They help this person fight depression by answering questions designed to implement CBT methods. Users then apply the concepts learned in the program to situations in their own lives. Responses of users, including work on all self-help exercises, are stored in a workbook feature of the software, and homework assignments are suggested for using the Workbook to practice CBT skills. Clinicians can access patient data to check on progress and to integrate the computer elements of therapy with the overall treatment plan.

Another computer program for CBT of depression (see the Climate Clinic Web site at http://www.climateclinic.tv) has recently been tested in a small preliminary study with 13 patients meeting *Diagnostic and Statistical Manual of Mental Disorders, Fourth Edition* (*DSM-IV*) criteria for MDD (Perini, Titov, & Andrews, 2008). The program includes six online lessons and weekly homework assignments. The focus of the 9-week program is on behavioral activation, cognitive restructuring, problem solving, and assertiveness skills. Patients are contacted by a clinical psychologist online via e-mail and also participate in an online discussion forum with other participants. The mean therapist time per patient was 208 minutes—an amount of time quite similar to time spent in CCBT with the Good Days Ahead software. Measures in this uncontrolled trial included the Patient Health Questionnaire (PHQ-9) and the Positive and Negative Affect Scale (PANAS). A large effect size was noted for the PHQ-9 (.98), and a moderate effect size of .60 was observed for the PANAS.

Clinical use of CAT for depression

Despite promising results of research on CCBT for depression, dating back to the 1980s, this method has not yet been widely used in clinical practice. Only two programs, Good Days Ahead and Beating the Blues are commercially available. Good Days Ahead can be obtained through the Mindstreet Web site at http://mindstreet.com/. Good Days Ahead has been used primarily in mental health practices such as psychiatric hospitals, outpatient clinics, and military behavioral health units. Beating the Blues has been used in general medical practice settings in the United Kingdom where it has been endorsed by the National Institute on Clinical Excellence (NICE) for this application. Beating the Blues can be obtained through the Wellness Shop Web site, http://www.thewellnessshop.co.uk/products/beating the blues/whatisbtb.html.

The barriers to more widespread dissemination of CCBT are not fully known, but lack of awareness of this technology and/or clinician resistance to change may be important factors. Health care economics may be another key issue in dissemination of CCBT. The success of Beating the Blues in the United Kingdom has hinged largely on National Health Service funding of use of the software in primary care settings. Currently, insurers in the United States do not cover costs of multimedia computer programs for depression, virtual reality therapy, or any other use of computer tools for treatment of mental disorders. Thus, clinicians or institutions who are interested in using CCBT must find alternate funding sources. Some of the options that have been used or could be used for funding of CCBT in the United States are institutional support from hospitals, military bases, or universities; self-payment by patients who are willing to incur out-of-pocket expenses; flexible health care spending accounts; and adoption of CCBT methods in capitated or health maintenance organization environments. Because one of the basic goals of CCBT is to reduce the overall cost of health care, it seems reasonable to expect that with further research and additional clinical experience, insurers will begin to fund CCBT methods.

Practitioners who wish to use this approach in their treatment settings need to learn how to integrate computer-assisted therapy into their daily work with patients. Table 1.1 lists some of our recommendations for using CCBT in clinical practice.

It is especially important for clinicians to learn about the programs they prescribe for patients. If clinicians use the software themselves, they can educate patients about what to expect from CCBT and also do a better job of guiding patients on effective use of the program. If manuals or clinician guides are available, it can be very helpful to read these materials in order to understand the mechanics of program registration, operation, data management and security, benefits and limitations, etc. For example, the Good Days Ahead Clinician Guide details ways to track patients' progress in using the program, facilitate learning of program material, pace sessions effectively, and combine CCBT with the work of the clinician in the overall treatment plan. Potential benefits and limitations of CCBT described in the Good Days Ahead Clinician Guide are shown in Table 1.2.

Chapter one: Mood disorders

Table 1.1 Suggestions for Using CCBT in Clinical Practice

- Learn about the computer application by using it yourself.
- Read manuals or clinician guides that educate clinicians on CCBT indications, benefits, limitations, and methods.
- Perform diagnostic assessment before recommending patient use of computer tools.
- Screen for conditions/problems that would interfere with CCBT.
- Evaluate and manage suicidal risk.
- Introduce and orient patient to computer program.
- Follow patient's progress in using computer-assisted therapy.
- Integrate CCBT into the comprehensive treatment plan.
- Customize the timing and pacing of CCBT to meet patient needs and capacities.
- Check homework from CCBT.

Table 1.2 Potential Benefits and Limitations of CCBT

What the program *can* do:

1. Teach users the basic concepts of CBT.
2. Show CBT principles in action.
3. Give feedback to users based on their responses to questions.
4. Coach users on applying CBT techniques in real-life situations.
5. Promote the use of CBT self-help procedures.
6. Encourage a hopeful attitude toward problem solving.
7. Provide multiple interactive exercises for practicing CBT skills.
8. Demonstrate the value of doing homework between sessions.
9. Give clinicians access to self-help exercises and homework completed during the program.
10. Measure patient's global levels of depression and anxiety.

What the program *cannot* do:

1. Take the place of clinician-administered therapy or eliminate need for clinician involvement in treatment.
2. Show the wisdom, empathy, and flexibility of an experienced clinician.
3. Elicit a psychiatric history or perform a psychiatric assessment.
4. Diagnose psychiatric disorders.
5. Assess suicidality or dangerousness.
6. Respond in a unique manner to each patient's particular set of problems.
7. Understand the nuances of patient communications.
8. Provide a useful learning experience for every patient.
9. Measure symptoms with standard rating scales.

As noted in Table 1.2, computer programs for CBT of depression do not typically provide a diagnostic assessment and do not evaluate or manage suicidality or dangerousness. Recommendations and warnings are given to users that a clinician should be contacted immediately if suicidal ideas are present. In current modes of CCBT delivery, the clinician must perform an assessment, make diagnoses, consider whether CCBT could be a valuable part of the treatment plan, and then supervise and manage the patient's use of the computer program. Even in primary care settings where there may be limited clinician support from a nurse or other health care professional, there is a minimum requirement for orientation to the computer-assisted therapy environment, brief supportive visits, answering of questions, and troubleshooting of any difficulties in using the software or understanding program content.

A fully integrated form of CCBT in which the clinician and computer program function as a human-technology team would usually include careful following of the patient's progress in using the computer program, suggestions for applying program content for specific problems in the patient's life, and customizing the pacing of use of the program to match the patient's symptoms, needs, and capacities. Also, program effectiveness can be enhanced when clinicians check homework from the computer program or make suggestions for targeting homework assignments on specific problems or issues.

We believe that effective use of CCBT is similar in some ways to effective implementation of psychopharmacology. Most clinicians would think that a doctor who recommends a medication but does not see the patient regularly to assess for side effects, positive outcomes, and needs for dosage adjustments or changes in the regimen, and who does not provide at least a modicum of support, is not doing the best job possible. In the same manner, a clinician who suggests a computer program, but then does nothing to monitor or support its use, is missing a large opportunity to help a patient benefit from CCBT.

Ethics of computer-assisted psychotherapy for depression

Ethical guidelines for CAT were suggested early in the history of this approach by Samson and Pyle (1983). These recommendations include (1) adequate safeguards for confidentiality, (2) assessment and determination of the appropriateness of the computer application, (3) introduction and orientation to the program, (4) up-to-date and accurate information, (4) well-functioning software, and (5) supervision of the treatment process by a clinician. Although these guidelines still seem appropriate, they were developed before the more recent emphasis on evidence-based medicine and conflict-of-interest considerations, and the widespread delivery of computer programs over the Internet. Thus, we believe an updated set of guidelines should include these types of statements: (1) program content should be derived from an evidence-based therapy approach and should accurately represent the basic features of this approach, (2) conflict-of-interest of any developers or investigators should be disclosed to users in research applications or if the program is being used with patients of the developers or investigators, and (3) if CCBT is being used in clinical practice but delivered over the Internet, all ethical standards for office-based administration of CCBT need to be applied (e.g., adequate safeguards for confidentiality, assessment, and determination of the appropriateness of the program; supervision by a clinician). Of course, self-help programs that are not prescribed or monitored by a health care professional would have somewhat different ethical standards. However, we believe that such programs should still meet ethical guidelines for confidentiality, program appropriateness, evidence-based content, and conflict of interest.

Self-help programs for depression

A number of self-help resources for depression have become available on the Internet. We give short descriptions of some of these resources here so that clinicians will be aware of programs that their patients may be accessing online. Clinicians may want to consider recommending some of these self-help materials or support groups as adjuncts to traditional, office-based treatment. Web sites that offer psychoeducational programs on the core principles of CBT and peer-to-peer support groups are listed in Table 1.3. One of the most active psychoeducational Web sites for CBT is MoodGym—a program that was developed in Australia. This Web site provides CBT training modules, a personal work log containing assessments, an interactive game, and a feedback form. Characters are used to illustrate important CBT concepts.

Table 1.3 Self-Help Resources for Depression Available on the Internet

Psycheducational Web sites for CBT
- Mood Gym, http://www.moodgym.anu.edu.au
- Overcoming Depression, http://www.calipso.co.uk

Support Groups
- Walkers in the Darkness, http://www.walkers.org
- Depression and Bipolar Support Alliance, http://www.dbsalliance.org
- National Alliance on Mental Illness (NAMI), http://www.nami.org
- Depression and Related Affective Disorders Association, http://www.drada.org

Several studies of MoodGym have reported significant usage and some evidence of symptom improvement (Christiansen & Griffiths, 2005; Christiansen, Griffiths, & Korten, 2002). There were 817,284 hits in the first month of operation (Christiansen et al., 2002). In an uncontrolled trial, scores on the Goldberg Depression Scale decreased if individuals had completed a significant number of the modules. However, it has been reported that the vast number of visitors to this Web site are "free range users and one-hit wonders" who view only a small portion of the program content (Christiansen & Griffiths, 2005; Christiansen et al., 2006).

Another psychoeducational program, Overcoming Depression, provides 10 structured workbooks for CBT for depression (Williams & Whitfield, 2001). The workbooks are offered in a range of delivery formats including a printed version and the Internet. The educational materials target dysfunctional thinking, poor problem solving, lack of assertiveness skills, reduced activity, sleep problems, and ability to take antidepressant medications. Six of the workbooks are available at the Calipso Web site at http://www.calipso.co.uk.

In addition to Internet-based self-help, there is a wide variety of support groups available. These include bulletin boards, chat rooms, news and discussion groups, and electronic mail lists in which each individual's message is copied and e-mailed to all subscribers. Support groups can offer a 24-hour availability of practical information, sharing of experiences, and provision of hope (Castelnuovo, Gaggioli, Mantorani, & Riva, 2003b). Depression tops the list of Internet support groups sought on the America Self-Help Clearinghouse (Lamberg, 2003). A support group of note is Walkers in Darkness—a nonprofit group for mood disorders with 500,000 visits annually. It has six current e-mail lists, posts 100 messages daily on living with mood disorders, has 18 Web-based forums, and 2,400 regular users (Lamberg, 2003).

The Depression and Bipolar Support Alliance had 14,000 users in February of 2003 and provided two weekly Internet support groups, one for individuals with mood disorders and one for families of those with mood disorders. The National Alliance on Mental Illness (NAMI) also has Web-based Internet support groups for MDD, anxiety, bipolar disorder, and other illnesses. DRADA (Depression and Related Affective Disorders Association) is a peer support program that links individuals with those who want to write or e-mail (Lamberg, 2003).

The first study of Internet support groups was performed with 103 users of the Walkers in Darkness Web site who were followed for one year (Houston, Cooper, & Ford, 2002). The investigators explored whether Internet support groups predicted changes in depression and social support. Of persons who used the Internet resource, 37.9% preferred the Internet support group, 50.5% preferred face-to-face contact with a clinician, and 11.7% had no preference. Users of the Internet group reported lower levels of emotional support compared with those who received conventional treatment (18 vs. 25%). The researchers concluded that use of this Internet support group is associated with improvements in depression in users who are socially isolated.

Andersson et al. (2005) studied 117 depressed subjects who were randomly assigned to receive either a CBT Web-based discussion group or a discussion group alone. Psychoeducational materials provided to participants included 89 pages of text in five modules: introduction, behavioral activation, cognitive restructuring, sleep and physical health, and relapse prevention and future goals. The mean reduction on BDI scores in this study was 5.2. ($p < .001$).

The overall effectiveness of peer-to-peer support groups was examined in a literature review of published studies (Eysenbach et al., 2004). The review failed to find any robust evidence of health benefits from peer-to-peer electronic support groups. Twelve of the studies were with depressed subjects. Only three of the depression studies produced a significant result. The authors concluded that there were significant problems with study design and many underpowered exploratory studies in this group of investigations. Regardless of these findings, Internet support groups continue to be a popular resource for persons with mood disorders and their families.

Although there has been little evidence that there is a negative impact of psychoeducational Web sites or support groups, there may be some concern that individuals who need or could benefit from professional treatment may use these resources instead of seeking help from a clinician. Also, conflicting or erroneous information from Web sites (e.g., recommendations for dietary supplements with unproven effectiveness or potential interactions with antidepressants, suggestions for using unhelpful remedies or therapies) could impair the reader's ability to overcome depression. Another possible disadvantage of Web support groups has been termed the Munchausen by Internet effect. Individuals can enter groups under false pretenses by using factitious information and end up disrupting well-established groups.

Telepsychiatry

"Telepsychiatry is the use of telecommunications technology to connect patients and health care providers, permitting effective diagnosis, education, treatment, consultation, transfer of medical data, research, and other health care activities" (Brown, 1998, p. 964). Areas that will be examined include use of teleconferencing, e-mail communications, and telephones to assist clinicians in making diagnoses or delivering treatment. We also briefly discuss issues in providing psychotherapy online.

Teleconferencing

Teleconferencing accounts for approximately 35% of all telemedicine and has frequently been used in remote areas (Norris, 2001). Specific uses are for case management, decision support, legal hearings, forensic evaluation, neuropsychological evaluations, and individual, group, and family therapy. Nearly all are conducted through videoconferencing (Hilty, Marks, Urness, Yellowlees, & Nesbitt, 2004). Examples of successful telepsychiatry programs are Rodeo Net, a telepsychiatry program in Oregon that has produced a 50% cost savings, and Kansas University Medical Center, which has opened 14 telemedicine sites throughout Kansas (Brown, 1998). Videoconferencing programs such as these have resulted in a high degree of patient acceptance, especially with high frame rates, which ensure more functional interaction (Castelnuovo, Gaggioli, Mantovani, & Riva, 2003a). Equipment for videoconferencing has been greatly reduced in cost and has wide availability. Clinicians may take advantage of this modality in areas that are remote and also when expertise from academic centers would improve management of difficult diagnostic or clinical cases.

E-mail

Use of e-mail as a therapeutic modality has been described by Mehta & Chalhoub (2006) as specifically useful in certain populations, e.g., adolescents. The authors describe 18 guidelines for the use of e-mail including use of encryption, obtaining consent, virus protection, establishing rules for what can and cannot be discussed via e-mail, how long until a reply occurs, and avoiding use of identifying information. Some of the benefits for use of e-mail in therapeutic activities are (1) having a "hard copy" of patient communications, (2) offering more choice, (3) easier to communicate with shy patients, and (4) offering a mechanism to administer and monitor homework. Also, e-mail applications can have features that can be quite useful for therapy. For instance, e-mails can be set up to be sent on a later date. This can serve as a reminder for patients to do therapeutic activities such as exercise. A helpful Web site, http://www.hassleme.co.uk, can be used to set up therapeutic reminders to patient's e-mail to complete activities agreed upon in the therapy session.

Some of the possible disadvantages of using e-mail include (1) clinician burden (clinician is available daily and may need to consume large amounts of time responding to e-mails), (2) lack of reimbursement for time spent on e-mail communication, (3) medical–legal issues (e.g., if advice is given via e-mail without full evaluation and adverse outcomes occur), (4) technical failures through viruses or other computer-based problems, (5) confidentiality issues, (6) problems in responding to time-sensitive communications, and (7) the potential for boundary violations. As a possible way of managing these potential problems, Oravec (2000) describes a type of therapeutic e-mail correspondence that is first contracted via http://www.therapyonline.ca. Agreements are made between the patient and therapist as to the rules and structure of their e-mail correspondence.

Whether e-mail is going to be a primary mode of communication between therapists and patients, or used as an adjunct to conventional therapy, it is important to come to an agreement with the patient about the following parameters or concerns:

1. The circumstances under which e-mail will and will not be an acceptable mode of communication.
2. That e-mail is never to be used for emergency communication.
3. Details on specific emergency communication procedures that are to be used.

Telephone-administered therapy

Simon, Ludman, Tutty, Operskalski, & Von Korff (2004) examined the effectiveness of telephone-administered CBT for primary care patients starting antidepressant treatment. Subjects were randomly assigned to telephone care management versus telephone care management plus 8 sessions of CBT provided over the telephone. Blind evaluations were done at 6 weeks, 3 months, and 6 months using the Hopkins Symptom Depression Scale. Those who received telephone CBT had decreased depression scores (p = .02), were more likely to identify themselves as "much improved" (80% vs. 55%, p < .001), and were very satisfied with treatment (59% vs. 29%, p < .001). Mean depression scores at follow-up in the therapy group demonstrated a significant mean reduction ($X^2 = 5.44$, p = .02). The authors conclude that telephone-administered CBT can be a new public health model for treatment of depression, but also note that at the present time, insurers do not reimburse for this type of treatment.

Another novel use of the telephone, in this case managed by a computerized interactive voice response (IVR) system, was described by Greist, Osgood-Hynes, Baer, & Marks (2000). Because this program is a self-help system for treatment of depression and does not involve a clinician, we only briefly outline findings of this study. The program included seven booklets combined with eleven IVR telephone calls and an introduction by videotape. Patients listened to recorded voice messages containing information or a question and responded by touching numbers on the telephone keypad. An uncontrolled, 12-week trial conducted in England and the United States (Greist et al., 2000) found significant reduction in

Hamilton Depression Rating Scale scores (41%). However, this program is not currently available.

In addition to standard telephone usage, smartphones and other cellular technologies are being examined for use as an adjunct to therapy. Boschen & Casey (2008) recently reviewed the proliferation of technological adjuncts to CBT. They describe the attributes of mobile phones that make them ideal for use as an adjunct to CBT. Some of these attributes include portability, being always on and connected, societal penetration and ubiquity, low initial cost, and probably the most important, acceptability. Horrigan (2008) reported that 62% of Americans have accessed the Internet wirelessly or used a nonvoice application of a cell phone. Given the widespread usage of mobile phones and their ever-growing features, it is not surprising that researchers have earmarked them as an ideal adjunct to CBT and other clinical populations. The primary usage, thus far, has been to gather ecological momentary assessment, to remind the patient to complete certain activities (e.g., gather information about blood glucose level), to facilitate homework completion, or to contact therapists between sessions. While mobile phone technology has not been specifically researched in the treatment of mood disorders, mood was often assessed in smoking and adolescent populations. Overall, the use of mobile phone technology was successful and patients reported a high degree of satisfaction (Boschen & Casey, 2008).

Even the simplest cell phone has an alarm feature. Clinicians may easily adapt their patient's cell phones to prompt them to complete therapeutic activities, use an activity schedule to monitor pleasurable or mastery activities, or set priorities.

Issues in the use of telepsychiatry

The various telepsychiatry modalities can have certain advantages such as accommodating different work schedules, ease of connection from home, and accommodating those who are distant from therapy services (Rehm, 2008). There are, however, obstacles that have to be overcome. The most commonly reported problem relates to nonverbal communication. Specifically, issues such as eye contact, posture, positioning, and voice tone can be obscured with teleconferencing approaches and, of course, the visual connection is completely lost when therapy is conducted only with online text or by telephone. Norman (2006) posits that these "distancing" effects may be useful for some patients who are intimidated by face-to-face contact.

A major concern with such therapies, specifically the issue of *working alliance*, was examined in a study of 15 patients using online therapy by e-mail. Only 20% of those studied were depressed, thus providing a very small sample of subjects for understanding how the therapeutic relationship may function in online therapy. Working alliance was measured at the third session using the Working Alliance Inventory (WAI) and was compared to a group of 25 patients receiving face-to-face therapy. The online group had higher means on the composite score of the WAI (t = 2.307, p < .05). These findings, if replicated in larger and better-controlled studies, may offer some reassurance that effective therapeutic relationships can be developed online.

Ethical and safety issues are also paramount in these alternative types of therapies. Haas, Benedict, & Kobos (1996) raises the concern that there is increased difficulty in providing for patient's safety in crisis situations and there are increased risks to privacy. Rehm (2008) expressed concern that telepsychiatry has left some questions unanswered or ambiguous. For instance, what are the ethical duties of someone treating an individual in another state or country? How much information is provided to the clinician? Where is a complaint lodged? What are the jurisdictions in relation to licensing? For an excellent discussion of the pros and cons of different teletherapy approaches, see Suler (2000).

Technology for self-monitoring in depression and related conditions

The importance of reporting symptoms in real time versus retrospective recall has been an important concept in relation to self-monitoring. If symptoms and related cognitions are recorded much after the fact, the reliability and validity of self-monitoring may be compromised. In a study of coping assessment that compared use of a palm top computer to a retrospective coping report, Stone et al. (1998) found discrepancies in reports of coping. They used two widely used instruments, the Daily Coping Inventory (Stone & Neale, 1984) and the Ways of Coping Scale (Folkman & Lazarus, 1980, 1985). An electronic diary with an audible alarm was used to collect real-time measures related to recent work/marital/other stressful events. When the device beeped they were required to complete the coping assessments. At the end of the two-day monitoring period, they were asked to select the most stressful event and to indicate with the two scales how they coped with this event. The results demonstrated overendorsement of the behavioral coping mechanisms (52.4%) and underendorsement of cognitive coping mechanisms (31.4%). The authors suggest that accurate recall of this type of information is impeded by many factors including the use of general schemas to reconstruct events as well as the influence of intervening events and psychological state at the time of recall.

Bolger, Davis, & Rafaeli (2003) observed that diary methods permit the examination of reported events and experiences in their natural, spontaneous context and reduce the likelihood of retrospective biases. Specifically, diary methods obtain reliable self-reported information, obtain reliable estimates of change over time, and strengthen the interpretation of those individual differences. Electronic diaries strengthen the assessment of self-reported information with the use of time stamps and date stamps. Additionally, data entry, management, and accuracy are assured with automated electronic diaries (Bolger et al.).

A federally funded freeware program, the Experience Sampling Program for electronic monitoring, is available for researchers and clinicians (Barrett & Feldman Barrett, 2000). The program can be accessed at http://www2.bc.edu/~barretli/esp/. Patient diaries are available commercially through invivodata.com. DiaryPro®, an electronic diary solution, offers configurable reminders and alerts, automatic data transfers, and patient accessibility features. There are many affordable choices that can be accessed on Amazon, Nextag, and other sites.

Research in electronic self-monitoring for mood disorders has been primarily conducted in patients with bipolar disorder. Scharer et al. (2002) reported on a pilot study of a palmtop device to collect information prospectively on the National Institute of Mental Health (NIMH) life chart. Patients used the palmtop device to collect data on mood, social function, impairment, duration of sleep, medication and use of drugs, life events, comorbid symptoms, hospitalizations, menses, and mood switches. Patients did not view use of this tool as a burden. Investigators concluded it was a viable, cost-effective alternative to traditional methods of collecting life chart information.

Bauer et al. (2004, 2008) published two validation studies of the home computer–based system ChronoRecord. The rationale for the use of such a system was the high degree of interindividual variation and heterogeneity with a wide range of phenotypes in bipolar disorder. The initial study (Bauer et al., 2004) reported on this computer-based system used outpatients with bipolar disorder. Self-reported information on mood, sleep, menstrual data, psychiatric medications, life events, and weight was collected on 96 subjects. Of those subjects, 83% (80/96) returned 8,662 days of data. The authors found that self-reported data was strongly correlated with clinician-rated HDRS ratings (–0.683, $p < .001$). Additionally, they noted a high acceptance of the use of ChronoRecord by patients.

Bauer et al. (2008) provided further validation data on ChronoRecord by including 27 inpatients with mania (57 ratings). Concurrent validation of ChronoRecord was established with clinician ratings of the Young Mania Rating Scale (YMRS; $r = .601$, $p < .001$). Additionally, an ROC (Receiver Operator Curve) analysis further demonstrated the ability of ChronoRecord to identify mania/hypomania. The ROC curve for normal moods and hypomania was .945 ($p < .001$, CI .923, .990) and .936 for normal and hypomania ($p < .001$, CI .867, 1.005). Agreement between

ChronoRecord and the YMRS occurred on 268/281 days (95.4%) without mania or hypomania, and 52/59 days (88.1%) with mania or hypomania using the YMRS (K = .80).

ChronoRecord is available through the ChronoRecord Association at http://www.chronorecord.org/providers.htm. Staff members enroll patients over the phone and send the required software and documentation. Each month, patients e-mail their mood data directly to the association. As soon as the data are received, mood charts are sent by e-mail to both the provider and the patient. All data and charts are encrypted for security.

Management of bipolar patients can be quite difficult due to the subtle changes in mood and behavior that may ultimately lead to dramatic mood shifts and/or hospitalization. Systematic collection of mood and behavioral information with the technologies reviewed here may assist clinicians to collaborate more effectively with bipolar patients in maintaining a more balanced mood and thereby improving their quality of life.

Summary

A principal theme that emerges throughout this chapter is that the role of technology in treatment of mood disorders is primarily adjunctive in nature. From an evidence-based perspective, programs that have attempted to "go it alone" without clinician involvement have not offered desirable clinical outcomes. For instance, full and accurate diagnostic assessment has not been possible via clinician-free software programs, but computerized assessment has been helpful in targeted applications such as completion of depression rating scales. Computer-assisted CBT with clinician screening, guidance, and support has been shown to be beneficial for depression. In contrast, CBT self-help for depression over the Internet has been marked with low completion rates and small effects on symptom reduction.

A number of different technologies have been developed to help clinicians in the provision of treatment for mood disorders. These range from fully developed programs for computer-assisted CBT to palmtop computers and smartphones that can be used for reminding, providing support, and gathering data for maintenance of a euthymic mood.

Most clinicians have not yet taken full advantage of the technological advances presented in this chapter. An important barrier is the lack of insurance reimbursement for the costs of the technology and the time of the clinician in administering some of the applications. For example, time spent using the telephone, e-mail, or data monitoring via smartphones is not typically covered by insurance plans. However, we predict that with growth of technology use by the general public and further evidence-based studies, there will be a substantial increase in the use of the treatment adjuncts outlined here. It seems unlikely that therapy for mood disorders will bypass the technological revolution and remain mired in the past.

References

Andersson, G., Bergstrom, J., Hollandare, F., Carlbring, P., Kaldo, V., & Ekselius, L. (2005). Internet-based self-help for depression: Randomised controlled trial. *British Journal of Psychiatry, 187,* 456–461.

Barrett, D. J., & Feldman Barrett, L. (2000). The Experience-Sampling Program (ESP). www.bc.edu. http://www2.bc.edu/~barretli/esp/

Bauer, M., Grof, P., Gyulai, L., Rasgon, N., Glenn, T., & Whybrow, P. C. (2004). Using technology to improve longitudinal studies: Self-reporting with ChronoRecord in bipolar disorder. *Bipolar Disorders, 6,* 67–74.

Bauer, M., Wilson, T., Neuhaus, K., Sasse, J., Pfennig, A., Lewitzka, U., et al. (2008). Self-reporting software for bipolar disorder: Validation of ChronoRecord by patients with mania. *Psychiatry Research, 159,* 359–366.

Bolger, N., Davis, A., & Rafaeli, E. (2003). Diary methods: Capturing life as it is lived. *Annual Review of Psychology, 54,* 579–616.

Boschen, M. J., & Casey, L. M. (2008). The use of mobile telephones as adjuncts to cognitive behavioral psychotherapy. *Professional Psychology: Research and Practice, 39,* 546–552.

Bowers, W., Stuart, S., MacFarlane, R., & Gorman, L. (1993). Use of computer-administered cognitive-behavior therapy with depressed inpatients. *Depression, 1,* 294–299.

Brown, F. W. (1998). Rural telepsychiatry. *Psychiatric Services, 49,* 963–964.

Castelnuovo, G., Gaggioli, A., Mantovani, F., & Riva, G. (2003a). From psychotherapy to e-therapy: The integration of traditional techniques and new communication tools in clinical settings. *CyberPsychology & Behavior, 6,* 375–382.

Castelnuovo, G., Gaggioli, A., Mantovani, F., & Riva, G. (2003b). New and old tools in psychotherapy: The use of technology for the integration of traditional clinical treatments. *Psychotherapy: Theory, Research, Practice, Training, 40,* 33–44.

Cavanaugh, K., Shapiro, D. A., Van Den Berg, S., Swain, S., Barkman, M., & Proudfoot, J. (2006). The effectiveness of computerized cognitive therapy in routine care. *British Journal of Clinical Psychology, 45,* 499–514.

Christiansen, H. & Griffiths, K. M. (2005). The prevention of depression using the Internet. *Medical Journal of Australia, 177,* 122–125.

Christiansen, H., Griffiths, K. M., Groves, C., & Korten, A. (2006). Free range users and one hit wonders: Community users of an Internet-based cognitive behaviour therapy program. *Australian NZ Journal of Psychiatry, 40,* 49–62.

Christiansen, H., Griffiths, K. M., & Korten, A. (2002). Web-based cognitive behavioral therapy: Analyses of site usage and changes in depression and anxiety scores. *Journal of Medical Internet Research, 4,* e3.

Colby, K. M., & Colby, P. M. (1990). Overcoming depression. *Malibu: Malibu Artificial Intelligence Works.*

Das, A. K. (2002). Computers in psychiatry: A review of past programs and an analysis of historical trends. *Psychiatric Quarterly, 73,* 351–365.

Emmelkamp, P. M. G. (2005). Technological innovations in clinical assessment and psychotherapy. *Psychotherapy and Psychosomatics, 74,* 336–343.

Eysenbach, G., Powell, J., Rizo, C., & Stern, A. (2004). Health related virtual communities and electronic support groups: Systematic review of the effects of online peer to peer interactions. *British Medical Journal, 328,* 1166–1170.

Farvolden, P., Denisoff, E., Selby, P., Bagby, M., & Rudy, L. (2005). Usage and longitudinal effectiveness of a web-based self-help cognitive behavioral therapy program for panic disorder. *Journal of Medical Internet Research, 7*(1), e7.

Folkman, S., & Lazarus, R. S. (1980). An analysis of coping in a middle-aged community sample. *Journal of Health and Social Behavior, 21,* 219–239.

Folkman, S., & Lazarus, R. S. (1985). If it changes, it must be a process: Study of emotion and coping during three stages of a college examination. *Journal of Personality and Social Psychology, 48,* 150–170.

Greenberg, P., Kessler, R., Birnbaum, H. G., Leong, S. A., Lowe, S. W., Berglund, P. A., & Corey-Lisle, P. K. (2003). The economic burden of depression in the United States: How did it change between 1990 and 2000? *Journal of Clinical Psychiatry, 64,* 1465–1475.

Greist, J. H., Osgood-Hynes, D. J., Baer, L., & Marks, I. M. (2000). Technology-based advances in the management of depression: Focus on the cope program. *Dis Manage Health Outcomes, 7,* 193–200.

Haas, L., Benedict, J. G., & Kobos, J. C. (1996). Psychotherapy by telephone: Risks and benefits for psychologists and consumers. *Professional Psychology: Research and Practice, 27,* 154–160.

Hilty, D. M., Marks, S. L., Urness, D., Yellowlees, P. M., & Nesbitt, T. S. (2004). Clinical and educational telepsychiatry applications: A review. *The Canadian Journal of Psychiatry, 49,* 12–23.

Horrigan, J. B. (2008). Mobile access to data and information. Pew Internet & American Life Project. http://www.pewinternet.org/press_release.asp?r=300

Houston, T. K., Cooper, L. A., & Ford, D. E. (2002). Internet support groups for depression: A 1-year prospective cohort study. *American Journal of Psychiatry, 159,* 2062–2068.

Houston, T. K., Cooper, L. A., Vu, H. T., Kahn, J., Toser, J., & Ford, D. E. (2001). Screening the public for depression through the Internet. *Psychiatric Services, 52,* 362–367.

Kessler, C., Berglund, P., Demler, O., Jin, R., & Walters, E. E. (2005). Lifetime prevalence and age-of-onset distributions of DSM-IV disorders in the national comorbidity survey replication. *Archives of General Psychiatry, 62,* 593–602.

Kessler, R. C., & Frank, R. G. (1997). The impact of psychiatric disorders on work loss days. *Psychological Medicine, 27,* 861–873.

Kobak, K. A., Mundt, J. C., Greist, J. H., Katzelnick, D. J., & Jefferson, J. W. (2000). Computer assessment of depression: Automating the Hamilton Depression Rating Scale. *Drug Information Journal, 34,* 145–156.

Lamberg, L. (2003). Online empathy for mood disorders: Patients turn to Internet support groups. *Journal of American Medical Association, 289,* 3073–3077.

Litz, B. T., Engel, C. C., Bryant, R. A., & Papa, A. (2007). A randomized, controlled proof-of-concept trial of an Internet-based, therapist-assisted self-management treatment for post-traumatic stress disorder. *American Journal of Psychiatry, 164*, 1676–1683.

Maris, R. W. (1992). Overview of the study of suicide assessment and prediction, pp. 3–24. In R. W. Maris, A. L. Berman, J. T. Maltsberger, & R. I. Yufit (Eds.). *Assessment and prediction of suicide*. New York: Guilford Press.

McCrone, P., Knapp, M., Proudfoot, J., Ryden, C., Cavanaugh, K., Shapiro, D. A., et al. (2004). Cost-effectiveness of computerized cognitive-behavioral therapy for anxiety and depression in primary care: Randomized controlled trial. *British Journal of Psychiatry, 185*, 55–62.

Mehta, S., & Chalhoub, N. (2006). An e-mail for your thoughts. *Child and Adolescent Mental Health, 11*, 168–170.

Mintz, J., Mintz, L. I., Arruda, M. J., & Hwang, S. S. (1992). Treatments of depression and the functional capacity to work. *Archives of General Psychiatry, 49*, 761–768.

Norman, S. (2006). The use of telemedicine in psychiatry. *Journal of Psychiatric and Mental Health Nursing, 13*, 771–777.

Norris, A. C. (2001). *Essentials of telemedicine and telehealth*. Chichester: John Wiley and Sons, Ltd.

O'Dell, J. W., & Dickson, J. (1984). ELIZA as "therapeutic" toll. *Journal of Clinical Psychology, 40*, 942–945.

Oravec, J. A. (2000). Online counselling and the Internet: Perspectives for mental health care supervision and education. *Journal of Mental Health, 9*, 121–135.

Perini, S., Titov, N., & Andrews, G. (2008). The Climate Sadness program of Internet-based treatment for depression: A pilot study. *Electronic Journal of Applied Psychology*.

Proudfoot, J., Goldberg, D., Mann, A., Everitt, B., Marks, I., & Gray, J. A. (2003). Computerized, interactive, multimedia cognitive behavioral therapy reduces anxiety and depression in general practice: A randomized controlled trial. *Psychological Medicine, 33*, 217–227.

Proudfoot, J., Ryden, C., Everitt, B., Shapiro, D. A., Goldberg, D., Mann, A., et al. (2004). Clinical efficacy of computerized cognitive-behavioral therapy for anxiety and depression in primary care: Randomized controlled trial. *British Journal of Psychiatry, 185*, 46–54.

Regier, D., Narrow, W., Rae, D., Manderscheid, D., Locke, B., & Goodwin, F. (1993). The de Facto US mental and addictive disorders service system. *Archives of General Psychiatry, 50*, 85–94.

Rehm, L. P. (2008). How far have we come in teletherapy? Comment on "Telephone-administered psychotherapy." *Clinical Psychology: Science and Practice, 15*, 259–261.

Sampson, J. P. & Pyle, K. R. (1983). Ethical issues involved with the use of computer-assisted counseling, testing, and guidance systems. *Personal and Guidance Journal, 61*, 283–287.

Scharer, L. O., Hartweg, V., Valerius, G., Graf, M., Hoern, M., Biedermann, C., et al. (2002). Life charts on a palmtop computer: First results of a feasibility study with an electronic diary for bipolar patients. *Bipolar Disorders, 4*, 107–108.

Selmi, P. M., Klein, M., Greist, J. H., & Harris, W. G. (1982). An investigation of computer-assisted cognitive behavior therapy in the treatment of depression. *Behavior Research Methods and Instruments, 14,* 181–185.

Selmi, P. M., Klein, M., Greist, J. H., Sorrell, S. P., & Erdman, H. P. (1990). Computer-administered cognitive behavioral therapy for depression. *American Journal of Psychiatry, 147,* 51–56.

Simon, G. E., Ludman, E. J., Tutty, S., Operskalski, B., & Von Korff, M. (2004). Telephone psychotherapy and telephone care management for primary care patients starting antidepressant treatment. *Journal of American Medical Association, 292,* 935–942.

Sletten, I. W., Altman, H., & Ulett, G. A. (1971). Routine diagnosis by computer. *American Journal of Psychiatry, 127,* 1147–1152.

Spek, V., Cuijpers, P., Nyklicek, I., Riper, H., Keyzer, J., & Pop, V. (2007). Internet-based cognitive behaviour therapy for symptoms of depression and anxiety: A meta-analysis. *Psychological Medicine, 37,* 319–328.

Stone, A. A., & Neale, J. M. (1984). New measure of daily coping: Development and preliminary results. *Journal of Personality and Social Psychology, 46,* 892–906.

Stone, A. A., Schwartz, J. E., Neale, J. M., Shiffman, S., Marco, C. A., Hickcox, M., et al. (1998). A comparison of coping assessed by ecological momentary assessment and retrospective recall. *Journal of Personality and Social Psychology, 74,* 1670–1680.

Stuart, S., & LaRue, S. (1996). Computerized cognitive therapy: The interface between man and machine. *Journal of Cognitive Psychotherapy, 10,* 181–191.

Suler, J. R. (2000). Psychotherapy in cyberspace: A 5-dimensional model of online and computer-mediated psychotherapy. *CyberPsychology & Behavior.*

Wang, L., LaBar, K. S., Smoski, M., et al. (2005). Amygdala activation to sad pictures during high-field (4 Tesla) functional magnetic resonance imaging. *Emotion, 5,* 12–22.

Weizenbaum, J. (1996). Computational linguistics. *Communications of the ACM, 9,* 36–45.

Wells, K. B., Stewart, A., Hays, R. D., Burnam, M. A., Rogers, W., Daniels, M., et al. (1989). The functioning and well-being of depressed patients. *Journal of American Medical Association, 262,* 914–919.

Whitfield, G., Hinshelwood, R., Pashely, A., Williams, C. J., & Campsie, L. (2006). The impact of a novel computerized CD-ROM (*Overcoming Depression*) offered to patients referred to clinical psychology. *Behavioral and Cognitive Psychotherapy, 34,* 1–11.

Williams, C., & Whitfield, G. (2001). Written and computer-based self-help treatments for depression. *British Medical Bulletin, 57,* 133–144.

Wilson, G. T. (1996). Treatment of bulimia nervosa: When CBT fails. *Behavioral Research Therapy, 34,* 197–212.

Wright, J. H. (2004). Computer-assisted cognitive-behavior therapy. In J. H. Wright (Ed.), *Cognitive-behavior therapy* (pp. 55–82). Washington, DC: American Psychiatric Publishing, Inc.

Wright, J. H. (2008). Computer assisted psychotherapy. *Psychiatric Times, 25,* 14–15.

Wright, J. H., Wright, A. S., Albano, A. M., Basco, M. R., Goldsmith, L. J., Raffield, T., & Otto, M. W. (2005). Computer-assisted cognitive therapy for depression: Maintaining efficacy while reducing therapist time. *American Journal of Psychiatry, 162,* 1158–1164.

Wright, J. H., Wright, A. S., & Beck, A. T. (2003). Good Days Ahead: The multimedia program for cognitive therapy. Mindstreet, Louisville.

Wright, J. H., Wright, A. S., Salmon, P., et al. (2002). Development and initial testing of a multimedia program for computer-assisted cognitive therapy. *American Journal of Psychotherapy, 56,* 76–86.

chapter two

Anxiety disorders

*Michelle G. Newman, Darryl F. Koif,
Amy Przeworski, and Sandra J. Llera*

Anxiety disorders

Cognitive-behavioral therapy (CBT) is considered the treatment of choice for anxiety disorders (Barlow, Gorman, Shear, & Woods, 2000; Hofmann & Spiegal, 1999; Newman, 2000). CBT is particularly well suited to delivery via interactive computer programs because it is highly structured with well-delineated procedures, it targets specific behaviors and symptoms, and it proceeds in a systematic fashion (Selmi, Klein, Greist, Sorrell, & Erdman, 1990). One of the most developed areas of research on the use of technology in conjunction with psychotherapy has been with respect to anxiety disorders. Technologies used for such treatment include palmtop computers, desktop computers, telephone-guided therapy, and virtual reality.

In the current chapter we will first discuss the reasons a therapist would want to implement technologies mostly tested to be used as self-help or minimal therapist-contact devices. Next, we will review the literature regarding the efficacy of these applications for anxiety disorders. Following the literature review, we will discuss the clinical skills needed to make use of the software and hardware tested for anxiety interventions. Next, we will review some of the pitfalls and ethical dilemmas surrounding these applications. Finally, we will discuss a model for making use of these technologies in private practice.

Why would a practitioner be interested in making use of a self-help or a minimal-contact device?

At first glance, it may not be obvious why a mental health provider would even want to learn about evidence-based self-help or minimal-contact applications. However, there are a variety of circumstances whereby such applications might be advantageous to a practitioner in the real world. For example, there are circumstances that a client can afford to pay for, or insurance has provided for, only a few sessions of therapy. Under those conditions, a therapist would want to maximize the use of his or her face-to-face contact time with a client. Similarly, clients often have multiple issues that they would like to work on or multiple comorbidities. Self-help and minimal-contact technologies provide an option for clients to work on one issue mostly at home and another issue mostly in the therapist's office. Moreover, working with a particular problem may not be within a therapist's scope of practice. Thus, the use of one of these devices might be preferable to referring a client to another therapist. In addition, most of the treatments tested with technological devices have been cognitive-behavioral in nature. Some tasks and techniques within CBT require a lot of repetition and such repetition can become arduous at times. The use of these devices allows for the freeing up of a therapist's time to work on other things with a client and can remove the burden of some of the less interesting CBT tasks.

In addition, few clients have the self-discipline to make use of self-help interventions in the absence of externally imposed structure (Newman, Erickson, Przeworski, & Dzus, 2003). In fact, many randomized controlled trials of self-help therapies have included weekly therapist check-ins to ensure that clients understand what they are reading and are adhering to homework (Newman et al., 2003). Furthermore, many evidence-based treatment manuals include client-focused workbooks to ensure that clients receive written descriptions of therapist-delivered instructions and rationales, thus reinforcing what a therapist has already covered in a particular session, such as self-monitoring, skills practice, and generalization. Technologies such as the Internet and handheld devices facilitate client practice and application of their tasks. For all of these reasons, many clients prefer to have a therapist available to work with on a weekly basis as they are working with technological devices, even devices meant to be used without therapist support.

How can technology enhance specific therapy techniques for the treatment of anxiety disorders?

A number of therapy components used in the treatment of anxiety disorders can be enhanced with technology. These components include assessment, psychoeducation, homework, and various cognitive-behavioral techniques. For example, computerized instruments can be used to assess disorder type and symptom severity as well as to monitor client performance throughout the course of treatment. Although structured interviews are the most reliable assessment method, these interviews are costly and require more detailed, consistent, and comprehensive assessment than is often

Chapter two: Anxiety disorders 29

performed in many clinical settings (Climent, Plutchik, Estrada, Gaviria, & Arevalo, 1975; Garb, 2005; Kiernan, McCreadie, & Flanagan, 1976; Miller, Dasher, Collins, Griffiths, & Brown, 2001). Moreover, even structured interviews are subject to unreliable administration. One study found that clinicians accidentally omit as much as 5% of required questions (Fairbairn, Wood, & Fletcher, 1959). Nonetheless, computerized structured interviews, such as the revised clinical interview schedule (Lewis, 1994), Hamilton Anxiety and Depression scales (Kobak, Reynolds, & Greist, 1994), and the Diagnostic Interview Schedule (Erdman et al., 1992; Greist et al., 1987), have shown high rates of reliability when compared with trained human interviewers administering structured assessments (Ancill, Rogers, & Carr, 1985; Carr & Ghosh, 1983; Greist et al., 1987; Lewis, 1994). Computer-based instruments also have the advantage of ensuring standardization and data completeness. Moreover, clients may feel less embarrassment when completing a computer-based assessment and therefore report more of their symptoms than they would in a face-to-face interview. In fact, studies have shown that clients are more open with a computer than with a person when discussing issues such as alcohol intake (Araya, Wynn, & Lewis, 1992; Erdman, Klein, & Greist, 1985), substance misuse (Supple, Aquilino, & Wright, 1999), suicidal ideation (Greist et al., 1973), and sexual experiences (Lapham, Henley, & Skipper, 1997; Romer et al., 1997).

Technological devices also create the option for clients to learn and master the psychoeducational components of a treatment in their own way and at their own pace. The interactive nature of technological packages allows access to information and treatment techniques in ways that are more personally relevant and engaging than would be the case with a traditional book format. Also, devices such as laptop computers, smartphones, or PDAs provide the additional benefit of portability, so clients can access elements of their psychoeducation while riding on the subway or sitting in a staff meeting or before entering a particularly fear-inducing situation. Because these devices have the option of connecting to the Internet, the information is easier to update than when it is presented in other ways.

Homework, a central feature of CBT for anxiety disorders, can also be enhanced using technological devices. Homework facilitates goal achievement by providing clients with opportunities to integrate and practice techniques introduced to them in session. However, the challenge of making homework effective is helping clients to complete their assigned tasks. Providing technological solutions may make completing homework less burdensome to the client. Of course, homework will always require client commitment and effort to be effective, so the goal is not to relieve clients of their responsibility to be fully engaged in their treatment. However, when clients are motivated but lack the self-structure to fully comply with assignments, technology can be used to assist them. Moreover, handheld devices provide the benefit of high portability of structured momentary assessment and therapy prompts. This also ensures greater accuracy of the information gathered as it reflects emotions, thoughts, and behaviors experienced in the moment. For example, Newman and colleagues (Kenardy, Dow, et al., 2003; Newman, Kenardy, Herman, & Taylor, 1997) developed two palmtop computer therapy programs to be used in the treatment of generalized anxiety disorder (GAD), social phobia, and panic disorder. These programs were equipped with audio alerts set to prompt clients several times per day to self-monitor their anxiety levels, to practice applied relaxation in response to anxiety cues, and to record the outcomes of behavioral experiments or exposure efforts. Using this device facilitated a client to self-monitor her thoughts, behaviors, and emotions, and to practice CBT skills in real time and led to treatment generalization (Przeworski & Newman, 2004). Recent advances in availability of Internet connections also mean that data can be transferred as soon as clients finish a task or complete a questionnaire.

In addition to facilitating homework compliance, technological devices can be used to facilitate exposure therapy. Virtual reality (VR) has been the technology most investigated as a delivery mechanism for exposure therapy. In virtual reality exposure (VRE), participants interact with a computer-generated, three-dimensional virtual world allowing them to experience a sense of presence and complete immersion in that world. Elements that can be incorporated into the virtual reality experience include audio cues, simulated manipulation of objects, and tactile augmentation in which the illusion is created that the client is actually touching, smelling, or feeling the feared object. Thus, VRE provides the therapist with the ability to immerse the client in an exposure exercise without the inconvenience and expense of in vivo exposure. VRE can be used for situations that are otherwise difficult to arrange and control (e.g., combat exposure) and can be made as mild or extreme as necessary. This technology can also be cost-effective when repeated exposure to the feared situation is expensive to create in real life (e.g., air travel with repeated takeoffs and landings). Other types of technologies can also be used to deliver exposure therapy in the real world. For example, videoconferencing and palmtop computers have been used to deliver interoceptive exposure and in vivo exposure to clients with panic disorder with agoraphobia (Bouchard et al., 2000; Newman et al., 1997).

Critical review of the literature

The study design and research methodology used to evaluate a technology application provides information about the generalizability of

Chapter two: Anxiety disorders

research results to clinical practice and the circumstances under which specific technologies may be helpful. Given the potential importance of the context in which the technologies are tested, we will include such information in our description of various studies. To do this, we will employ four categorical descriptors of therapist contact modified from those used by Glasgow and Rosen (1978). These descriptors are as follows: (a) self-administered therapy (SA; therapist contact for assessment, at most), (b) predominantly self-help (PSH; therapist contact beyond assessment is for periodic check-ins, teaching clients how to use the self-help tool, and/or for providing the initial therapeutic rationale), (c) minimal-contact therapy (MC; active involvement of a therapist, though to a lesser degree than traditional therapy for this disorder; includes any treatment in which the therapist helps to train the client in the application of specific therapy techniques, and (d) predominantly therapist-administered treatments (PTA; clients have regular contact with a therapist for a typical number of sessions, but the study attempts to determine whether the use of a self-help tool augments the impact of the standard therapy). Also, because some studies have required clients to make weekly appointments to come to a lab to use a particular application, we have included this information in our review even though the actual use of the software may or may not involve any therapist contact.

In studies of panic disorder treatment, controlled trials found that SA Internet cognitive therapy was superior to self-monitoring (Klein & Richards, 2001). In addition, PSH Internet CBT plus e-mail and phone check-in (averaging 90 to 240 minutes contact) was superior to wait list (Carlbring, Westling, Ljungstrand, & Andersson, 2001; Carlbring, Bohman, et al., 2006), but comparable to 10 sessions of therapist-delivered CBT (Carlbring et al., 2005). Also, Internet delivered CBT plus e-mail contact (averaging 332.5 minutes of contact) was better than a self-help manual plus phone calls (averaging 245.3 minutes of contact) at reducing visits to a general practitioner and at reducing negative health ratings, and both of these treatments were superior to an information-only Internet site (Klein, Richards, & Austin, 2006). Moreover, Internet CBT plus stress management (averaging 300.3 minutes therapist contact) was superior at post-assessment to Internet CBT alone (averaging 376.3 minutes therapist contact) on panic severity and general anxiety, although there were no longer any differences between the two active treatments at 3-month follow-up (Richards, Klein, & Austin, 2006). PSH desktop computer-assisted exposure delivered in the lab (averaging 3.2 contact hours) was equivalent to 3 to 10 sessions of therapist-delivered exposure (Ghosh & Marks, 1987). Also, a palmtop computer plus 6 hours of therapist-delivered CBT was equivalent to 12 hours of therapist-delivered CBT (Kenardy, Dow, et al., 2003; Newman et al., 1997). Further, 8 sessions of therapist-assisted VRE was significantly better than wait list (North, North, & Coble, 1996; Vincelli et al., 2003) and equivalent to 12 sessions of therapist-delivered exposure.

Studies failing to find support for active computer-delivered therapies found that PSH Internet CBT plus e-mail and phone check-in (averaging 90 to 240 minutes contact) was equivalent to PSH Internet-applied relaxation (Carlbring, Ekselius, & Andersson, 2003). In addition, desktop computer-assisted exposure delivered in the lab (averaging 3.2 contact hours) was equivalent to a self-help exposure manual used at home (averaging 1.5 contact hours; Ghosh & Marks, 1987). Also, a palmtop computer plus 6 hours of therapist-delivered CBT was comparable to 6 hours of therapist-delivered CBT without the computer (Kenardy, Dow, et al., 2003).

Studies of computer-treated obsessive-compulsive disorder (OCD) were fewer and somewhat less promising than was found with panic disorder. For example, an uncontrolled trial showed nonsignificant improvement in symptoms as a result of vicarious exposure and response prevention (ERP) delivered in a lab using a desktop computer (Clark, Kirkby, Daniels, & Marks, 1998). Additional studies employing a manual plus a computer response system accessed by telephone found that although it led to statistically significant improvement (Bachofen et al., 1999; Marks et al., 1998), as a PSH intervention it was not as effective as therapist-delivered ERP (Greist et al., 2002). On the other hand, when this intervention was used as an adjunct to therapist-delivered treatment, participants required fewer sessions with the therapist but did as well (Nakagawa et al., 2000). Also, an open trial of a self-help Internet intervention for trichotolomania showed significant reductions of self-reported hair pulling (Mouton-Odum, Keuthen, Wagener, Stanley, & DeBakey, 2006).

In contrast to treatment of OCD, computer studies of social phobia appear more promising. In a case study, 6 sessions of group CBT plus a palmtop computer led to symptom improvement (Przeworski & Newman, 2004). Moreover, 6 to 10 sessions of therapist-guided cognitive therapy, breathing retraining, and VR exposure decreased social phobia symptoms (Anderson, Rothbaum, & Hodges, 2003; Anderson, Zimand, Hodges, & Rothbaurn, 2005). In controlled trials, a CBT Internet site in conjunction with therapist e-mail contact (averaging 2 to 3 hours) was superior to no treatment (Andersson et al., 2006; Carlbring, Furmark, Steczkó, Ekselius, & Andersson, 2006; Carlbring et al., 2007). In addition, 8 sessions of group CBT plus a palmtop computer was equivalent to 12 sessions without the computer (Gruber, Moran, Roth, & Taylor, 2001), and twelve 45-minute sessions of VRE was equivalent to twelve 2-hour sessions of group CBT (Klinger et al., 2005). Also, in the treatment of public speaking anxiety, 4 to 5 sessions of VR active exposure was superior to VR trivial exposure (North, North, & Coble, 1997) and wait list (Harris, Kemmerling, & North, 2002).

There are only a few studies of generalized anxiety disorder (GAD) or posttraumatic stress disorder (PTSD). A case study of a person with GAD found that 6 sessions of group treatment plus the use of a palmtop computer led to diminished symptoms (Newman, Consoli, & Taylor, 1999). In the treatment of PTSD, 2 uncontrolled studies found that therapist-assisted (TA) VR exposure (VRE) lasting from eight to ten 90-minute sessions led to clinically meaningful and statistically significant reductions in combat veterans' symptomatology (Ready, Pollack, Rothbaum, & Alarcon, 2006; Rothbaum, Hodges, Anderson, Price, & Smith, 2002; Rothbaum, Hodges, Ready, Graap, & Alarcon, 2001). Studies of the impact of MC Internet treatment on trauma victims found that the site led to clinically significant change (Lange et al., 2000) and was superior to wait list (Lange et al., 2003; Lange, van de Ven, Schrieken, & Emmelkamp, 2001).

By far, the greatest number of computer studies has been conducted on the treatment of simple phobias. However, only uncontrolled studies examined the treatment of fear of driving or claustrophobia. Uncontrolled studies found that PTA VRE led to change in self-reported and physiological symptoms of driving phobia (Wald & Taylor, 2003; Walshe, Lewis, Kim, O'Sullivan, & Wiederhold, 2003) but no change in driving behavior (Wald & Taylor, 2003). On the other hand, a case study series found that TA VRE was effective in the treatment of claustrophobia (Botella, Banos, Villa, Perpina, & Garcia Palacios, 2000).

In terms of fear of flying or flight phobia, minimal-contact VRE was superior to progressive muscle relaxation (Mühlberger, Herrmann, Wiedemann, Ellgring, & Pauli, 2001). In addition, minimal-contact VRE plus cognitive therapy was superior to cognitive therapy alone and to a wait list (Mühlberger, Wiedemann, & Pauli, 2003). Also, minimal-contact anxiety management training plus VRE (simulated flights including takeoff and landing) was equivalent to minimal-contact anxiety management training plus in vivo exposure (to an airport and parked airplane), and both were superior to wait list (Anderson et al., 2006; Rothbaum et al., 2002, 2006). Interestingly, 6 sessions of vicarious exposure to images of flights on a computer screen alone was superior to 11 sessions of computer exposure plus relaxation and information, and both were superior to wait list (Bornas, Tortella Feliu, Llabres, & Fullana, 2001). However, 6 sessions of computer-aided vicarious exposure plus flight sounds was equivalent to 6 sessions of nonexposure CBT (Bornas, Tortella-Feliu, & Llabrés, 2006). Moreover, minimal-contact cognitive therapy plus VRE plus motion simulation was equivalent to cognitive therapy plus VRE without motion simulation (Mühlberger et al., 2003). However, MC VRE plus physiological feedback was superior to VRE without physiological feedback, and both were better than imaginal exposure (Wiederhold, Jang, Kim, & Wiederhold, 2002; Wiederhold & Wiederhold, 2003). Less positive results showed that MC VRE led to a flight in 67% of participants after treatment, but only 23% at 1 year follow-up in an uncontrolled study (Kahan, Tanzer, Darvin, & Borer, 2000). Also, VRE was not superior to an attention placebo (Maltby, Kirsch, Mayers, & Allen, 2002). However, the majority of studies suggested that VRE and computer-based exposure are effective in the treatment of fear of flying.

Studies of acrophobia treatment found that MC VRE was equivalent to MC in vivo exposure (Emmelkamp, Bruynzeel, Drost, & Van Der Mast, 2001; Emmelkamp et al., 2002). In addition, MC VRE using a more expensive system with greater presence was equivalent to a less expensive system with less presence (Krijn et al., 2004). Also, MC and TA VRE were superior to wait list (Krijn et al., 2004; Lamson, 1994; Rothbaum et al., 1995) and TA VRE plus D-cycloserine was superior to VRE plus placebo (Ressler et al., 2004).

In the treatment of spider phobia, studies found that 3 sessions were equivalent to 6 sessions of computer-aided vicarious exposure. (Fraser, Kirkby, Daniels, Gilroy, & Montgomery, 2001), but 3 sessions were superior to wait list (Dewis et al., 2001). However, when compared to live graded exposure, computer-aided vicarious exposure was inferior in three studies (Dewis et al., 2001; Heading et al., 2001; Nelissen, Muris, & Merckelbach, 1995), and equivalent in one (Gilroy, Kirkby, Daniels, Menzies, & Montgomery, 2000, 2003). Moreover, although vicarious exposure was superior to progressive muscle relaxation at 3-month follow-up (Gilroy et al., 2000), it was equivalent at 33-month follow-up (Gilroy et al., 2003). Also, relevant vicarious exposure plus feedback was equivalent to irrelevant vicarious exposure with feedback (Smith, Kirkby, Montgomery, & Daniels, 1997). On the other hand, studies of VRE for spider phobia found that it was superior to wait list (Garcia-Palacios, Hoffman, Carlin, Furness, & Botella, 2002) and no treatment (Hoffman, Garcia Palacios, Carlin, Furness, & Botella Arbona, 2003). Moreover, VRE plus tactile augmentation was superior to VRE without tactile augmentation (Hoffman et al., 2003).

In sum, more studies are needed on computer treatment of OCD, GAD, PTSD, social phobia, claustrophobia, and driving phobia. Studies of mixed anxiety, panic disorder, and social phobia treatment are promising. However, no mixed anxiety or panic disorder study found differences across active treatments, and none compared active treatments to attention placebo. The same pattern emerges for studies of GAD and PTSD. Within the domain of simple phobias, treatment for flying phobia and acrophobia are particularly promising, although these studies relied mostly on VRE technology and are therefore difficult to generalize to other types of computer treatments. In addition, in the treatment of OCD at least minimal therapist contact appears optimal. Finally, vicarious exposure is not as helpful as therapist-directed exposure in the treatment of OCD or spider phobia.

Problems with these studies include small sample sizes, limiting power to detect differences between conditions (Carlbring et al., 2003; Chandler, Burck, & Sampson, 1986; Chandler, Burck, Sampson, & Wray, 1988; Clark et al., 1998; Kirkby et al., 2000; Klein & Richards, 2001; Newman et al., 1999; Vincelli et al., 2003); lack of control group (Bachofen et al., 1999; Chandler et al., 1988; Harcourt, Kirkby, Daniels, & Montgomery, 1998; Kirkby et al.; Marks et al., 1998; Newman, Kenardy, Herman, & Taylor, 1996; Newman et al., 1999; Richards & Alvarenga, 2002; Shaw, Marks, & Toole, 1999; White, Jones, & McGarry, 2000); nonrandomized assignment (Kenwright, Liness, & Marks, 2001; Kenwright & Marks, 2004); large number of dropouts (Bachofen et al.; Kenwright et al., 2001; Kenwright & Marks; Richards & Alvarenga); failure to include a structured interview diagnosis (Kenwright et al., 2001) or use of an analogue sample (North et al., 1996); no follow-up assessment (Bachofen et al.; Carlbring et al.; Clark et al., 1998; Harcourt et al., 1998; Kenardy, McCafferty, & Rosa, 2003; Kirkby et al.; Klein & Richards; Marks et al.; North et al.; Vincelli et al., 2003); failure to assess compliance (Carlbring et al., 2001; Ghosh & Marks, 1987; Klein & Richards); low compliance (Bachofen et al.; Carlbring et al., 2005; Greist et al., 2002; Marks et al.); and very few studies included an attention placebo control. In addition, most studies relied solely on self-report outcome measures (Bachofen et al.; Harcourt et al.; Klein & Richards; Marks et al.)

Clinical skills needed

Effectively integrating information technology into the treatment of anxiety disorders requires first and foremost that the mental health care provider have a working knowledge of the change principles and the techniques used in CBT. CBT emphasizes the role of faulty information processing, or the way a person thinks about a situation, as well as maladaptive behaviors as precursors to painful emotions. Altering the client's information processing strategy will lead to changes in how the client feels and behaves. Conversely, attending to the client's feeling states and behaviors can result in changes to the client's thought processes. Providers also need to have knowledge of, and skills with, the specific CBT techniques that a particular program or device is addressing. For example, in using virtual reality to do exposure, the provider must be aware of the necessary and sufficient conditions that will maximize the exposure experience (Foa & Kozak, 1986). Similarly, when using technology to augment cognitive restructuring, the provider must be trained in the use of cognitive restructuring. In addition, the provider should understand that to be successful, clients must learn how to gain awareness of automatic thought habits, to challenge their thoughts, and then to replace the maladaptive thoughts with realistic thoughts and beliefs. This level of knowledge of CBT will allow the provider to guide the client through the process of change being enhanced with information technology. All of this requires at least a moderate working knowledge of CBT as well as a willingness to learn how to use new software and computer devices.

Pitfalls and ethical dilemmas

An advantage of computer-assisted therapy is its cost-effectiveness. Studies have estimated a savings of $540 to $630 per client when compared with standard individual CBT (Newman et al., 1997, 1999). The cost advantages are most pronounced over long-term usage of the technology in the clinician's practice. The initial cost of implementing a technical solution may discourage use of technology if clinicians do not think of those initial costs as an investment in their practice. Initial costs may include the price of purchasing items such as the necessary hardware, the software packages, or technical assistance in setting up the system. For Internet-based solutions, the server costs need to be considered and a decision made as to whether to purchase and maintain the server within the practice or to purchase server space through a service provider. It should be noted that some technologies will take longer to realize cost benefits. For instance, virtual reality is a costly technology to implement and its cost-effectiveness may not be realized for several years after its introduction into the practice and only after significant use with a large enough client population. Finally, in considering the future cost benefits of any technological solution, clinicians should remember to account for long-term maintenance and upgrade costs. These are "hidden" costs that people often fail to take into consideration when deciding to integrate a technology solution in their practices.

Despite their promise, technology-based interventions have been received with skepticism and resistance by some researchers and psychotherapists and continue to be used by only a small portion of therapists (Whitfield & Williams, 2004). Privacy concerns are the biggest issues concerning the use of technology in behavioral health. This issue is especially important with the use of technologies such as the Internet or wireless communication. Access to programs can be password protected and data records must follow federal privacy guidelines. However, even when following these security measures and privacy guidelines strictly, it is not possible for a clinician to guarantee the confidentially of client records over the Internet or wireless communication. Clients must be clearly informed of the possible risk inherent in such technologies. If using a service provider to provide server access or data collection, the clinician should be certain that the provider will do all that is possible to protect sensitive client information. The methods that the provider will use to ensure the confidentiality of data and communication traffic should be enumerated in writing in the contractual agreement as well as any compensatory measures that will be followed in case of a loss of data or a breach in the security of the system.

Implementing technology in practice

To maximize the effectiveness of computer- or Internet-based applications in the treatment of anxiety disorders, mental health providers should begin with a comprehensive implementation plan. Sampson and colleagues (2003) have developed an implementation model for computer applications that can be adapted to the mental health care setting. The model consists of seven steps including (1) program evaluation, (2) software selection, (3) software integration, (4) staff training, (5) trial use, (6) operation, and (7) evaluation. Each step in the model consists of a number of substeps. Depending on your needs and goals for technology use in your practice, not all of the substeps may be necessary. The critical substeps enumerated here correspond with the enumeration of the seven primary steps above: (1) review the needs of the clients; (2) select the computer application that best balances features, quality, and costs; (3) decide how the computer application will be used in delivering treatment; (4) develop a plan for staff training to provide assistance to help clients make effective use of the application; (5) observe and interview trial users to identify the strengths and limitations of resources and services that support computer application use; (6) collect evaluation data; and (7) refine resources and services that support computer application use based on evaluation results. Note that evaluation is not limited to steps six and seven. Evaluation should occur throughout the implementation process, and the data from any evaluation activities should inform execution of any relevant step in the process.

The implementation model should be applicable whether the provider is implementing a purchased application or an application created within the organization. Additional considerations for providers developing their own applications include acquiring the appropriate knowledge of computer development, debugging, and testing. Rapid application development (RAD) software packages are available to assist the programming effort. RAD is a programming approach designed to minimize the time it takes to develop a computer application. There are RAD software development tools that require minimal to no previous programming knowledge. RAD development tools exist for creating applications for desktop computers, handheld devices, and the Internet.

In summary, there are advantages and disadvantages to using technological devices in the treatment of anxiety disorders. These advantages and disadvantages should be weighed carefully by a practitioner before employing technological devices in practice. The data suggests that for many anxiety disorders these devices can be helpful when used appropriately.

References

Ancill, R. J., Rogers, D., & Carr, A. C. (1985). Comparison of computerised self-rating scales for depression with conventional observer ratings. *Acta Psychiatrica Scandinavica, 71*, 315–317.

Anderson, P. L., Jacobs, C. H., Lindner, G. K., Edwards, S., Zimand, E., Hodges, L. L., et al. (2006). Cognitive behavior therapy for fear of flying: Sustainability of treatment gains after September 11. *Behavior Therapy, 37*, 91–97.

Anderson, P. L., Rothbaum, B. O., & Hodges, L. F. (2003). Virtual reality exposure in the treatment of social anxiety. *Cognitive and Behavioral Practice, 10*, 240–247.

Anderson, P. L., Zimand, E., Hodges, L. F., & Rothbaurn, B. O. (2005). Cognitive behavioral therapy for public-speaking anxiety using virtual reality for exposure. *Depression and Anxiety, 22*, 156–158.

Andersson, G., Carlbring, P., Holmström, A., Sparthan, E., Furmark, T., Nilsson-Ihrfelt, E., et al. (2006). Internet-based self-help with therapist feedback and in vivo group exposure for social phobia: A randomized controlled trial. *Journal of Consulting and Clinical Psychology, 74*, 677–686.

Araya, R. I., Wynn, R., & Lewis, G. (1992). Comparison of two self-administered psychiatric questionnaires (GHQ-12 and SRQ-20) in primary care in Chile. *Social Psychiatry and Psychiatric Epidemiology, 27*, 168–173.

Bachofen, M., Nakagawa, A., Marks, I. M., Park, J. M., Greist, J. H., Baer, L., et al. (1999). Home self-assessment and self-treatment on obsessive-compulsive disorder using a manual and a computer-conducted telephone interview: Replication of a U.K.–U. S. study. *Journal of Clinical Psychiatry, 60*, 545–549.

Barlow, D. H., Gorman, J. M., Shear, M. K., & Woods, S. W. (2000). Cognitive-behavioral therapy, imipramine, or their combination for panic disorder: A randomized controlled trial. *Journal of the American Medical Association, 283*, 2529–2536.

Bornas, X., Tortella-Feliu, M., & Llabrés, J. (2006). Do all treatments work for flight phobia? Computer-assisted exposure versus a brief multicomponent nonexposure treatment. *Psychotherapy Research, 16*, 41–50.

Bornas, X., Tortella-Feliu, M., Llabrés, J., & Fullana, M. A. (2001). Computer-assisted exposure treatment for flight phobia: A controlled study. *Psychotherapy Research, 11*, 259–273.

Botella, C. A., Banos, R. M., Villa, H., Perpina, C., & Garcia Palacios, A. (2000). Virtual reality in the treatment of claustrophobic fear: A controlled, multiple-baseline design. *Behavior Therapy, 31*, 583–595.

Bouchard, S., Payeur, R., Rivard, V., Allard, M., Paquin, B., Renaud, P., et al. (2000). Cognitive behavior therapy for panic disorder with agoraphobia in videoconference: Preliminary results. *CyberPsychology and Behavior, 3*, 999–1007.

Carlbring, P., Bohman, S., Brunt, S., Buhrman, M., Westling, B. E., Ekselius, L., et al. (2006). Remote treatment of panic disorder: A randomized trial of Internet-based cognitive behavioral therapy supplemented with telephone calls *American Journal of Psychiatry, 163*, 2119 – 2125.

Carlbring, P., Ekselius, L., & Andersson, G. (2003). Treatment of panic disorder via the Internet: A randomized trial of CBT vs. applied relaxation. *Journal of Behavior Therapy and Experimental Psychiatry, 34*, 129–140.

Carlbring, P., Furmark, T., Steczkó, J., Ekselius, L., & Andersson, G. (2006). An open study of Internet-based bibliotherapy with minimal therapist contact via email for social phobia. *Clinical Psychologist, 10*, 30–38.

Carlbring, P., Gunnarsdóttir, M., Hedensjö, L., Andersson, G., Ekselius, L., & Furmark, T. (2007). Treatment of social phobia: Randomised trial of Internet-delivered cognitive-behavioural therapy with telephone support. *British Journal of Psychiatry, 190*, 123–128.

Carlbring, P., Nilsson-Ihrfelt, E., Waara, J., Kollenstam, C., Buhrman, M., Kaldo, V., et al. (2005). Treatment of panic disorder: Live therapy vs. self-help via the Internet. *Behaviour Research and Therapy, 43*, 1321–1333.

Carlbring, P., Westling, B. E., Ljungstrand, P., & Andersson, G. (2001). Treatment of panic disorder via the Internet: A randomized trial of a self-help program. *Behavior Therapy, 32*, 751–764.

Carr, A. C., & Ghosh, A. (1983). Accuracy of behavioural assessment by computer. *British Journal of Psychiatry, 142*, 66–70.

Chandler, G. M., Burck, H. D., & Sampson, J. P. (1986). A generic computer program for systematic desensitization: Description, construction and case study. *Journal of Behavior Therapy and Experimental Psychiatry, 17*, 171–174.

Chandler, G. M., Burck, H., Sampson, J. P., & Wray, R. (1988). The effectiveness of a generic computer program for systematic desensitization. *Computers in Human Behavior, 4*, 339–346.

Clark, A., Kirkby, K. C., Daniels, B. A., & Marks, I. M. (1998). A pilot study of computer-aided vicarious exposure for obsessive-compulsive disorder. *Australian and New Zealand Journal of Psychiatry, 32*, 268–275.

Climent, C. E., Plutchik, R., Estrada, H., Gaviria, L. F., & Arevalo, W. (1975). A comparison of traditional and symptom-checklist-based histories. *The American Journal of Psychiatry, 132*, 450–453.

Dewis, L. M., Kirkby, K. C., Martin, F., Daniels, B. A., Gilroy, L. J., & Menzies, R. G. (2001). Computer-aided vicarious exposure versus live graded exposure for spider phobia in children. *Journal of Behavior Therapy and Experimental Psychiatry, 32*, 17–27.

Emmelkamp, P. M. G., Bruynzeel, M., Drost, L., & Van Der Mast, C. A. P. G. (2001). Virtual reality treatment in acrophobia: A comparison with exposure in vivo. *CyberPsychology and Behavior, 4*, 335–339.

Emmelkamp, P. M. G., Krijn, M., Hulsbosch, A. M., de Vries, S., Schuemie, M. J., & van der Mast, C. (2002). Virtual reality treatment versus exposure in vivo: A comparative evaluation in acrophobia. *Behaviour Research and Therapy, 40*, 509–516.

Erdman, H. P., Klein, M. H., & Greist, J. H. (1985). Direct patient computer interviewing. *Journal of Consulting and Clinical Psychology, 53*, 760–773.

Erdman, H. P., Klein, M. H., Greist, J. H., Skare, S. S., Husted, J., Robins, L. N., et al. (1992). A comparison of two computer-administered versions of the NIMH Diagnostic Interview Schedule. *Journal of Psychiatric Research, 26*, 85–95.

Fairbairn, A. S., Wood, C. H., & Fletcher, C. M. (1959). Variability in answers to a questionnaire on respiratory symptoms. *British Journal of Preventive and Social Medicine, 13*, 175–193.

Foa, E. B., & Kozak, M. J. (1986). Emotional processing of fear: Exposure to corrective information. *Psychological Bulletin, 99*, 20–35.

Fraser, J., Kirkby, K. C., Daniels, B., Gilroy, L., & Montgomery, I. M. (2001). Three versus six sessions of computer-aided vicarious exposure treatment for spider phobia. *Behaviour Change, 18*, 213–223.

Garb, H. N. (2005). Clinical judgment and decision making. *Annual Review of Clinical Psychology, 1*, 67–89.

Garcia-Palacios, A., Hoffman, H., Carlin, A., Furness, T. A., III, & Botella, C. (2002). Virtual reality in the treatment of spider phobia: A controlled study. *Behaviour Research and Therapy, 40*, 983–993.

Ghosh, A., & Marks, I. M. (1987). Self-treatment of agoraphobia by exposure. *Behavior Therapy, 18*, 3–16.

Gilroy, L. J., Kirkby, K. C., Daniels, B. A., Menzies, R. G., & Montgomery, I. M. (2000). Controlled comparison of computer-aided vicarious exposure versus live exposure in the treatment of spider phobia. *Behavior Therapy, 31*, 733–744.

Gilroy, L. J., Kirkby, K. C., Daniels, B. A., Menzies, R. G., & Montgomery, I. M. (2003). Long-term follow-up of computer-aided vicarious exposure versus live graded exposure in the treatment of spider phobia. *Behavior Therapy, 34*, 65–76.

Glasgow, R. E., & Rosen, G. M. (1978). Behavioral bibliotherapy: A review of self-help behavior therapy manuals. *Psychological Bulletin, 85*, 1–23.

Greist, J. H., Gustafson, D. H., Strauss, F. F., Rowse, G. L., Laughren, T. P., & Chiles, J. A. (1973). A computer interview for suicide-risk prediction. *American Journal of Psychiatry, 130*, 1327–1332.

Greist, J. H., Klein, M. H., Erdman, H. P., Bires, J. K., Bass, S. M., Machtinger, P. E., et al. (1987). Comparison of computer- and interviewer-administered versions of the Diagnostic Interview Schedule. *Hospital and Community Psychiatry, 38*, 1304–1311.

Greist, J. H., Marks, I. M., Baer, L., Kobak, K. A., Wenzel, K. W., Hirsch, M., et al. (2002). Behavior therapy for obsessive-compulsive disorder guided by a computer or by a clinician compared with relaxation as a control. *Journal of Clinical Psychiatry, 63*, 138–145.

Gruber, K., Moran, P. J., Roth, W. T., & Taylor, C. B. (2001). Computer-assisted cognitive behavioral group therapy for social phobia. *Behavior Therapy, 32*, 155–165.

Harcourt, L., Kirkby, K., Daniels, B., & Montgomery, I. (1998). The differential effect of personality on computer-based treatment of agoraphobia. *Comprehensive Psychiatry, 39*, 303–307.

Harris, S. R., Kemmerling, R. L., & North, M. M. (2002). Brief virtual reality therapy for public speaking anxiety. *CyberPsychology and Behavior, 5*, 543–550.

Heading, K., Kirkby, K. C., Martin, F., Daniels, B. A., Gilroy, L. J., & Menzies, R. G. (2001). Controlled comparison of single-session treatments for spider phobia: Live graded exposure alone versus computer-aided vicarious exposure. *Behaviour Change, 18,* 103–113.

Hoffman, H. G., Garcia Palacios, A., Carlin, A., Furness, T. A., III, & Botella Arbona, C. (2003). Interfaces that heal: Coupling real and virtual objects to treat spider phobia. *International Journal of Human Computer Interaction, 16,* 283–300.

Hofmann, S. G., & Spiegal, D. A. (1999). Panic control treatment and its applications. *Journal of Psychotherapy Practice and Research, 8,* 3–11.

Kahan, M., Tanzer, J., Darvin, D., & Borer, F. (2000). Virtual reality-assisted cognitive-behavioral treatment for fear of flying: Acute treatment and follow-up. *CyberPsychology and Behavior, 3,* 387–392.

Kenardy, J., McCafferty, K., & Rosa, V. (2003). Internet-delivered indicated prevention for anxiety disorders: A randomized controlled trial. *Behavioural and Cognitive Psychotherapy, 31,* 279–289.

Kenardy, J. A., Dow, M. G. T., Johnston, D. W., Newman, M. G., Thomson, A., & Taylor, C. B. (2003). A comparison of delivery methods of cognitive-behavioral therapy for panic disorder: An international multicenter trial. *Journal of Consulting and Clinical Psychology, 71,* 1068–1075.

Kenwright, M., Liness, S., & Marks, I. (2001). Reducing demands on clinicians by offering computer-aided self-help for phobia/panic: Feasibility study. *British Journal of Psychiatry, 179,* 456-459.

Kenwright, M., & Marks, I. M. (2004). Computer-aided self-help for phobia/panic via Internet at home: A pilot study. *British Journal of Psychiatry, 184,* 448–449.

Kiernan, W. E., McCreadie, R. G., & Flanagan, W. L. (1976). Trainees' competence in psychiatric case writing. *British Journal of Psychiatry, 129,* 167–172.

Kirkby, K. C., Berrios, G. E., Daniels, B. A., Menzies, R. G., Clark, A., & Romano, A. (2000). Process-outcome analysis in computer-aided treatment of obsessive-compulsive disorder. *Comprehensive Psychiatry, 41,* 259–265.

Klein, B., & Richards, J. C. (2001). A brief Internet-based treatment for panic disorder. *Behavioural and Cognitive Psychotherapy, 29,* 113–117.

Klein, B., Richards, J. C., & Austin, D. W. (2006). Efficacy of Internet therapy for panic disorder. *Journal of Behavior Therapy and Experimental Psychiatry, 37,* 213–238.

Klinger, E., Bouchard, S., Legeron, P., Roy, S., Lauer, F., Chemin, I., et al. (2005). Virtual reality therapy versus cognitive behavior therapy for social phobia: A preliminary controlled study. *CyberPsychology and Behavior, 8,* 76–88.

Kobak, K. A., Reynolds, W. M., & Greist, J. H. (1994). Computerized and clinician assessment of depression and anxiety: Respondent evaluation and satisfaction. *Journal of Personality Assessment, 63,* 173–180.

Krijn, M., Emmelkamp, P. M. G., Biemond, R., de Wilde de Ligny, C., Schuemie, M. J., & van der Mast, C. A. P. G. (2004). Treatment of acrophobia in virtual reality: The role of immersion and presence. *Behaviour Research and Therapy, 42,* 229–239.

Lamson, R. (1994). Virtual therapy of anxiety disorders. *CyberEdge Journal, 4,* 1–28.

Lange, A., Rietdijk, D., Hudcovicova, M., van de Ven, J. P., Schrieken, B., & Emmelkamp, P. M. G. (2003). Interapy: A controlled randomized trial of the standardized treatment of posttraumatic stress through the Internet. *Journal of Consulting and Clinical Psychology, 71,* 901–909.

Lange, A., Schrieken, B., van de Ven, J. P., Bredeweg, B., Emmelkamp, P. M. G., van der Kolk, J., et al. (2000). "Interapy": The effects of a short protocolled treatment of posttraumatic stress and pathological grief through the Internet. *Behavioural and Cognitive Psychotherapy, 28,* 175–192.

Lange, A., van de Ven, J. P., Schrieken, B., & Emmelkamp, P. M. G. (2001). Interapy. Treatment of posttraumatic stress through the Internet: A controlled trial. *Journal of Behavior Therapy and Experimental Psychiatry, 32,* 73–90.

Lapham, S. C., Henley, E., & Skipper, B. J. (1997). Use of computerized prenatal interviews for assessing high-risk behaviors among American Indians. *American Indian and Alaska Native Mental Health Research, 8,* 11–23.

Lewis, G. (1994). Assessing psychiatric disorder with a human interviewer or a computer. *Journal of Epidemiology and Community Health, 48,* 207–210.

Maltby, N., Kirsch, I., Mayers, M., & Allen, G. J. (2002). Virtual reality exposure therapy for the treatment of fear of flying: A controlled investigation. *Journal of Consulting and Clinical Psychology, 70,* 1112–1118.

Marks, I. M., Baer, L., Greist, J. H., Park, J. M., Bachofen, M., Nakagawa, A., et al. (1998). Home self-assessment of obsessive-compulsive disorder: Use of a manual and a computer-conducted telephone interview: Two U.K.–U.S. studies. *British Journal of Psychiatry, 172,* 406–412.

Miller, P. R., Dasher, R., Collins, R., Griffiths, P., & Brown, F. (2001). Inpatient diagnostic assessments: 1. Accuracy of structured vs. unstructured interviews. *Psychiatry Research, 105,* 255–264.

Mouton-Odum, S., Keuthen, N. J., Wagener, P. D., Stanley, M. A., & DeBakey, M. E. (2006). StopPulling.com: An interactive, self-help program for trichotillomania. *Cognitive and Behavioral Practice, 13,* 215–226.

Mühlberger, A., Herrmann, M. J., Wiedemann, G., Ellgring, H., & Pauli, P. (2001). Repeated exposure of flight phobics to flights in virtual reality. *Behaviour Research and Therapy, 39,* 1033–1050.

Mühlberger, A., Wiedemann, G., & Pauli, P. (2003). Efficacy of a one-session virtual reality exposure treatment for fear of flying. *Psychotherapy Research, 13,* 323–336.

Nakagawa, A., Marks, I. M., Park, J. M., Bachofen, M., Baer, L., Dottl, S. L., et al. (2000). Self-treatment of obsessive-compulsive disorder guided by manual and computer-conducted telephone interview. *Journal of Telemedicine and Telecare, 6,* 22–26.

Nelissen, I., Muris, P., & Merckelbach, H. (1995). Computerized exposure and in vivo exposure treatments of spider fear in children: Two case reports. *Journal of Behavior Therapy and Experimental Psychiatry, 26,* 153–156.

Newman, M. G. (2000). Recommendations for a cost-offset model of psychotherapy allocation using generalized anxiety disorder as an example. *Journal of Consulting and Clinical Psychology, 68,* 549–555.

Newman, M. G., Consoli, A. J., & Taylor, C. B. (1999). A palmtop computer program for the treatment of generalized anxiety disorder. *Behavior Modification, 23,* 597–619.

Newman, M. G., Erickson, T., Przeworski, A., & Dzus, E. (2003). Self-help and minimal-contact therapies for anxiety disorders: Is human contact necessary for therapeutic efficacy? *Journal of Clinical Psychology, 59,* 251–274.

Newman, M. G., Kenardy, J., Herman, S., & Taylor, C. B. (1996). The use of handheld computers as an adjunct to cognitive-behavior therapy. *Computers in Human Behavior, 12,* 135–143.

Newman, M. G., Kenardy, J., Herman, S., & Taylor, C. B. (1997). Comparison of palmtop-computer-assisted brief cognitive-behavioral treatment to cognitive-behavioral treatment for panic disorder. *Journal of Consulting and Clinical Psychology, 65,* 178–183.

North, M. M., North, S. M., & Coble, J. R. (1996). Effectiveness of virtual environment desensilizalion in the treatment of agoraphobia. *PRESENCE: Teleoperators and Virtual Environments, 5,* 346–352.

North, M. M., North, S. M., & Coble, J. R. (1997). Virtual reality therapy: An effective treatment for the fear of public speaking. *International Journal of Virtual Reality, 3,* 2–7.

Przeworski, A., & Newman, M. G. (2004). Palmtop computer-assisted group therapy for social phobia. *Journal of Clinical Psychology, 60,* 179–188.

Ready, D. J., Pollack, S., Rothbaum, B. O., & Alarcon, R. D. (2006). Virtual reality exposure for veterans with posttraumatic stress disorder. *Journal of Aggression, Maltreatment & Trauma, 12,* 199–220.

Ressler, K. J., Rothbaum, B. O., Tannenbaum, L., Anderson, P., Graap, K., Zimand, E., et al. (2004). Cognitive enhancers as adjuncts to psychotherapy: Use of D-cycloserine in phobic individuals to facilitate extinction of fear. *Archives of General Psychiatry, 61,* 1136–1144.

Richards, J. C., & Alvarenga, M. E. (2002). Extension and replication of an Internet-based treatment program for panic disorder. *Cognitive Behaviour Therapy, 31,* 41–47.

Richards, J. C., Klein, B., & Austin, D. W. (2006). Internet cognitive behavioural therapy for panic disorder: Does the inclusion of stress management information improve end-state functioning? *Clinical Psychologist, 10,* 2–15.

Romer, D., Hornik, R., Stanton, B., Black, M., Li, X., Ricardo, I., et al. (1997). "Talking" computers: A reliable and private method to conduct interviews on sensitive topics with children. *Journal of Sex Research, 34,* 3–9.

Rothbaum, B. O., Anderson, P., Zimand, E., Hodges, L. F., Lang, D., & Wilson, J. (2006). Virtual reality exposure therapy and standard (in vivo) exposure therapy in the treatment of fear of flying. *Behavior Therapy, 37,* 80–90.

Rothbaum, B. O., Hodges, L. F., Anderson, P. L., Price, L., & Smith, S. (2002). Twelve-month follow-up of virtual reality and standard exposure therapies for the fear of flying. *Journal of Consulting and Clinical Psychology, 70,* 428–432.

Rothbaum, B. O., Hodges, L. F., Kooper, R., Opdyke, D., Williford, J. S., & North, M. (1995). Effectiveness of computer-generated (virtual reality) graded exposure in the treatment of acrophobia. *American Journal of Psychiatry, 152,* 626–628.

Rothbaum, B. O., Hodges, L. F., Ready, D., Graap, K., & Alarcon, R. D. (2001). Virtual reality exposure therapy for Vietnam veterans with posttraumatic stress disorder. *Journal of Clinical Psychiatry, 62,* 617–622.

Sampson, J., James P., Carr, D. L., Panke, J., Arkin, S., Minvielle, M., et al. (2003). An implementation model for web site design and use in counseling and career services. Retrieved October 17, 2008, from Florida State University, Center for the Study of Technology in Counseling and Career Development Web Site: http://www.career.fsu.edu/documents/implementation/Implementation%20Model%20for%20Web%20Site%20Design%20and%20Use.DOC.

Selmi, P. M., Klein, M. H., Greist, J. H., Sorrell, S. P., & Erdman, H. P. (1990). Computer-administered cognitive-behavioral therapy for depression. *American Journal of Psychiatry, 147,* 51–56.

Shaw, S. C., Marks, I. M., & Toole, S. (1999). Lessons from pilot tests of computer self-help for agora/claustrophobia and panic. *MD Computing, 16,* 44–48.

Smith, K. L., Kirkby, K. C., Montgomery, I. M., & Daniels, B. A. (1997). Computer-delivered modeling of exposure for spider phobia: Relevant versus irrelevant exposure. *Journal of Anxiety Disorders, 11,* 489–497.

Supple, A. J., Aquilino, W. S., & Wright, D. L. (1999). Collecting sensitive self-report data with laptop computers: Impact on the response tendencies of adolescents in a home interview. *Journal of Research on Adolescence, 9,* 467–488.

Vincelli, F., Anolli, L., Bouchard, S., Wiederhold, B. K., Zurloni, V., & Riva, G. (2003). Experiential cognitive therapy in the treatment of panic disorders with agoraphobia: A controlled study. *CyberPsychology and Behavior, 6,* 321–328.

Wald, J., & Taylor, S. (2003). Preliminary research on the efficacy of virtual reality exposure therapy to treat driving phobia. *CyberPsychology and Behavior, 6,* 459–465.

Walshe, D. G., Lewis, E. J., Kim, S. I., O'Sullivan, K., & Wiederhold, B. K. (2003). Exploring the use of computer games and virtual reality in exposure therapy for fear of driving following a motor vehicle accident. *CyberPsychology and Behavior, 6,* 329–334.

White, J., Jones, R., & McGarry, E. (2000). Cognitive behavioural computer therapy for the anxiety disorders: A pilot study. *Journal of Mental Health (UK), 9,* 505–516.

Whitfield, G., & Williams, C. (2004). If the evidence is so good—Why doesn't anyone use them? A national survey of the use of computerized cognitive behaviour therapy. *Behavioural & Cognitive Psychotherapy, 32,* 57–65.

Wiederhold, B. K., Jang, D. P., Kim, S. I., & Wiederhold, M. D. (2002). Physiological monitoring as an objective tool in virtual reality therapy. *CyberPsychology and Behavior, 5,* 77–82.

Wiederhold, B. K., & Wiederhold, M. D. (2003). Three-year follow-up for virtual reality exposure for fear of flying. *CyberPsychology and Behavior, 6,* 441–445.

chapter three

Posttraumatic stress disorder

**Carmen P. McLean, Maria M. Steenkamp,
Hannah C. Levy, and Brett T. Litz**

Most people who suffer from posttraumatic stress disorder (PTSD) do not receive care (e.g., Kessler, Sonnega, Bromet, Hughes, & Nelson, 1995; Kulka et al., 1990). How can technology be used to make PTSD care more efficiently delivered and widely available and can the technology be exploited in an evidence-based manner? In this chapter, after briefly reviewing the epidemiology of PTSD and evidence-based psychological treatments for PTSD, we provide a summary and synthesis of the available evidence supporting the use of technology-assisted interventions for PTSD, including Web-based therapist-assisted interventions and virtual reality exposure therapy (VRET). Next, we discuss the advantages and challenges of using this approach with a PTSD population and offer a detailed description of a Web-based early intervention for Marines at risk for PTSD as a case example. We end with some practical suggestions for clinicians interested in technology-assisted intervention strategies.

The phenomenology and epidemiology of PTSD

Exposure to potentially traumatizing events (PTEs) puts anyone at risk for developing posttraumatic adjustment problems. An event or context is considered potentially traumatizing if it is unpredictable, is uncontrollable, and involves a severe or catastrophic violation of fundamental beliefs and expectations about safety, physical integrity, trust, and justice (Everly & Lating, 2004). Examples of PTEs include direct life threats, physical injury, observing violence or extreme suffering, and sexual assault.

Although PTEs are extraordinary, they are not rare. Epidemiological research suggests that 60% of men and 51% of women endure exposure to at least one PTE in their lifetime (Kessler et al. 1995). Despite the ubiquity of exposure to PTEs over the life span, most individuals exposed to PTEs experience an immediate traumatic stress reaction but do not go on to develop an acute stress disorder that greatly interferes with their ability to return to their normal family, social, and work routines (e.g.,

Koopman, Classen, Cardeña, & Spiegel, 1995). However, while most individuals exposed to trauma will recover on their own, a salient and often silent minority do go on to develop PTSD.

PTSD is characterized by intrusive reexperiencing symptoms, avoidance behaviors, and elevated arousal (American Psychiatric Association, 1994). Epidemiological research indicates that approximately 1 in 4 individuals exposed to trauma develop PTSD (Bryant & Harvey, 1998). PTSD is often chronic, disabling, and resistant to treatment (e.g., Kessler et al., 1995; Kulka et al., 1990). Thus, it is important to provide effective interventions to reduce the incidence of chronic PTSD (e.g., Litz, Gray, Bryant, & Adler, 2002).

According to the National Comorbidity Survey, the lifetime prevalence rate of PTSD is 8% (Kessler et al., 1995). In another large epidemiological study, 24% of women exposed to trauma reported current PTSD (Breslau, Davis, Andreski, & Peterson, 1991). Certain groups are at greater risk for developing PTSD: service members exposed to a war zone (e.g., Kulka et al., 1990), emergency medical technicians, police, firefighters, and members of communities or geographical regions affected by natural and man-made disasters (e.g., Davidson & Baum, 1986; Green, 1991). For example, the National Vietnam Veterans Readjustment Study found prevalence rates of current PTSD to be 15.2% and 8.5% for male and female war-exposed veterans, respectively (Kulka et al., 1990). Similarly, recent studies have shown that between 10% and 20% of military personnel returning from Iraq or Afghanistan screened positive for PTSD (e.g., Hoge et al., 2004). Primary and secondary prevention of PTSD is especially critical for these special at-risk groups.

Unique stigma of PTSD

The stigma of PTSD remains one of the most formidable barriers to effective care. Prejudice toward people with mental disorders is prevalent across diverse nationalities and cultures (e.g., Chung, Chen, & Liu, 2001; Hamre, Dahl, & Malt, 1994). Sadly, some mental health and medical professionals hold stigmatizing attitudes toward people with mental illness (Keane, 1990; Lyons & Ziviani, 1995; Scott & Philip, 1985).

Research among military personnel clearly suggests that stigma acts as a potent dissuasion from seeking treatment. Studies have found that military personnel are often unwilling to seek treatment and believe that admitting to a psychological problem would be highly stigmatizing (Britt, 2000; Hoge et al., 2004) and could potentially damage their military careers (Britt, 2000).

Fear of stigmatization is not limited to military populations, however. A recent study (Davis, Ressler, Schwartz, Stephens, & Bradley, 2008) among low-income African American women showed that despite high rates of PTSD (22%) and a clear desire for mental health services, most participants had not received trauma-focused treatment due to personal and institutional barriers to care. Fear of stigmatization and perceived family and community disapproval were particularly highly endorsed as reasons for not seeking care, a finding confirmed in other studies (Atdjian & Vega, 2005; Gary, 2005).

Efficacy of CBT treatment for PTSD

All evidence-based psychological interventions for PTSD involve a combination of exposure to trauma-related memories and contexts, cognitive restructuring, and negative affect and arousal management techniques (e.g., Foa et al., 1999), a set of strategies known as cognitive-behavioral therapy (CBT). Numerous studies, including randomized controlled trials, indicate that CBT is effective in reducing the array of PTSD symptoms (e.g., Bryant, 2000; Bryant & Friedman, 2001; Foa & Meadows, 1997). CBT is effective for acute and chronic PTSD (Bryant, Sackville, Dang, Moulds, & Guthrie, 1999; Foa et al., 1999) and gains are generally maintained at follow-ups of a year or more (see Taylor, 2004). Furthermore, exposure-based CBT such as prolonged exposure has been consistently associated with rapid change and shown to maintain large effect sizes over time (e.g., Foa et al., 2005; Taylor et al., 2003).

Meta-analyses show that CBT is associated with lower dropout rates than pharmacotherapy, and that CBT is equally effective as selective serotonin reuptake inhibitors in the short term (van Etten & Taylor, 1998), although long-term data is sparse. Similar research suggests that CBT is more effective than supportive counseling and short-term psychodynamic therapy. Although CBT has been found to have comparative efficacy as eye movement desensitization and reprocessing (EMDR; Hembree et al., 2003; van Etten & Taylor, 1998), research has shown that the eye movements in EMDR do not contribute to treatment outcome (Davidson & Parker, 2001), which suggests that imaginal exposure plays the most important role in the efficacy of EMDR (Devilly, 2002). These results are consistent with expert consensus that exposure therapy is the most effective and fastest acting psychotherapeutic technique (Ballenger et al., 2000; Foa et al., 1999).

A number of dismantling studies have been conducted in order to gauge the efficacy of various components of cognitive-behavioral interventions. These studies have shown that imaginal exposure, situational exposure, cognitive restructuring, and their combination are more effective than no treatment, supportive counseling, and some forms of relaxation training (e.g., Foa, Rothbaum, Riggs, & Murdock, 1991; Marks, Lovell, Noshirvani, Livanou, & Thrasher, 1998; Taylor et al., 2003).

Barriers to effective care

There is sufficient scientific evidence to justify the widespread routine use of systematic CBT to target PTSD whenever possible. This conclusion is consistent with practice guidelines published by the International Society for Traumatic Stress (Foa, Keane, Friedman, & Cohen, 2008), the American Psychiatric Association (Ursano et al., 2004), and the Departments of Veterans Affairs and Defense (2004). Unfortunately, however, CBT is not widely available or routinely employed outside of specialty clinics and research settings, despite recommendations by numerous practice guidelines that CBT treatments be standardized as the treatment of choice for trauma (e.g., Rosen et al., 2004; Zayfert et al., 2005). Because CBT requires significant professional training and expertise to administer (e.g., Becker, Zayfert, & Anderson, 2004) and the availability of professionals trained in CBT procedures is limited, sufficient resources to provide individual, multisession CBT are scarce.

In addition to being relatively therapist intensive, CBT also requires considerable patient time and resources. As a result, the majority of people who suffer from PTSD are unable to access this form of expert specialty care or they are provided interventions that have little to no evidence supporting their use. The latter is a particularly pernicious problem leading to unnecessary suffering, wasted time and resources, and decreased motivation to seek care in the future.

Given the enormous public health and societal costs associated with chronic PTSD (e.g., Kessler et al., 1995), efficiency of care delivery and the dissemination of evidence-based practices present some of the biggest current challenges for the field. Thus, CBT interventions that can be efficiently delivered and readily received by individuals who are otherwise reluctant to seek mental health care in a specialty setting need to be developed.

Computer-assisted and Internet-assisted treatments for PTSD

The use of computer-based therapeutic interventions for PTSD has increased considerably in the past decade. The development of new methods of treatment delivery has been motivated, in part, by the needs of returning veterans from the Iraq and Afghanistan wars. This influx of veterans has prompted clinicians to explore alternative ways of providing treatment to a population where barriers to care are particularly high (e.g., Bauer, Williford, McBride, McBride, & Shea, 2004; Rosenheck & Stolar, 1998).

One of the alternatives to traditional psychotherapy made possible by modern technology is Internet-based CBT. To examine the efficacy of this treatment delivery method, Lange, Rietdijk, van de Ven, Schrieken, and

Emmelkamp (2003) randomly assigned individuals with mild to severe PTSD symptoms to either an Internet-based CBT program ("Interapy") or a wait-list control group that received psychoeducational materials about PTSD. After 5 weeks of Internet-based treatment (ten 45-minute sessions twice per week) using written exposure therapies supervised by a clinician, participants enrolled in Interapy demonstrated significant improvement in trauma-related symptoms (e.g., avoidance and intrusions) and general psychopathology (e.g., mood and anxiety) as compared to participants in the control condition (Lange et al., 2003).

In a related study, a community sample of 18 trauma-exposed individuals reporting re-experiencing and avoidance symptoms were randomly assigned to an Internet-based CBT program (Hirai & Clum, 2005). The Internet-based CBT condition consisted of psychoeducational materials, breathing and relaxation exercises, cognitive restructuring, and written exposure modules. Participants improved significantly in coping and self-efficacy as compared to a wait-list control group of 18 participants with comparable PTSD symptoms. The intervention involved considerably less clinician involvement relative to previous Internet-based interventions: therapists were available to participants for technological assistance, information on the program's timeline, and encouragement to complete the online assessments, but they did not provide any feedback on the written exposure exercises.

Litz, Engel, Bryant, and Papa (2007) randomly assigned service members with PTSD to one of two therapist-assisted Internet-based conditions: CBT or supportive counseling. All 45 participants had an initial face-to-face meeting with a clinician, learned relaxation exercises, obtained psychoeducation on the benefits of stress management, and had access to therapists via e-mail and telephone. Participants in the CBT intervention were asked to complete a series of self-guided in vivo exposure exercises and also completed seven trauma writing sessions. In contrast, participants in supportive counseling were asked to self-monitor daily nontrauma-related concerns and experiences and to write about them on the Web site. Consistent with conventional treatment outcome research, results showed that participants in the CBT intervention exhibited a sharper decline in total PTSD and depressive symptoms compared to participants in the supportive counseling condition. This suggests that the superior effects of CBT relative to supportive therapy generalize to alternate methods of delivery.

There is encouraging initial support for the efficacy of Internet-based CBT for the treatment of posttraumatic stress disorder. In all of the studies reviewed, participants enrolled in the Internet-based CBT conditions improved significantly in PTSD symptomology as compared to participants in the control conditions. However, no research has compared Internet-based CBT to traditional individual psychotherapy for the treatment of

PTSD. As such, it is unclear whether Internet-based CBT is more, less, or equally efficacious as compared with traditional face-to-face CBT to treat posttraumatic stress. However, worth mentioning is a study conducted by Jacobs and colleagues (2001) that compared a computer-based therapeutic learning program (TLP; Gould, 1989) that combined psychodynamic and CBT strategies to traditional individual psychotherapy. Although participants in the individual psychotherapy condition showed greater improvement in level of pathology and clinical change, participants enrolled in the computer-based TLP program did evidence clinical change across target complaints and decreased pathology as a result of the treatment. With these results in mind, there is thus evidence to suggest that Internet-based CBT may perform well compared with traditional face-to-face CBT. Further research is needed to determine how the degree of human support provided within Internet-based CBT (ranging from self-help to predominately therapist-administered treatment) relates to treatment outcome (see Benight, Ruzek, & Waldrep, 2008).

Virtual reality exposure therapy for PTSD

Virtual reality technology has also been used as a supplement to traditional CBT for the treatment of PTSD. Because imaginal exposure can be difficult to engage in, VRET can help the clinician immerse the patient in a highly controlled computer-generated virtual environment that is specific to the patient's feared situation, with full clinician control of the introduction of new stimuli to move the patient along the fear hierarchy. In this way, virtual reality technology provides an efficacious alternative to traditional imaginal exposure therapy, and could be especially beneficial for those patients who have difficulty engaging in imaginal exposure.

A number of case studies have examined the efficacy of VRET for PTSD. Rothbaum and colleagues (1999) demonstrated that 14 sessions of VRET lead to a significant reduction in the total score on the Clinician-Administered PTSD Scale (CAPS; Blake et al., 1995) among a Vietnam veteran with combat-related PTSD at posttreatment and further improvement at 6-month follow-up. Using only six sessions of VRET, Reger and Gahm (2008) found significant reductions in PTSD Checklist–Military Version (PCL-M; Weathers, Litz, Huska, & Keane, 1994) among active-duty Iraq War infantrymen. A similar case study by Wood and colleagues (2007) found that 10 sessions of VRET given to a military officer with chronic PTSD resulted in lower PCL-M scores and reduced arousal during trauma recall (as measured by skin conductance, peripheral temperature, and heart rate). After six sessions of VRET, Difede and Hoffman (2002) showed that a patient with acute PTSD as a result of the September 11 attacks no longer met criteria for PTSD, major depression, or any other psychiatric disorder as determined by an independent evaluator. The authors note that a previous course of imaginal exposure had not been successful with this patient and suggest that VRET may be particularly useful for patients who have difficulty engaging in traditional exposure therapy.

Although there have been only a handful of larger studies examining the efficacy of VRET for PTSD to date, the results appear promising. A study by Rothbaum, Hodges, Ready, Graap, and Alarcon (2001) examined the effectiveness of VRET among 16 patients with combat-related PTSD who had not responded to imaginal exposure and reported difficulty with emotional engagement during imaginal exposure. The results showed significant reductions in PTSD symptoms, as measured by the CAPS, both at posttreatment and 6-month follow-up. These results are consistent with preliminary work by Hodge and colleagues showing that VRET can effectively facilitate emotional engagement during exposure (Hodges, Rothbaum, Alarcon, Ready, et al., 1999).

Two studies have examined the efficacy of VRET among motor vehicle accident victims. Using VRET that involved real-time driving scenarios, Beck, Paylo, Winer, Schwagler, and Ang (2007) demonstrated significant reductions in re-experiencing, avoidance, and emotional numbing symptoms among six victims of motor vehicle accidents with PTSD. Walshe and colleagues (2003) enrolled 14 victims of motor vehicle accidents who met criteria for either specific phobia of driving or PTSD in a combined VRET plus relaxation and cognitive restructuring program. After 12 sessions, all participants showed significant reductions in subjective distress, driving anxiety, CAPS total score, and depression ratings.

There has been only one randomized controlled trial of VRET for PTSD. This study compared the effects of 14 sessions of VRET among a sample of individuals with PTSD (n = 9) with a wait-list control group (n = 8) following the World Trade Center attacks. The treatment group showed significantly greater decline in CAPS scores at posttreatment compared with wait list (Difede, Cukor, Patt, Giosan, & Hoffman, 2006). Despite the small sample size, this study found large effect size and provides the most rigorous test of the effects of VRET for PTSD to date. Two upcoming studies are designed to examine the specific treatment effects of VRET. The first of these studies will compare VRET, medication (setraline), and combined VERT plus medication (Gamito, Pacheco, Ribeiro, Pablo, & Saraiva, 2005). The second study involves two blinded comparisons: CBT/VRET compared with supportive psychotherapy and relaxation in a virtual environment and CBT/VRET compared with CBT/imaginal exposure (Roy et al., 2006). Findings from these studies will help clarify the unique contribution of VRET relative to standard pharmacologic treatments and CBT techniques (imaginal exposure).

Current developments in virtual reality exposure therapy

Josman and colleagues (2006) are currently developing BusWorld, a virtual reality exposure therapy program for victims of terrorist bus bombings in Israel. The program will consist of 10 VRET sessions of 90 to 120 minutes each, and include other components of CBT such as psychoeducation about common reactions to trauma, breathing skills training, and homework assignments.

In addition to BusWorld, the United States Office of Naval Research is currently funding a project to develop a virtual reality exposure program for returning veterans of Operation Iraqi Freedom and Operation Enduring Freedom with combat-related PTSD (Rizzo, Rothbaum, & Graap, 2007). This virtual reality program will feature both olfactory and tactile stimuli, and will be adjustable to fit weather, time of day, and light conditions. Additionally, the software will include six scenarios relevant to the environment in Iraq and Afghanistan: city scenes, checkpoints, city building interiors, small rural villages, desert bases, and desert road convoys.

Advantages of incorporating technology into the treatment of PTSD

Using technology to increase access to trauma services and overcoming barriers to care

With the majority of North Americans now using the Internet (Porter & Donthu, 2006), computer- and Internet-based treatments substantially increase access to mental health care. These treatments could reach people who are geographically isolated and would otherwise be unlikely to receive services. Rural areas in particular may be underserved, often only provided access to mental health services through general medical practitioners, rather than specialized mental health providers such as psychologists and psychiatrists (Przeworski & Newman, 2006). Given that general practitioners often do not correctly identify PTSD (Taubman-Ben-Ari, Rabinowitz, Feldman, & Vaturi, 2001) and little is often known about PTSD in such settings (Lecrubier, 2004), increased availability of specialized care is needed.

Computer- and Internet-based interventions have great value in circumventing logistical constraints, such as traveling large distances to attend therapy sessions, which would otherwise prevent patients from easily accessing care. For example, veterans who must travel farther and who report travel difficulties are less likely to use VA mental health services (Bauer et al., 2005; Rosenheck & Stolar, 1998). Computer-based treatments increase access to care among individuals who cannot easily travel to weekly sessions due to physical injuries and disabilities stemming from combat or motor vehicle accidents, a population that may have particularly high rates of PTSD (Grieger et al., 2006; Norris, 1992).

The option of computer- and Internet-based treatment is a cost-effective alternative to traditional therapy. Indeed, some of the fastest rates of growth in Internet use have been among people with lower incomes (Porter & Donthu, 2006) because potentially traumatic and other stressful events tend to occur more frequently in low socioeconomic groups (Hatch & Dohrenwend, 2007); the Internet may be an especially effective mode of treatment to high risk individuals. Over and above therapist fees, costs associated with transportation, taking time off work to attend therapy, and paying for child care during therapy sessions often dissuade individuals from seeking care. In one study of mental health care use in veterans returning from the current war, over half of participants reported that being unable to take time off from work to attend therapy sessions was a significant obstacle to receiving treatment (Hoge et al., 2004). The convenience of therapist-assisted computer- and Internet-based therapy may thus be particularly attractive for individuals juggling multiple life and family demands and who do not have the resources—financial or otherwise—to participate in traditional therapy.

Even in situations where quality care is available, affordable, and easily accessible, traumatized individuals may be reluctant to seek help due to shame and fear of stigmatization. Feelings of shame are often associated with traumas such as abuse and sexual assault (Dutra et al., 2008; Ginzburg et al., 2006; Vidal & Petrak, 2007; Wilson, Drozdek, & Turkovic, 2006) and can serve as a significant barrier to care (Lee, Scragg, & Turner, 2001). In addition to shame, other psychological consequences of trauma such as fear, guilt, social impairment, mistrust, and avoidance of thinking and talking about the trauma may in themselves also prevent trauma survivors from seeking face-to-face therapy (Bremner, Quinn, Quinn, & Veledar, 2006). Combined, these factors result in only a minority of trauma survivors seeking care (Koss et al., 1991; Schwarz & Kowalski, 1992). For example, only 12% of crime victims seek mental health services within the first three months of a crime (Norris, Kanisasty, & Scheer, 1990).

The relative anonymity of computer- and Internet-based interventions may be especially attractive to these populations. People feel more "psychologically protected" when doing therapy anonymously (Schultze, 2006) and may be more willing to disclose traumatic events in writing than face to face (Leibert, Archer, Munson, & York, 2006). Given that positive appraisals of coping efficacy have been shown to be important predictors of posttraumatic adjustment (Benight, Cieslak, Molton, & Johnson, 2008), computer- and Internet-based interventions hold the additional benefit of increasing patients' capacity to help themselves, thereby allowing them to regain feelings of mastery, self-efficacy, and control that may have been lost following the trauma.

Using technology to support and enhance the delivery of evidence-based care

Computer-based and Internet-based interventions can enhance and streamline traditional therapy. Therapists can make use of technological advances to maximize the effects of behavioral interventions such as exposure. VRET can be particularly useful in enhancing prolonged exposure for patients who characteristically engage in emotional detachment (Reger & Gahm, 2008). It has been argued, for example, that exposure to trauma cues through VRET can heighten the emotional engagement of the exposure and help activate the underlying fear structure (see Foa & Kozak, 1986), which are important factors in treatment outcome (Reger & Gahm, 2008). In addition to enhancing the visual experience, recent advances in VRET introduce sensory stimuli such as odor and temperature to further heighten the salience of the exposure. More broadly, traditional treatment delivery can also be augmented with DVDs, CD-ROMs, and the Internet. Psychoeducational information provided to patients can be augmented through the use of visuals such as movies, pictures, animated graphics, and interactive programs to enhance understanding of the disorder and promote engagement in treatment (Ritterband et al., 2003).

Another advantage is that technology allows patients to complete the more tedious aspects of CBT, such as self-monitoring, psychoeducation, and skills training, via computer, DVD, or the Internet. This frees up therapist time for providing therapy, increases the cost-effectiveness of standard care, and promotes patient participation in clinical activities outside of session (Taylor & Luce, 2003). An additional administrative advantage is the automation of assessment to help clinicians screen and diagnose cases of PTSD, which requires significantly fewer personnel resources. In circumstances where there is a large need for mental health care and limited therapist resources, as is currently the case in many VA mental health clinics that have seen an influx of returning veterans in recent years, this may be particularly helpful.

More broadly, beyond increasing the accessibility and quality of trauma services in general, computers and the Internet may be used to specifically disseminate evidence-based treatments that have been shown to be most effective in reducing symptoms of posttraumatic stress. CBT is especially well suited to computer and Internet delivery because it is a highly structured, systematic treatment that focuses on concrete behaviors and symptoms (Prezowski & Newman, 2006). As discussed above, there is growing evidence that CBT treatments continue to maintain their efficacy when delivered with limited therapist intervention, and they are significantly more effective when compared to supportive online counseling (e.g., Litz et al., 2007).

Using the Internet to provide education on traumatic stress and to encourage individuals to seek care

The Internet is a particularly valuable tool to provide information on PTSD and trauma to the general public, and to increase knowledge, self-care, and help-seeking behaviors. Through specialized Web sites, people can educate themselves about the psychological consequences of trauma and the treatments available. For example, the National Center for PTSD's Web site (http://www.ncptsd.va.gov) provides the public with comprehensive information on PTSD and other reactions to trauma, including videos, fact sheets, and printable guides on PTSD, as well as information on finding therapists and support groups. There is evidence that traumatized individuals make use of the Internet to look up mental health–related information and high rates of traumatization and PTSD have been found in users of mental health Web sites: a study of users of the Anxiety Disorders Association of America (ADAA) Web site found that of the 1,558 participants, 87% had a history of trauma, and 38% met criteria for current PTSD (Nicholls, Abraham, Connor, Ross, & Davidson, 2006).

Through the use of e-mail, chat rooms, online support groups, and discussion boards, the Internet may also provide a means for social support. In fact, Internet and e-mail use has been associated with decreased risk for PTSD and prolonged grief reactions after bereavement, possibly due to an increased sense of connection with the outside world (Vanderwerker & Prigerson, 2004).

A related final benefit of computer- and Internet-based treatments is that the text of the therapy (be it e-mail exchanges with the therapist, the content of sessions, or ad hoc reading materials) is permanently available to the patient to review and reread (Murphy & Mitchell, 1998). This information can be accessed whenever the patient wishes, and can be digested at his or her own pace, facilitating learning and retention of information and key concepts.

Challenges to incorporating technology into trauma care

Challenges associated with decreased face-to-face interaction

Many of the challenges associated with computer- and Internet-based treatments pertain to decreased contact with the therapist. First, encouraging patient adherence to CBT in the absence of a therapeutic alliance may be challenging. In traditional therapy, the working relationship with the therapist can encourage the patient to stay in treatment in the face of a difficult and painful healing process; this relationship is not present in computer- and Internet-based treatments. Successfully completing a computer- or Internet-based CBT program for trauma would thus require significant commitment and self-discipline on the part of the patient. However, it is encouraging to note that dropout rates from Internet-based therapist-assisted treatments for PTSD are similar to those of regular CBT interventions (30%; Litz et al., 2007), and studies of other anxiety disorders have shown that it is possible to obtain high rates of adherence with Internet-based treatments when adding even minimal therapist contact, such as weekly telephone calls (Carlbring et al., 2007). Treatment adherence can be maximized through initial phone calls with the patient, by scheduling phone calls or e-mails before difficult parts of the treatments (such as exposure), and by encouraging patients to contact the therapist via phone or e-mail when feeling discouraged.

Second, certain therapeutic techniques associated with CBT may be difficult to apply successfully without the presence of the clinician. For example, even traditional, face-to-face exposure therapy requires great clinical skill in order to facilitate the necessary emotional engagement with raw and painful emotions such as fear, horror, loss, guilt, and shame (Jaycox, Foa, & Morral, 1998). Computer- and Internet-based interventions employing exposure-based techniques will need to pay careful attention to how exposures are explained and framed in order to maximize the likelihood of attaining sustained emotional engagement. Being mindful of potential pitfalls, such as avoidance, and addressing them in explanations of the exercise will be important.

Other exercises may be aimed at meaning making rather than extinction of fear responses. In such exercises, the patient is encouraged to find meaning in the trauma and to think about it in different, more adaptive ways. In the case of trauma stemming from the loss of a buddy in combat, for example, the patient may be asked to write a letter to the deceased friend conveying how his or her death has affected the patient, or the patient may be encouraged to write a letter to his or her friend's relatives, expressing feeling for the deceased friend. In a similar vein, cognitive restructuring exercises consisting of prewritten Socratic questions can encourage patients to seek out different points of view on bothersome situations. Overall, computer- and Internet-based interventions should not be shackled to the techniques of traditional therapy; there is much room for creative and novel ways of achieving the overarching aims of emotional processing and adaptive thinking.

It should be noted that the "human element" of therapy is arguably more salient in PTSD treatments than for most other disorders. Treating trauma requires the patient to self-disclose and engage with painful memories and emotions that have been avoided (Litz & Salters-Pedneault, 2007). The intimate human context and the therapist's bearing witness to poignant, raw, and compelling stories of trauma is particularly important in cases where the trauma involved interpersonal malice, abuse, or violence. Empathic, caring, and nonjudgmental reactions during the retelling and reliving of the trauma are in themselves corrective, as is the therapist's continued positive regard for the patient (Litz & Salters-Pendeault, 2007). However, there is evidence that disclosure does not need to be face to face to yield positive effects—experiments show that when writing narratives of traumatic events, simply knowing that the narrative will be read by someone else is sufficient to result in decreases in cognitive intrusions and avoidance (Radcliffe, Lumley, Kendall, Stevenson, & Beltran, 2007).

Studies of patients' perceptions of online counseling confirm that although the absence of a relationship with the therapist is considered a disadvantage by patients, for many this is outweighed by greater disinhibition and ease with self-disclosure (Leibert et al., 2006). Online relationships may thus result in higher levels of self-disclosure than in face-to-face therapy and have been shown to equal face-to-face relationships in depth and breadth of the relationship (Parks & Roberts, 1998). As such, a therapeutic alliance can be adequately established within online treatments: comparisons between Internet-based treatments and traditional treatments show comparable levels of working alliance between patient and therapist across both modalities (Cook & Doyle, 2002). In addition, Knaevelsrud and Maercker (2006) examined the quality of the working alliance in online therapy with traumatized adults and found that it was possible to establish a positive and stable therapeutic relationship online. However, unlike traditional therapy, patients' perceptions of the working alliance did not predict treatment outcome. This suggests that the working alliance may be a less-relevant predictor of treatment outcome than in face-to-face approaches.

The challenge of creating flexible computer- and Internet-based treatments

In traditional face-to-face therapy, the clinician tailors the intervention in an idiographic, iterative, and evolving process. Because the content of protocol-driven computer- and Internet-based interventions is more static than that of traditional therapy, where the content of the session can be changed from moment to moment, the fit between the patient and the treatment in Internet-based programs may be more rudimentary. Nonverbal cues such as facial expressions, tone of voice, and level of eye contact, which provide the therapist with important information on the patient's emotional state in the moment, are not available in online treatments. However, there is some flexibility in the ways that computer- and Internet-based treatments can be delivered, and these treatments are becoming increasingly nuanced and sophisticated. For example, computer- and Internet-based treatments may begin with an assessment and then, based on the results of the assessment, be tailored according to the level of distress and symptoms present for that particular individual. A number of suggestions have also been made for overcoming the lack of nonverbal cues in Internet or e-mail-based therapy (see Murphy & Mitchell, 1998), such as *emotional bracketing* in which, during e-mail exchanges with the patient, the therapist puts the intended emotional material in brackets (such as "concern" or "worry") after sentences. It is also possible to individualize computer- and Internet-based treatments according to the progress the patient has made and the activities that they have completed. For example, pop-up messages may be used to steer patients toward pertinent reading materials and exercises, based on symptom scores.

Challenges regarding quality control of information and services

Although the Internet is an excellent resource for people to educate themselves about the psychological effects of trauma, a disadvantage is that anyone can post information or treatments for PTSD online: there is no vetting, oversight, or control over the quality of the information provided. A recent study of the quality of trauma-related Web sites found that psychiatrists and psychologists are involved in the development of only 20% of the sites available on psychological trauma, and that only a minority of sites (18%) draw from peer-reviewed literature, with most of the sites not providing a source for their information. Of the sites reviewed, 42% provided unhelpful information, and 6% of sites provided harmful information about adjustment to psychological trauma (Bremner et al., 2006).

Ethical concerns

Finally, another challenge associated with computer- and Internet-based interventions relates to protecting patient privacy and confidentiality. This is especially pertinent in trauma-focused work, where issues of perpetration and culpability are often at stake. During the course of computer- and Internet-based treatments for PTSD, patients may be asked to disclose very private information about themselves, for example, when providing a written narrative account detailing the trauma. Despite advances in encryption technology, it must be made clear to patients that confidentiality cannot be fully guaranteed (Taylor & Luce, 2003). Internet communication such as e-mail and Web sites visited is available to system administrators who monitor customer use, and many companies and government agencies monitor the use of the Internet by their employees (Shaw & Shaw, 2006). It is the mental health provider's responsibility to ensure that the technology being used is as secure as possible, and that the possibility of third-party interception of information and unauthorized disclosure of patient information is minimized. Information should also be clearly provided regarding circumstances under which confidentiality will be breached for legal reasons, for example, in the case of ongoing child or elder abuse.

Case example: DE-STRESS

As an example of an Internet-based treatment of PTSD, we will discuss DE-STRESS, a therapist-assisted self-management intervention that uses the Internet as a means of promoting, prompting, and monitoring applied stress management. DE-STRESS is short for DElivery of Self-TRaining and Education for Stressful Situations, a title aimed at reducing stigma and emphasizing the self-management aspect of cognitive behavioral therapy. DE-STRESS uses CBT techniques in a self-management framework to facilitate adaptive coping in the face of negative affect and arousal. In particular, DE-STRESS is designed to teach patients strategies to help them cope with and manage their reactions to situations that trigger recall of traumatic experiences. Overall, the goal is to reduce PTSD symptoms and to promote greater self-efficacy and confidence in coping capacities. The 18-session treatment involves only one face-to-face session with a therapist (or "coach"), followed by scheduled and as-needed e-mail and telephone contact with the coach. Patients are provided with a password to a secure Web site, which they are asked to log on to three times a week for 6 weeks. Logons typically require about 15 minutes, followed by homework exercises of varying duration. Patients may log on to the Web site anytime and as many times as they wish. At each logon, they are asked to make daily ratings of their PTSD symptoms (using a modified PTSD checklist [PCL]; Weathers, Litz, Herman, Huska, & Keane, 1993) and a global rating of their level of depression. If a patient endorses severe depression, coaches are automatically notified via e-mail to contact the patient. Patients can also request a phone call from a coach or e-mail their coaches at any time with questions or comments. Coaches have access to a special administrative interface of the DE-STRESS Web site, where they can monitor compliance and symptom levels.

A large component of the intervention involves psychoeducation. Patients are provided ad lib access to educational information about PTSD, stress, and trauma on the Web site, as well as information on common comorbid problems and symptoms they might experience post-trauma (for example, depression or survivor guilt). Patients are also provided unrestricted access to information on strategies to manage their anger and sleep hygiene, as well as in-depth information on how to perform and practice two easy stress management strategies, diaphragmatic breathing and progressive muscle relaxation. Patients also obtain information about cognitive reframing techniques, such as how to challenge unhelpful thought patterns and alter self-talk to effectively manage demanding situations.

Upon logging into the Web site, patients see a To-Do List for that day. At each logon, the list will include daily symptom reporting, questions about homework experiences (and compliance), and a description of the next homework assignment. After the first day, each time patients log on to the DE-STRESS site, they will be prompted to enter responses from the previous day's homework assignment. If patients have not completed a given homework assignment, they are encouraged to do so before continuing to the next assignment. The site is designed to proceed to the next assignment only after patients enter data from the previous day's homework.

The specific components of DE-STRESS are as follows:

1. Self-monitoring of situations that trigger trauma-related distress (first 2 weeks)
2. Generating a hierarchy of these trigger contexts in terms of their degree of threat or avoidance (starting week 3)
3. Didactics on stress management strategies that, once practiced (starting day 1), are used for:
4. Graduated, self-guided, in vivo exposure to items from the personalized hierarchy (starting with the least threatening or least avoided item in week 3)
5. Seven online trauma writing sessions (week 7)
6. A review of progress (charts of daily symptom reports are presented), a series of didactics on relapse prevention, and the generation of a personalized plan for future challenges (week 8)

In the initial face-to-face meeting with the therapist, a preliminary hierarchy of stressful situations is generated collaboratively. It is assumed that most patients with PTSD are not sufficiently aware of what triggers the recall of trauma memories and, as a result, in the first two weeks the patient is asked to monitor his or her reactions to various stressful and demanding situations. After this period, the therapist assists the patient in generating a final personalized stress hierarchy via e-mail. In the initial session, the therapist also provides training in two stress-management strategies (diaphragmatic breathing and progressive muscle relaxation) and initial training in simple cognitive reframing techniques (how to challenge unhelpful thought patterns and alter self-talk to effectively manage demanding situations). Subsequently, homework assignments are given online to further build upon these skill sets.

As part of the treatment protocol, therapists telephone participants at the end of week six of treatment (at which time they have learned relaxation strategies and completed their hierarchy of exposure to avoided situations) to determine if they are ready to do the trauma narrative portion of intervention. If deemed ready, in week seven patients are asked to write (type) a detailed first-person, present-tense account of a particularly salient and troubling traumatic experience. They are then asked to read this trauma narrative and are encouraged to experience any emotions or memories that they may have been avoiding, while using coping skills acquired earlier in the program. In the subsequent six logons, patients rewrite and process their traumatic memory, including any other memories that might not have occurred to them initially. The writing task is a variant of techniques employed in cognitive behavior therapy and cognitive processing therapy to target PTSD. The goal is to promote mastery of the memory (the person learns that he can allow himself to fully remember the event without any catastrophic outcomes occurring) and to reduce avoidance, as opposed to maximizing emotional processing to promote extinction, as is the case in exposure therapy. Coping skills learned within the first two weeks are applied to minimize distress while writing and reading the narrative.

Early randomized controlled trials have demonstrated the efficacy of DE-STRESS (see Litz, Bryant, Williams, Wang, & Engel, 2004, and Litz et al., 2007; see above) and studies testing DE-STRESS in active-duty military personnel and veterans returning from the current wars are currently ongoing.

Conclusion

Only a minority of individuals with PTSD receive evidence-based treatment. Untreated or undertreated, PTSD is associated with significant adverse health and life consequences. CBT with exposure therapy is the preferred nonpharmacologic therapy, but there are significant barriers to care including lack of resources, scarcity of trained clinicians, and fear of stigmatization that limit the efficiency of evidence-based treatment delivery. Technology-assisted mental health interventions, including computer- and Internet-based treatment and virtual reality technology can greatly reduce these barriers.

There are a number of significant advantages to Internet-based treatments for PTSD including overcoming barriers to receiving trauma-related care (such as cost and time constraints associated with traditional therapy), supporting and enhancing evidence-based interventions (for example, augmenting exposures through VRET), increasing the accessibility and quality of trauma services, and providing education on traumatic stress to the general public and encouraging people to seek needed care. Challenges to Internet-based PTSD treatment include difficulties associated with decreased face-to-face interaction between the patient and therapist (such as possible difficulty maintaining patient adherence or creating emotional engagement in exposures), challenges in creating flexible interventions that will allow CBT principles and techniques to be optimally delivered, challenges regarding lack of oversight and quality control in the information and services provided over the Internet, and lastly, challenges regarding ethical concerns, particularly privacy and confidentiality. As was described, however, these obstacles are not insurmountable and there is evidence for the efficacy of Internet-based treatments of PTSD in spite of the unique challenges associated with this form of treatment.

Virtual reality technology provides clinicians with full control over the selection of feared scenarios and introduction of stimuli to increase the realism of the exposure. These modifications all help to reduce clinicians' administrative role, make evidence-based treatments more accessible to individuals with PTSD, and therefore increase the efficiency of evidence-based care delivery.

Research examining the utility and efficacy of computer- and Internet-based interventions for PTSD is beginning to accumulate. Results suggest that VRET is effective in reducing PTSD symptomology and may be especially useful for patients who have difficulty engaging in imaginal exposure. Data from research examining combat PTSD are particularly encouraging given that the wars in Iraq and Afghanistan have exposed active-duty service members to high rates of potentially traumatic events and PTSD. Although further controlled trials are needed, these preliminary findings offer encouraging initial support for the efficacy of treatments that use innovative technology to supplement well-established CBT techniques in treating PTSD.

References

American Psychiatric Association (1994). *Diagnostic and statistical manual of mental disorders*, 4th edition. Washington DC: Author.

Atdjian, S., & Vega, W. A. (2005). Disparities in mental health treatment in U.S. racial and ethnic minority groups: Implications for psychiatrists. *Psychiatric Services, 56*(12), 1600–1602.

Ballenger, J. C., Davidson, J. R. T., Lecrubier, Y., Nutt, D. J., Foa, E. B., Kessler, R. C., et al. (2000). Consensus statement on posttraumatic stress disorder from the international consensus group on depression and anxiety. *Journal of Clinical Psychiatry, 61*(Suppl 5), 60–66.

Bauer, M., Williford, W., McBride, L., McBride, K., & Shea, N. (2005). Perceived barriers to health care access in a treated population. *International Journal of Psychiatry in Medicine, 35*(1), 13–26.

Becker, C. B., Zayfert, C., & Anderson, E. (2004). A survey of psychologists' attitudes towards and utilization of exposure therapy for PTSD. *Behaviour Research and Therapy, 42*(3), 277–292.

Benight, C. C., Cieslak, R., Molton, I. R., & Johnson, L. E. (2008). Self-evaluative appraisals of coping capability and posttraumatic stress following motor vehicle accidents. *Journal of Consulting and Clinical Psychology, 76,* 677–685.

Benight, C., Ruzek, J., & Waldrep, E. (2008). Internet interventions for traumatic stress: A review and theoretically based example. *Journal of Traumatic Stress, 21*(6), 513–520.

Bremner, J., Quinn, J., Quinn, W., & Veledar, E. (2006). Surfing the Net for medical information about psychological trauma: An empirical study of the quality and accuracy of trauma-related websites. *Medical Informatics and The Internet in Medicine, 31*(3), 227–236.

Breslau, N., Davis, G. C., Andreski, P., & Peterson, E. (1991). Traumatic events and posttraumatic stress disorder in an urban population of young adults. *Archives of General Psychiatry, 48*(3), 216–222.

Britt, T. W. (2000). The stigma of psychological problems in a work environment: Evidence from the screening of service members returning from Bosnia. *Journal of Applied Social Psychology, 30*(8), 1599–1618.

Bryant, R. A. (2000). Cognitive behavioral therapy of violence-related posttraumatic stress disorder. *Aggression and Violent Behavior, 5*(1), 79–97.

Bryant, R., & Harvey, A. (1998). Relationship between acute stress disorder and posttraumatic stress disorder following mild traumatic brain injury. *American Journal of Psychiatry, 155*(5), 625–629.

Bryant, R. A., & Friedman, M. (2001). Medication and non-medication treatments of post-traumatic stress disorder. *Current Opinion in Psychiatry, 14*(2), 119–123.

Bryant, R. A., Harvey, A. G., Dang, S., Sackville, T., & Basten, C. (1998). Treatment of acute stress disorder: A comparison of cognitive-behavioral therapy and supportive counseling. *Journal of Consulting and Clinical Psychology, 66,* 862–866.

Bryant, R. A., Sackville, T., Dang, S. T., Moulds, M., & Guthrie, R. (1999). Treating acute stress disorder: An evaluation of cognitive behavior therapy and supporting counseling techniques. *American Journal of Psychiatry, 156*(11), 1780–1786.

Carlbring, P., Gunnarsdottir, M., Hedensjo, L., Andersson, G., Ekselius, L., & Furmark, T. (2007). Treatment of social phobia: Randomised trial of Internet-delivered cognitive-behavioural therapy with telephone support. *The British Journal of Psychiatry, 190,* 123–128.

Chung, K. F., Chen, E. Y. H., & Liu, C. S. M. (2001). University students' attitudes towards mental patients and psychiatric treatment. *International Journal of Social Psychiatry, 47*(2), 63–72.

Cook, J. E., & Doyle, C. (2002). Working alliance in online therapy as compared to face-to-face therapy: Preliminary results. *CyberPsychology & Behavior, 5,* 95–105.

Davidson, L. M., & Baum, A. (1986). Chronic stress and posttraumatic stress disorders. *Journal of Consulting and Clinical Psychology, 54*(3), 303–308.

Davidson, P. R., & Parker, K. C. H. (2001). Eye movement desensitization and reprocessing (EMDR): A meta-analysis. *Journal of Consulting and Clinical Psychology, 69*(2), 305–316.

Davis, R. G., Ressler, K. J., Schwartz, A. C., Stephens, K. J., & Bradley, R. G. (2008). Treatment barriers for low-income, urban African Americans with undiagnosed posttraumatic stress disorder. *Journal of Traumatic Stress, 21*(2), 218–222.

Devilly, G. J. (2002). Eye movement desensitization and reprocessing: A chronology of its development and scientific standing. *The Scientific Review of Mental Health Practice, 1*(2), 113–138

Difede, J., Cukor, J., Patt, I., Giosan, C., & Hoffman, H. (2006). The application of virtual reality to the treatment of PTSD following the WTC attack. *Psychobiology of posttraumatic stress disorders: A decade of progress* (Vol. 1071) (pp. 500–501). Malden, MA US: Blackwell Publishing.

Difede, J., & Hoffman, H. (2002, December). Virtual reality exposure therapy for World Trade Center post-traumatic stress disorder: A case report. *CyberPsychology & Behavior, 5*(6), 529–535.

Everly, G., & Lating, J. (2004). The defining moment of psychological trauma: What makes a traumatic event traumatic? In *Personality-guided therapy for posttraumatic stress disorder* (pp. 33–51). Washington, DC: American Psychological Association.

Foa, E. B., Dancu, C. V., Hembree, E. A., Jaycox, L. H., Meadows, E. A., & Street, G. P. (1999). A comparison of exposure therapy, stress inoculation training, and their combination for reducing posttraumatic stress disorder in female assault victims. *Journal of Consulting and Clinical Psychology, 67,* 194–200.

Foa, E. B., Hembree, E. A., Cahill, S. P., Rauch, S. A. M., Riggs, D. S., Feeny, N. C., et al. (2005). Randomized trial of prolonged exposure for posttraumatic stress disorder with and without cognitive restructuring: Outcome at academic and community clinics. *Journal of Consulting and Clinical Psychology, 73*(5), 953–964.

Foa, E. B., Keane, T. M., & Friedman, M. J. (2008). *Effective treatments for PTSD: Practice guidelines from the International Society for Traumatic Stress Studies.* New York: Guilford Press.

Foa, E. B., & Kozak, M. J. (1986). Emotional processing of fear: Exposure to corrective information. *Psychological Bulletin, 99*(1), 20–35.

Foa, E. B., & Meadows, E. A. (1997). Psychosocial treatments for posttraumatic stress disorder: A critical review. *Annual Review of Psychology, 48,* 449–480.

Foa, E. B., Rothbaum, B. O., Riggs, D. S., & Murdock, T. B. (1991). Treatment of posttraumatic stress disorder in rape victims: A comparison between cognitive-behavioral procedures and counseling. *Journal of Consulting and Clinical Psychology, 59*(5), 715–723.

Gamito, P., Pacheco, J., Ribeiro, C., Pablo, C., & Saraiva, T. (2005). Virtual war PTSD: A methodological thread. *Annual Review of CyberTherapy and Telemedicine, 3,* 173–178.

Gary, F. A. (2005). Stigma: Barrier to mental health care among ethnic minorities. *Issues in Mental Health Nursing, 26*(10), 979–999.

Gould, R. L. (1989). Therapeutic Learning Program (Version 5.0) [Computer software]. Santa Monica, CA: Interactive Health Systems.

Green, B. (1991). Evaluating the effects of disasters. *Psychological Assessment: A Journal of Consulting and Clinical Psychology, 3*(4), 538–546.

Grieger, T. A., Cozza, S. J., Ursano, R. J., Hoge, C., Martinez, P. E., Engel, C. C., et al. (2006). Posttraumatic stress disorder and depression in battle-injured soldiers. *American Journal of Psychiatry, 163,* 1777–1783.

Hamre, P., Dahl, A. A., & Malt, U. F. (1994). Public attitudes to the quality of psychiatric treatment, psychiatric patients, and prevalence of mental disorders. *Nordic Journal of Psychiatry, 48*(4), 275–281.

Hatch, S. L., & Dohrenwend, B. P. (2007). Distribution of traumatic and other stressful life events by race/ethnicity, gender, SES and age: A review of the research. *American Journal of Community Psychology, 40,* 313–332.

Hembree, E. A., Foa, E. B., Dorfan, N. M., Street, G. P., Kowalski, J., & Xin T. (2003). Do patients drop out prematurely from exposure therapy for PTSD? *Journal of Traumatic Stress, 16*(6), 555–562

Hirai, M., & Clum, G. A. (2005). An Internet-based self-change program for traumatic event–related fear, distress, and maladaptive coping. *Journal of Traumatic Stress, 18,* 631–636.

Hodges, L., Rothbaum, B., Alarcon, R., Ready, D., Shahar, F., Graap, K., et al. (1999). A virtual environment for the treatment of chronic combat-related post-traumatic stress disorder. *CyberPsychology & Behavior, 2*(1), 7–14.

Hoge, C. W., Castro, C. A., Messer, S. C., McGurk, D., Cotting, D. I., & Koffman, R. L. (2004). Combat duty in Iraq and Afghanistan, mental health problems, and barriers to care. *New England Journal of Medicine, 351,* 13–22.

Jacobs, M. K., Christensen, A., Huber, A., Snibbe, J. R., Dolezal-Wood, S., & Polterak, A. (2001). A comparison of computer-based versus traditional individual psychotherapy. *Professional Psychology: Research and Practice, 32,* 92–96.

Jaycox, L. H., Foa, E. B., & Morral, A. R. (1998). Influence of emotional engagement and habituation on exposure therapy for PTSD. *Journal of Consulting and Clinical Psychology, 66,* 185–192.

Josman, N., Somer, E., Reisberg, A., Weiss, P. L., Garcia-Palacios, A., & Hoffman, H. (2006). BusWorld: Designing a virtual environment for post-traumatic stress disorder in Israel: A protocol. *CyberPsychology and Behavior, 9,* 241–244.

Keane, M. (1990). Contemporary beliefs about mental illness among medical students: Implications for education and practice. *Academic Psychiatry, 14*(3), 172–177.

Kessler, R. C., Sonnega, A., Bromet, E., Hughes, M., & Nelson, C. B. (1995). Posttraumatic stress disorder in the National Comorbidity Survey. *Archives of General Psychiatry, 52,* 1048–1060.

Knaevelsrud, C., & Maercker, A. (2006). Does the quality of the working alliance predict treatment outcome in online psychotherapy with traumatized patients? *Journal of Medical Internet Research, 8,* 1–9.

Koopman, C., Classen, C., Cardeña, E., & Spiegel, D. (1995). When disaster strikes, acute stress disorder may follow. *Journal of Traumatic Stress, 8*(1), 29–46.

Kulka, R. A., Schlenger, W. E., Fairbank, J. A., Hough, R. L., Jordan, B. K., Marmar, C. R., & Weiss, D. S. (1990). *Trauma and the Vietnam War generation.* New York: Brunner/Mazel.

Lecrubier, Y. (2004). Posttraumatic stress disorder in primary care: A hidden diagnosis. *Journal of Clinical Psychiatry, 65,* 49–54.

Lee, D. A, Scragg, P., & Turner, S. (2001). The role of shame and guilt in traumatic events: A clinical model of shame-based and guilt-based PTSD. *British Journal of Medical Psychology, 74,* 451–466.

Leibert, T., Archer, J., Munson, J., & York, G. (2006). An exploratory study of client perceptions of Internet counseling and the therapeutic alliance. *Journal of Mental Health Counseling, 28*(1), 69–83.

Litz, B. T., Bryant, R., Williams, L., Wang, J., & Engel Jr., C. C. (2004). A therapist-assisted Internet self-help program for traumatic stress. *Professional Psychology: Research and Practice, 35,* 628–634.

Litz, B. T., Engel, C. C., Bryant, R. A., & Papa, T. (2007). A randomized, controlled proof-of-concept trial of an Internet-based, therapist-assisted self-management treatment for posttraumatic stress disorder. *American Journal of Psychiatry, 164,* 1676–1683.

Litz, B. T., Gray, M. J., Bryant, R., and Adler, A. B. (2002). Early intervention for trauma: Current status and future directions. *Clinical Psychology: Science and Practice, 9,* 112–134.

Litz, B. T., & Salters-Pedneault, K. (2007). The art of evidence-based treatment of trauma survivors. In S. G. Hofmann & J. Weinberger (Eds.), *The art and science of psychotherapy.* New York: Routledge/ Taylor & Francis.

Lyons, M., & Ziviani, J. (1995). Stereotypes, stigma, and mental illness: Learning from fieldwork experiences. *American Journal of Occupational Therapy, 49*(10), 1002–1008.

Marks, I., Lovell, K., Noshirvani, H., Livanou, M., & Thrasher, S. (1998). Treatment of posttraumatic stress disorder by exposure and/or cognitive restructuring: A controlled study. *Archives of General Psychiatry, 55*(4), 317–325.

Murphy, L. J., & Mitchell, D. L. (1998). When writing helps to heal: E-mail as therapy. *British Journal of Guidance & Counselling, 26,* 21–32.

Nicholls, P. J., Abraham, K., Connor, K. M., Ross, J., & Davidson, J. R. T. (2006). Trauma and posttraumatic stress in users of the Anxiety Disorders Association of America Web site. *Comprehensive Psychiatry, 47,* 30–34.

Norris, F. (1992). Epidemiology of trauma: Frequency and impact of different potentially traumatic events on different demographic groups. *Journal of Consulting and Clinical Psychology, 60,* 409–418.

Norris, F. H., Kanisasty, K. Z., & Scheer, D. A. (1990). Use of mental health services among victims of crime: Frequency, correlates and subsequent recovery. *Journal of Consulting and Clinical Psychology, 58,* 538–547.

Parks, M. R., & Roberts, L. D. (1998). Making MOOsic: The development of personal relationships on line and a comparison to their offline counterparts. *Journal of Social and Personal Relationships, 15,* 517–537.

Porter, C. E., & Donthu, N. (2006). Using the technology acceptance model to explain how attitudes determine Internet usage: The role of perceived access barriers and demographics. *Journal of Business Research, 59*, 999–1007.

Przeworski, A., & Newman, M. G. (2006). Efficacy and utility of computer-assisted cognitive behavioral therapy for anxiety disorders. *Clinical Psychologist, 10*, 43–53.

Radcliffe, A. M., Lumley, M. A., Kendall, J., Stevenson, J. K., & Beltran, J. (2007). Written emotional disclosure: Testing whether social disclosure matters. *Journal of Social and Clinical Psychology, 26*, 362–384.

Reger, G. M., & Gahm, G. A. (2008). Virtual reality exposure therapy for active duty soldiers. *Journal of Clinical Psychology: in session, 64*, 940–946.

Ritterband, L., Gonder-Frederick, L., Cox, D., Clifton, A., West, R., & Borowitz, S. (2003). Internet interventions: In review, in use, and into the future. *Professional Psychology: Research and Practice, 34*(5), 527–534.

Rizzo, A., Rothbaum, B. O., & Graap, K. (2007). Virtual reality applications for the treatment of combat-related PTSD. In C. R. Figley & W. P. Nash, W. P. (Eds.), *Combat Stress Injury: Theory, Research, and Management* (pp. 183–204). New York: Taylor and Francis Books.

Rosen, C. S., Chow, H. C., Finney, J. F., Greenbaum, M. A., Moos, R. H., Sheikh, J. I., et al. (2004). VA practice patterns and practice guidelines for treating posttraumatic stress disorder. *Journal of Traumatic Stress, 17*(3), 213–222.

Rosenheck, R., & Stolar, M. (1998). Access to public mental health services: Determinants of population coverage. *Medical Care, 36*(4), 503–512.

Rothbaum, B. O., Hodges, L., Alarcon, R., Ready, D., Shahar, F., Graap, K., et al. (1999). Virtual reality exposure therapy for PTSD Vietnam veterans: A case study. *Journal of Traumatic Stress, 12*, 263–271.

Rothbaum, B. O., Hodges, L. F., Ready, D., Graap, K., & Alarcon, R. D. (2001). Virtual reality exposure therapy for Vietnam veterans with posttraumatic stress disorder. *Journal of Clinical Psychiatry, 62*(8), 617–622.

Roy, M. J., Law, W., Patt, I., Difede, J., Rizzo, A., Graap, K., et al. (2006). Randomized controlled trial of CBT with virtual reality exposure therapy for PTSD. *Annual Review of CyberTherapy and Telemedicine, 4*, 39–44.

Schultze, N. (2006). Success factors in Internet-based psychological counseling. *CyberPsychology & Behavior, 9*(5), 623–626.

Schwarz, E., & Kowalski, J. (1992). Personality characteristics and posttraumatic stress symptoms after a school shooting. *Journal of Nervous and Mental Disease, 180*(11), 735–737.

Scott, D. J., & Philip, A. E. (1985). Attitudes of psychiatric nurses to treatment and patients. *British Journal of Medical Psychology, 58*(2), 169–173.

Shaw, H. E., & Shaw, S. F. (2006). Critical ethical issues in online counseling: Assessing current practices with an ethical intent checklist. *Journal of Counseling & Development, 84*, 41–53.

Taubman-Ben-Ari, O., Rabinowitz, J., Feldman, D., & Vaturi, R. (2001). Posttraumatic stress disorder in primary-care settings: Prevalence and physicians' detection. *Psychological Medicine, 31*, 555–60.

Taylor, C. B., & Luce, K. H. (2003). Computer and Internet-based psychotherapy interventions. *Current Directions in Psychological Science, 12*, 18–22.

Taylor, S. (2004). *Advances in the treatment of posttraumatic stress disorder: Cognitive-behavioral perspectives*. New York: Springer Publishing Co.

Taylor, S., Thordarson, D. S., Maxfield, L., Fedoroff, I. C., Lovell, K., & Ogrodniczuk, J. (2003). Comparative efficacy, speed, and adverse effects of three PTSD treatments: Exposure therapy, EMDR, and relaxation training. *Journal of Consulting and Clinical Psychology, 71*(2), 330–338.

Ursano, R. J., Bell, C., Eth, S., Friedman, M., Norwood, A., et al., for the Work Group on ASD and PTSD. (2004). *Practice guideline for the treatment of patients with acute stress disorder and posttraumatic stress disorder.* Arlington, VA: American Psychiatric Association, 1–95.

U.S. Department of Veterans Affairs & U.S. Department of Defense. (2004). *VA/DoD clinical practice guidelines for the management of post-traumatic stress.* U.S. Department of Veterans Affairs & U.S. Department of Defense, Washington, DC.

Vanderwerker, L. C., & Prigerson, H. G. (2004). Social support and technological connectedness as protective factors in bereavement. *Journal of Loss & Trauma, 9,* 45–57.

van Etten, M. L., & Taylor, S. (1998). Comparative efficacy of treatments for post-traumatic stress disorder: A meta-analysis. *Clinical Psychology & Psychotherapy, 5*(3), 126–144.

Walshe, D. G., Lewis, E. J., Kim, S. I., O'Sullivan, K., & Wiederhold, B. K. (2003). Exploring the use of computer games and virtual reality in exposure therapy for fear of driving following a motor vehicle accident. *CyberPsychology and Behavior, 6,* 329–334.

Weathers, F. W., Litz, B. T., Herman, J. A., Huska, J. A., & Keane, T. M. (1993, October). *The PTSD Checklist (PCL): Reliability, validity and diagnostic utility.* Paper presented at the 9th annual conference of the International Society for Traumatic Stress Studies, San Antonio, TX.

Wilson, J., Drozdek, B., & Turkovic, S. (2006). Posttraumatic shame and guilt. *Trauma, Violence, & Abuse, 7*(2), 122–141.

Wood, D. P., Murphy, J., Center, K., McLay, R., Reeves, D., Pyne, J., et al. (2007). Combat-related post-traumatic stress disorder: A case report using virtual reality exposure therapy with physiological monitoring. *CyberPsychology and Behavior, 10,* 309–315.

Zayfert, C., DeViva, J. C., Becker, C. B., Pike, J. L., Gillock, K. L., & Hayes, S. A. (2005). Exposure utilization and completion of cognitive behavioral therapy or PTSD in a "real world" clinical practice. *Journal of Traumatic Stress, 18*(6), 637–645.

chapter four

Schizophrenia

Armando J. Rotondi

Overview

There is a growing use of electronic technologies in the provision of health services. This includes electronic medical records, e-mail reminders, client access to components of their medical records often with the ability to create their own personal records, home telemonitoring systems that can assess and transmit physical symptoms such as blood pressure to health care professionals, and so forth. A somewhat smaller area, though with considerable activity, is the use of telehealth technologies in the delivery of mental and psychiatric health services. This is an area of research and practice that is accelerating quite rapidly. This chapter presents a Web-based intervention that was used to provide multifamily psychoeducational therapy to persons with schizophrenia and their family members via home computers. The chapter reviews the need for telehealth approaches to mental health services delivery, presents the theoretical foundations upon which the intervention was designed, explains the process that was used to design the Web site interface to make it accessible to persons who may have cognitive impairments, and summarizes the results from a randomized clinical trial of the intervention.

Background

In the United States an estimated 8.3% of the adult population has a serious mental illness (SMI; Substance Abuse and Mental Health Services Administration [SAMHSA], 2004). The social and economic burden of mental illness in America is estimated to be more than $148 billion a year in direct and indirect costs (National Institute of Mental Health [NIMH], 2003). An international study of the indirect costs of illness and injury revealed that mental disorders account for more than 10% of the global burden of disease (Murray & Lopez, 1996). Four mental disorders (unipolar major depression, schizophrenia, bipolar disorder, and obsessive-compulsive disorder) rank among the 10 leading causes of disability worldwide.

However, in the United States only about half of those with an SMI receive any treatment (Kessler et al., 2001; SAMHSA, 2003), with only 4.1% receiving even minimally adequate treatment in any given year (Wang, Demler, & Kessler, 2002). Additionally, the landmark Schizophrenia PORT study found that 77% of persons with schizophrenia have ongoing contact with family (Lehman et al., 1998). The only evidence-based practice (EBP) that includes families is family psychoeducation (FPE), which has been shown in over 30 randomized clinical trials to dramatically reduce relapse rates and family burden (Falloon, Held, Roncone, Coverdale, & Laidlaw, 1998). The compelling evidence for the effectiveness of FPE programs has established it as the standard of care for all families in contact with their relatives with schizophrenia. Nevertheless, fewer than 9% of families received any educational or supportive services (Dixon, 1999), and an even smaller proportion of consumers and families received FPE. Thus, for persons with SMI, the number who receive mental health treatments, the conformity of those treatments to accepted guidelines and EBPs, and the provision of services to families fall substantially short of the established standards of care.

One of the reasons for the gap between what is needed and what is received is that health care systems have traditionally been designed to provide a symptom-driven response to acute illnesses. This has been found to be a poor configuration to meet the needs of those with chronic conditions (Wagner, Austin, & Von Korff, 1996b) such as diabetes (Funnell & Anderson, 2004) and SMI. There is a growing recognition that management of chronic conditions can be improved by systems of care that are patient- or consumer-centered, and are designed to meet the needs of those with chronic illnesses and provide long-term support (Funnell & Anderson, 2004). Increasingly, patient participation has been shown to be a critical element for successful chronic illness management (Lorig et al., 1999; Wagner, Austin, & Von Korff, 1996a). Systems that support the development of informed and activated consumers have demonstrated positive chronic illness outcomes (Glasgow et al., 2002; Lorig et al., 1999). Further, chronic disease self-management strategies structured around enhancement of self-efficacy have resulted in improvements in health status and reductions in unnecessary use of health services (Loring et al., 1999, 2001). Consumer-centered care, characterized by an effective partnership between health care professionals and consumers that promotes active participation by consumers in their care, results in better health outcomes and satisfaction (Anderson et al., 1995; Greenfield, Kaplan, & Ware, 1985). In order to improve outcomes, systems to treat chronic illnesses are recognizing that while health professionals are experts on evidence-based care, consumers are experts on their own lives and living with their illnesses. The day-to-day behaviors, supports, interactions, and decisions that occur in the community and at home are more important determinants of patient and family long-term outcomes and treatment efficacy than what occurs in a clinic visit (Von Korff, Gruman, Schaefer, Curry, & Wagner, 1997). Thus, for management of chronic illness to be successful, professionals must be in partnership with consumers, and consumer self-management must be successful (Funnell & Anderson, 2004). Providers, therapies, and services need to facilitate self-efficacy and self-management. Both empirical evidence and theory indicate that services provided in clinics have only limited success in changing the long-term outcomes for persons with chronic conditions (Von Korff et al., 1997). Therapies should thus be designed specifically to promote patient empowerment and to facilitate behavior change (Funnell & Anderson, 2004). This approach empowers consumers to be responsible for their self-management and provides for self-management support. The nature of the clinician–consumer therapeutic alliance is thus a key dimension of effective chronic illness management (Stewart et al., 2003).

An approach that puts consumers at the center of their care is consistent with a framework for health care that addresses the struggle to live life to its fullest. This aspect of care, which addresses a subjective attitude about chronic mental illness, is often called *recovery*. The term *recovery* is used to acknowledge that people can successfully contend with severe and persistent psychiatric illnesses, function well, and create positive lives. Recovery is grounded in resiliency—the personal and environmental characteristics that enable people to surmount crises and stress and live fulfilling lives. The intervention presented in this chapter is seen as a tool to promote recovery.

Weaknesses of the current mental health care delivery system targeted by the intervention

There are several features of many treatment delivery systems where improvements could be made to increase access and effectiveness for persons with SMI. At least five have the potential to lead to significant improvements, and the intervention described in this chapter has been specifically designed to address these. The first is that services are provided in person. This requires clients to travel to clinics, which can add considerable burden to those with cognitive impairments, and to family if they help them get to clinics. Given that the symptoms of SMI may be cyclical, travel to services may be periodically interrupted when exacerbations occur. For those in suburban communities, or in the case of rural residents where fewer or no services may be available, it may simply be too difficult to travel frequently enough to receive adequate psychosocial services. A second issue is that families by and large do not receive any services, not even disease-related education, and are even less likely to be involved in the treatment process. Third, services may not be designed to facilitate a partnership between professionals and consumers in a way that empowers consumers to play a greater role in their care and promotes self-efficacy and self-management. A fourth issue is the lack of appropriate peer and social support. Yet again, such resources are often not available to consumers or family members despite their proven value. Finally, services that are provided may not conform to evidence-based practices and standards of care. Improvements in these five areas could have marked effects on treatment effectiveness. Online delivery of services offers distinct advantages in each of these cases. There are certainly obstacles that must be addressed before online mental health services can be an easily accessible and commonly provided, but if these are appropriately addressed, the benefits should be pronounced.

The intervention

To help improve the access to, and effectiveness of, services for persons with SMI, we created a Web-based intervention that was specifically designed to address the five problem areas mentioned above. The Web-based intervention was developed to provide therapeutic services to persons with schizophrenia and their family members, as well as other important informal support persons. The intervention was designed based on the following primary therapeutic foundations.

1. *Multifamily Psychoeducational Therapy.* The primary foundation of this intervention, which is based on the groundbreaking work of Dr. Carol Anderson and her colleague Gerald Hogarty, is family psychoeducational therapy (Anderson, Hogarty, & Reiss, 1980). Approximately 25 years ago clinicians and researchers began to recognize that a partnership with the families of persons with schizophrenia could contribute to better outcomes for both. Efforts were undertaken to engage families in the treatment process by sharing illness information, identifying behaviors that decrease stimulation and promote recuperation, and teaching families and persons with schizophrenia pragmatic coping strategies to reduce distress (Anderson et al., 1980; Falloon & Liberman, 1983; McFarlane & Lukens et al., 1995). The group of therapies that emerged are known collectively as family psychoeducation (FPE). Based on the stress-vulnerability-coping-competence model of schizophrenia (Nuechterlein et al.,

1992; Zubin & Spring, 1977), which recognizes that schizophrenia is a brain disorder that is influenced by environmental factors, and that families can have a significant effect on their relative's recovery, FPE programs are designed to reduce family distress and improve consumer well-being by teaching what is known about the illness and its management, teaching coping strategies to minimize crises, providing coaching in problem-solving skills to solve day-to-day problems, providing emotional support, and providing crisis management (Anderson, Reiss, & Hogarty, 1986; Lehman et al., 2004). An overall goal of FPE is to promote social functioning and integration of the person with schizophrenia. In addition, these models are designed to reduce the high levels of burden and distress in relatives by providing help in adjusting to the illness (Barrowclough, Tarrier, & Johnston, 1996). Research indicates that long-term improvement in outcomes may derive a substantial benefit from FPE programs that have an ongoing multifamily group component as part of the treatment. The significant main effect of the group environment component indicates that there are factors present in the social context of multifamily groups that add to the therapeutic effects of single-family FPE (McFarlane, Link, Dusbay, Marchal, & Crilly, 1995).

2. *Empowerment: Self-Management & Self-Efficacy.* Perceived self-efficacy is the belief in one's abilities to influence events that affect one's life. Self-efficacy leads to greater empowerment, which is a critical dimension of recovery. Self-management is engaging in activities that protect and promote health (e.g., monitoring and managing illness symptoms and signs; monitoring the impact of illness on functioning, emotions, and interpersonal relationships; and monitoring treatment regimes). Studies have found that self-management facilitates recovery from mental illnesses (Allott, Loganathan, & Fulford, 2002).

The Web site is designed to promote self-efficacy and self-management. It provides the opportunity for users to play an active role in directing resources to meet their self-identified needs. It allows users to learn from others in a similar situation, and obtain information and services directed by their needs and at their convenience. The Web site provides educational materials on topics and needs that consumers and family members identified as important to them, and includes bibliotherapy—*self-help* materials and tutorials are provided to help users master skills (e.g., problem solving) to both improve illness-related knowledge and develop skills to cope with stressors; and guidance is provided on available community services (e.g., financial aid), their eligibility requirements, and how to obtain those services. Via the online therapy and peer groups, social learning occurs, and skills are taught, modeled, and reinforced (e.g., problem solving, communication skills). Participants are able to test their abilities to use a skill successfully and obtain feedback on skill usage to facilitate adjustment when appropriate and improve skill mastery; and peers have an opportunity to act as models for other participants. For example, peers provide suggestions for how to solve problems, how to cope with symptoms, and their experiences with solutions they have tried. This provides examples of similar others, rather than professionals, solving problems. It helps participants learn that their own experiences have value, and they are put in the role of being the helper versus being helped. These characteristics of the Web site are designed to help foster a greater sense of control and self-determination (Bandura, 1994; Onken, Dumont, Ridgway, Dorman, & Ralph, 2002).

3. *Involvement of Family and Other Significant Support Persons.* The intervention allows families as well as other informal support persons to receive services and to be involved in the treatment when appropriate. This recognizes that there is a reciprocal relationship between the health of persons with chronic illnesses, such as SMI, and the health of their family members (Lefley, 1992). The health and well-being of each impacts the other, and thus both, when present, can benefit from services and involvement in treatment.

4. *Peer Support.* The Web site provides mechanisms to support social learning and engage in peer support and social networking. Consumers and family members often open up more quickly and speak more candidly to others in a similar situation than to health care professionals (Long, 1989). They accommodate suggestions for change more easily as they recognize in others what they have experienced, and they may be more engaged in the presence of others who are managing circumstances similar to their own (Bandura, 1981). The value to consumers and families of peer networking and knowing that others are in the same situation and have found methods of coping can be tremendous. Peer support may be a quite important element given that over time social networks can become quite limited for persons with SMI and their families. Through offering "support, empathy, sharing, and assistance ... feelings of loneliness, rejection, discrimination, and frustration" can be countered (Flexer & Solomon, 1993, p. 48).

In addition to the practical and psychological value of appropriate peer support, there are indirect health benefits as well. There is substantial evidence of a significant independent association between social support and health status (Cohen & Wills, 1985; Wortman, 1984). There is an increased risk of mortality, physical deterioration, psychological burden, and psychological breakdown among persons with a low quantity and quality of social support (Biegel, 1991; Schulz, Tompkins, & Rau, 1988). Social support is a modifiable vulnerability factor that has been targeted by this intervention.

5. *Problem Solving.* Problem solving, in the context of this intervention, refers to the process by which people employ effective means of coping with stressful events and situations and meet perceived needs (D'Zurilla & Nezu, 1982; McFarlane, 2002). The ability to recognize situations as modifiable and realize that one can take actions to improve a situation or change events to make them less stressful was seen as a modifiable vulnerability. A cornerstone of the intervention was to help participants recognize these two characteristics of day-to-day problems and stressors, and to teach and promote active problem-solving skills as a way to reduce stresses and meet personal needs.

6. *Help Users to Meet Self-Perceived Needs.* The ability of those with chronic illnesses to meet needs affects their well-being. There is extensive theoretical and empirical work to indicate that meeting an individual's needs will reduce stress, promote better adaptation to injury-related difficulties, and improve outcomes (McCubbin, Thompson, & McCubbin, 1996). Additionally, when needs go unmet, distress increases and outcomes worsen (Allen & Mor, 1997). This intervention was specifically designed to help users meet and better formulate personal needs and goals in order to facilitate their attainment.

The above therapeutic elements can have a dramatic impact on the effectiveness of clinic-based treatments. The intervention is designed to allow its capabilities to be tailored to a given patient, patient group, illness, clinic, clinician, or environment (e.g., public, private, large medical center, or provider network). That is, given the particular circumstances and goals, all or only some of these capabilities could be made available by any given clinic or clinician. This would allow for a partial or a progressive implementation of these services as the need indicates or the ability to support features is developed by a clinic or clinician. Providing informational and educational materials via a Web site that consumers and families can access may be a very simple approach to improving comprehension of treatment regimens and support for treatments. Consumers of services often benefit from additional explanation of illness conditions and treatment recommendations, and answers to important questions once they return home, such as what to do if a dose of medication is missed. Providing these via a Web site could be a cost-effective approach to addressing basic information needs of patients and families and exploring the strengths of online services. Given the availability of low-cost Web site hosting, it is a relatively simple task for individual clinicians and clinics to offer basic services online. There is also great potential to team with local consumer groups, either de novo or via established groups such as the National Alliance on Mentally Illness (NAMI).

Designing the intervention to address the accessibility needs of persons with cognitive impairments

There is, on average, a 15- to 20-year time lag between discovering effective forms of treatment and incorporating them into routine patient care (Institute of Medicine [IOM], 2001). As an example, FPE is past the 25-year marker and is still not close to being routinely available. Clearly, a major public health challenge is to devise methods for more rapidly translating and disseminating mental health treatments and services to diverse community settings. To this end, the President's New Freedom Commission on Mental Health (2003) has, as one of its goals, the use of technology to access mental health care and information. This goal states, "Consumers and families will be able to regularly communicate with agencies and personnel that deliver treatment and support services. Health technology and telehealth will offer a powerful means to improve access to mental health care. . . . To address this technological need in the mental health care system, this goal envisions . . . a robust telehealth system to improve access to care" (2003, p. 79). In terms of providing psychosocial treatments to persons with SMI, online services are potentially less costly than face-to-face methods, which would support their dissemination to more settings. Clinics would not have to set up their own programs, but could use a centralized program. Although the potential usefulness of computer and Web-based applications for persons with SMI is considerable, research on the design needs of this population has been almost nonexistent to date.

Given that the use of technologies such as computers, Web sites, and the Internet can involve complex cognitive activities (Stephanidis, 2000), the cognitive deficits associated with SMI are likely to influence the accessibility of these technologies. For example, impairments have been found in the working memory system, particularly in spatial abilities and language (Green, 1998) and episodic memory (Sharma & Harvey, 2001). Deficits have been found in visual–spatial processing, psychomotor skills, and executive functioning, which encompass the cognitive abilities necessary for complex goal-directed behavior, planning, problem solving, task sequencing, concept formation, and anticipating outcomes (Sharma

& Harvey, 2001). Studies have found deficits in vigilance, or the ability to sustain attention (Green, 1998). Deficits exist in information processing, including the use of erroneous logic (Arieti, 1955) and the propensity for personal meanings to dominate interpretation of themes and categories (Steffy, 1993). The potential for cognitive deficits in many with SMI could affect their use of Web tools, and it is likely that interfaces and guidelines will need to be developed for use by persons with SMI that are tailored to their specific cognitive challenges and needs.

Unfortunately, a dearth of research has been conducted on how to design usable interfaces for persons with SMI (Edwards, 1995; Rotondi et al., 2007) or others who may have reduced cognitive abilities, such as those with traumatic brain injury or posttraumatic stress disorder, or the elderly. Understanding the limits of current technology interface designs for persons with SMI and developing appropriate guidelines to overcome them is essential for enhancing the positive impact of these important technologies.

Findings from usability studies of persons with physical impairments, and our studies on persons with SMI (Rotondi et al., 2007), indicate that people with different types of disabilities (e.g., vision, hearing, cognitive) have different design needs. To identify these needs and determine the best way to meet them, usability studies need to use persons who have a particular condition. There has been a dearth of research on how to design Web sites for persons with SMI and cognitive deficits (Bulger, 2002; Rotondi et al., 2007). As a result, Web designers do not have guidance on how to make their sites usable by persons with SMI. At the time that the current intervention was designed, a search of 16 publication databases (e.g., Medline, PsycINFO, Information Science & Technology Abstracts, Applied Science and Technology Abstracts) and several accessibility Web sites (e.g., World Wide Web Consortium [W3C], Usability. gov, UniversalUsability.org) revealed no studies that assessed Web site usability performance with any persons who have SMI and/or cognitive impairments, or any research-based Web design guidelines for such persons. Although there are several prominent published guidelines for designing Web sites, they are based on research with non-SMI users only (Architectural and Transportation Barriers Compliance Board, 2000; IBM, 2001; Koyani, Bailey, & Nall, 2003; Microsoft, 2005; W3C, 1999).

As a result, we conducted our own usability testing to identify the design features that influence the ability of persons with SMI to use a Web-based application. On the basis of our findings, we designed the Web site intervention presented here, which was named Schizophrenia On-line Access to Resources (SOAR; Rotondi et al., 2007). The usability testing involved 98 persons with SMI and consisted of several components, including testing alternatives, vocabulary, link and heading designs, ways to organize information, design elements to present Web site content, and complete Web site designs. We identified several tendencies on the part of participants that made a standard Web site design inappropriate, leading us to experiment with alternative Web site designs. These tendencies included the following:

Difficulty understanding geometric conventions. Many of the common geometric forms and their interpretation were not fully understood by some participants. For example, some users did not understand that the columns in a table are intended to divide a row into discrete cells of separate contents. They would read the contents across a row with no recognition of the column separators.

Difficulty understanding complex phrasing. Compound sentences and complex phrasing were often not understood. For example, the sentence "I need to take my medications" may not be understood, though its two components were understood (i.e., *I need*, and *take my medications*).

Using personal meanings to interpret content and guide navigational decision making. A tendency that emerged was the use of personal experiences as the basis for interpreting Web site content. As a result, when presented with the following navigation task of having to choose between "Getting help to pay your bills" or "Information for families" in order to find programs to help a consumer pay the rent, some chose the latter. Their rationale was that when they needed money they went to their families.

Difficulty with abstract reasoning. Virtually any situation that required users to abstract a meaning from the Web site's contents proved difficult for some. One of the most important situations involved instances where participants had to make navigational decisions. Errors occurred, especially when trying to interpret the meanings of one- or two-word links and labels.

Easily confused. Our usability testing showed that some users had considerable difficulties when they were asked to navigate through multiple levels of a Web site. They had difficulty "descending" accurately, and "retreating" to find their way back to previous contents.

Being unable to scan a page. Usability testers observed that some participants seemed to read a Web page from beginning to end without scanning it. That is, by beginning at the upper left and reading every line until they reached either the contents that they were looking for or the end (i.e., lower right). We did not collect retinal scan data that could confirm this impression, however.

Easily distracted. Our testing indicated that some users could be sidetracked by content or design elements that were irrelevant to their task. As a result, we kept the final design of the Web site free of "enhancing" content such as decorative images.

Having a low level of reading comprehension. Though it was necessary to be able to read at a sixth grade level to participate in our testing, many of the participants had reading comprehension that was below their assessed reading level.

Distinctive design features of the SOAR Web site

Based on the above findings, several alternative designs for the Web site were created and their performances evaluated. The final Web site contained several design features that were used to accommodate unique needs we identified for persons with SMI, and are not necessarily consistent with published Web site design guidelines.

Flat, two-level hierarchy. We created a flat hierarchical structure by listing all categories of the Web site's resources on the homepage and using a unique pop-up menu structure to list the resources for each category. A user could survey each of the 12 or so resources in a given category and then choose a resource without leaving the homepage. Making a Web site that is only one-page deep reduced the chances that a user would become lost or disoriented.

Homepage presents low-level modular concepts. Modules were organized around lower-level concepts, resulting in relatively more modules on the homepage. Each choice takes users to a smaller amount of relatively focused content.

Explicit versus inferential link labels (i.e., navigational links). Instead of using one or two words to represent links and labels, long phrases were used (up to 12 words), providing more information and requiring less interpretation on the part of users.

One constant navigational dimension. One constant (i.e., present on every page) navigational tool bar was used to help users navigate the site.

Minimal superfluous content. Given the potential vulnerability to overstimulation, bright colors, icons, and graphics were avoided when possible.

Our usability findings support several inferences about how Web sites should be designed to meet the needs of persons with SMI:

1. Shallow hierarchies are better than deeper structures.
2. Web sites that use "memory aids" to support navigation (e.g., pop-up menus, pull-down menus, lists) should be more usable
3. There should be minimal need to think abstractly in order to use the Web site
4. Decorative features that may distract users should be avoided.
5. Modular designs that use brief labels and increase the depth of a site should not be used.
6. Web sites should present the simplest and most accessible design and allow users to add complexity and functions if they are capable.

These characteristics should reduce the need for users to (1) rely on working memory stores and the visual–spatial sketchpad to learn and create a mental model of, memorize, or explore the layers of a site; (2) identify the concrete meanings of abstract links and labels; (3) grasp the logic of the concepts used by designers to organize the contents of a Web site; and (4) use executive functions to plan and execute multiple-step searches to find information. It allowed users with cognitive impairments, who made highly personal associations and had difficulty thinking abstractly, to understand the presentation, locate desired information, and explore the site effectively, while not requiring them to learn the hierarchy. A user could effectively navigate the site each time as though it were the first visit. This design model is appropriate for persons with SMI, who may have a reduced ability to search for information, recall information, learn the hierarchy of a Web site, think abstractly, quickly scan to differentiate relevant from irrelevant information, or execute multistep sequences, all of which are required in a Web site designed in accordance with standard guidelines, but are minimized in this design.

Components of the intervention Web site

Therapy groups. The site provided a Web-based environment for each of three separate and private therapy groups: (a) family members and support persons only, (b) persons with schizophrenia only, and (c) a multifamily therapy group for all intervention participants. For each of the groups, bulletin board formats were provided for Internet communication among group members. The groups were facilitated by experienced mental health professionals (Master of Social Work and PhD clinicians) who were trained in the monitoring and management of Web-based interventions. Intervention implementation was guided by a standardized facilitation protocol that included an outline for facilitating the groups and the process for solving problems, based on the manuals of Dr. Anderson and Dr. McFarlane (Anderson et al., 1986; McFarlane et al., 1991).

Ask Our Experts Your Questions. This module allowed users to anonymously ask questions and receive responses. Members of the research team answered these questions or contacted outside experts when additional information was needed.

Questions and Answers Library. The questions and answers from the Ask Our Experts Your Questions module, with personal information eliminated, were added to this library. Initially, the library was seeded with answers to commonly asked questions.

Educational and Tutorial Reading Materials (Articles). The Web site contained educational materials, including articles on typical family responses to schizophrenia, emotional problems experienced by families, stress management strategies, stages of schizophrenia, side effects of medication, obtaining financial assistance, and other topics of interest.

Community Resources. This module identified community resources and state and federal programs that might be of interest to the participants. It also had a financial assessment component that helped locate possible financial resources based on a user's eligibility characteristics.

What's New. This module identified activities in the community or news items that were relevant to persons with schizophrenia and their support persons.

Participants in the project

The participants in the project (Table 4.1) had to meet the following inclusion criteria: age 14 years or older, DSM-IV diagnosis of schizophrenia or schizoaffective disorder, more than one psychiatric hospitalization or emergency department visit within the previous two years, ability to speak English and read English at a sixth-grade reading level, and the absence of physical limitations that would preclude the use of a computer even with commonly available accessibility accommodations. No previous computer experience was required. The intervention group had 15 persons with schizophrenia (9 had no prior computer experience) and 11 support persons (i.e., 11 persons with schizophrenia were joined by a support person). The treatment as usual (TAU) control group had 14 persons with schizophrenia and 10 support persons (Table 4.1). As needed, one

Table 4.1 Demographic Characteristics (Intervention & Control)

Demographic	Persons with schizophrenia/ schizoaffective disorder	Family/support persons
Age (mean (*SD*))	37.5 (10.7)	51.52 (12.4)
	n (%)	n (%)
Female	21 (72)	14 (66.7)
Race	n = 29	
Caucasian	14 (48)	11 (52.4)
African American	14 (45)	8 (38.1)
Other	2 (7)	2 (9.6)
Intervention/Web site group only		
No previous computer experience	9 (60)	4 (40)

Table content provided by the SOAR Web site.

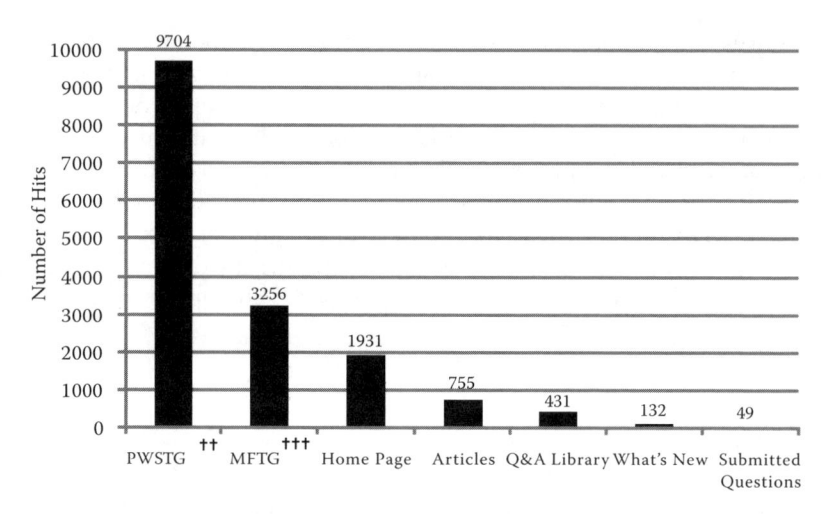

82 Using technology to support evidence-based behavioral health practices

[Bar chart showing Number of Hits by component: PWSTG†† = 9704, MFTG††† = 3256, Home Page = 1931, Articles = 755, Q&A Library = 431, What's New = 132, Submitted Questions = 49]

† Persons with schizophrenia
†† Persons with schizophrenia therapy group
††† Multi-family therapy group

Figure 4.1 Consumer Web site usage by component.

home computer was provided to each dyad, or when no family member was present, to each person with schizophrenia, as well as access to the Internet via a dial-up modem.

Uses and outcomes from the intervention

Use of the SOAR Web site by persons with schizophrenia. In the first three months of the study, the total number of pages accessed (i.e., page hits) on the Web site was 17,292 for persons with schizophrenia assigned to the intervention condition. The average number of hits per person was 1,080.7 (SD = 1812.0, range = 86 – 6,325). The most frequently used component of the Web site was the persons with schizophrenia therapy group (PWSTG; Figure 4.1; Rotondi et al., 2005). Figure 4.2 displays the total usage of the Web site, including all of its components, over the first 12 months of the project.

Use of the SOAR Web site by family and support persons. In the first three months of the study, family members and support persons had a total of 2,527 pages accessed, with an average total hits per person of 229.7 (SD = 164.5, range = 55 – 362). The most frequently used components were the two therapy groups (Figure 4.3). The usage of the Web site's components by family members over the first 12 months of the project is shown in Figure 4.4.

Outcomes for consumers and family members. In the first three months, when compared to the TAU control group, persons with schizophrenia in

Chapter four: Schizophrenia 83

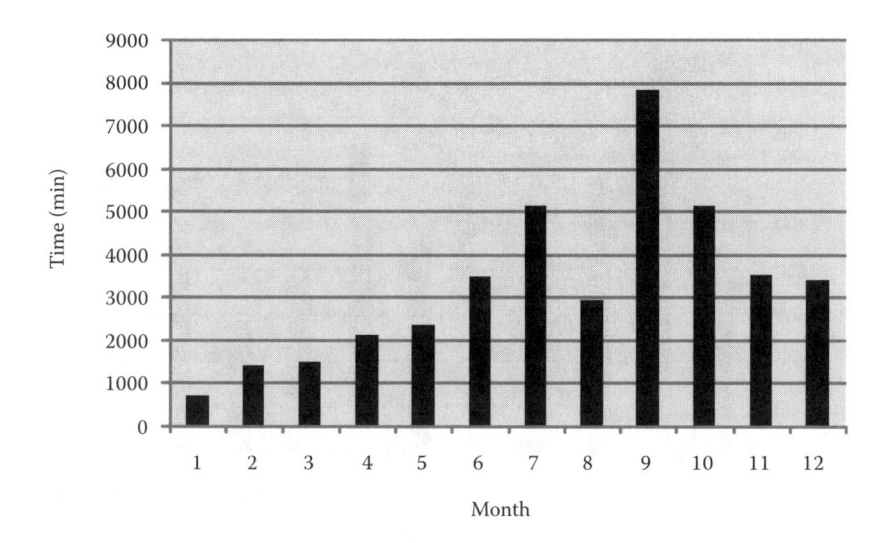

Figure 4.2 Consumer Web site usage first 12 months.

†† Family Support Person therapy group
††† Multi-family therapy group

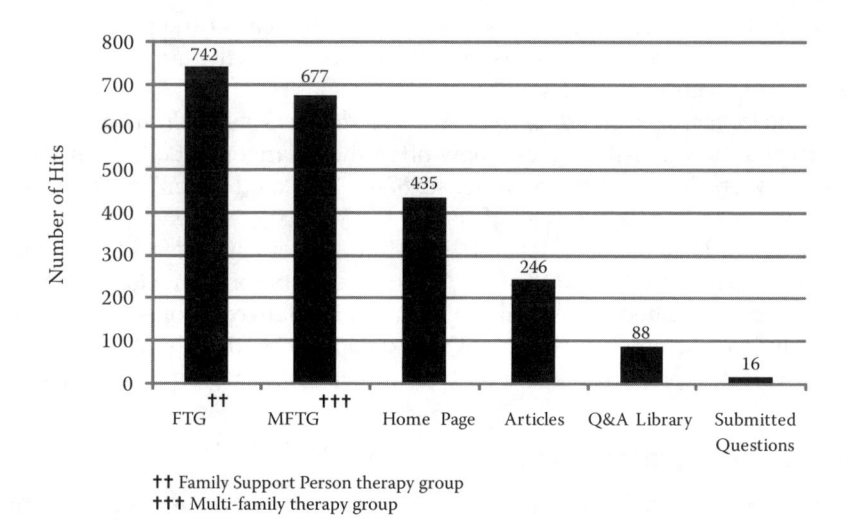

Figure 4.3 Family Support (F/S⁺) Web site usage by component.

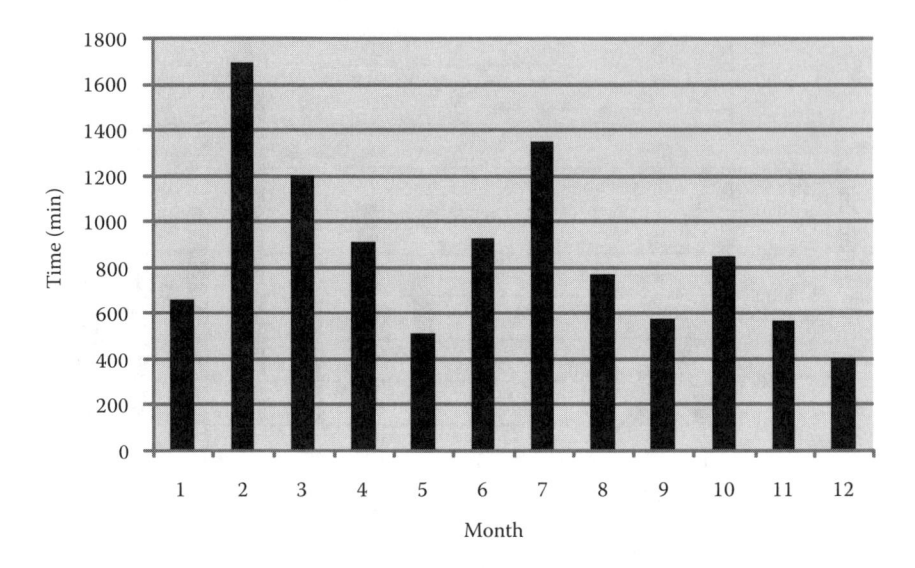

Figure 4.4 Family member total Web site usage first 12 months.

the telehealth group had significantly less perceived stress (p = 0.044) and showed a trend toward greater perceived social support (p = 0.062). Over the year of the study, persons with schizophrenia in the telehealth group had a significant and large reduction in positive symptoms (differential effect size = 0.98, p = 0.013), and improved knowledge of schizophrenia, compared with the TAU control group.

At the three-month time period and at one year of the project, there were no significant differences in outcomes between the family members in the intervention and TAU groups.

Subjective evaluations of the Web site by persons with schizophrenia. Participants were asked to rate how often the arrangement of information on the Web site's screens made sense: 87.6% indicated "often" or "always." When asked to rate the ease of use of the Web site and each of its components, at least 62.5% indicated "very" or "extremely" easy. Participants were asked to rate the value of the Web site and its components. Over 80% of participants rated the Web site overall and its main components as "moderately" to "extremely" valuable. These data indicate that the Web site was relatively comprehensible and usable, and was valuable to participants.

Future directions

An overriding goal of this intervention is to provide consumers and their support persons with an environment, including tools and professional partnerships, that will improve their ability to take a more active role in their illness management, well-being, and recovery. Additional capabilities that have been considered for this intervention include (1) the ability for each user to develop, track, and manage personal goals; (2) the ability to identify and track side-effects of medications and unwanted symptoms to aid in illness management and discussions with health care professionals; and (3) provision of a more formal integration of the intervention with a clinic's standard of care procedures.

Summary and conclusions

The use of telehealth systems to provide treatment and supportive services has several advantages, especially when these systems can be accessed from users' homes. Receiving services in the home provides convenience and the potential for a less stressful environment than clinical settings. It increases access and allows certain resources to be available 24 hours a day, 7 days a week. These characteristics should increase service utilization and effectiveness, and allow users a greater ability to take an active role by choosing when and which resources to access based on their needs. This is intended to create a greater sense of control and self-determination. This in turn may lead to better use of services, such as medication, psychosocial treatments, and employment, and ultimately to improvements in quality of life. Telehealth psychosocial services are potentially less costly than in-person services, which would support their dissemination to more settings. Clinics would not have to expend the resources needed to set up their own online treatment programs, but could use a centralized program offered over the Internet.

While some lower income groups are likely to have less computer and Internet access than those with more resources, studies have found that when these technologies are made available and are designed to be accessible, these individuals become avid computer system users (Eng et al., 1998; Rotondi et al., 2005; Rotondi, Sinkule, & Spring, 2005). Since underserved populations often lack access to many health care resources, including illness-specific information (Eng et al., 1998) and are thus likely to have more unmet health needs, it is likely that providing the opportunity to access educational resources and other online services may be cost effective and produce significant benefits.

The findings from the study presented here suggest that (1) when telehealth services are designed specifically to accommodate the needs and cognitive requirements of persons with SMI, they can and will be used; (2) family members, support persons, and persons with schizophrenia will use a Web site for psychoeducational therapy frequently and will value it highly; and (3) the use of psychoeducational telehealth services is associated with improved outcomes in persons who have schizophrenia. To date, however, outside of our own work, there has been a dearth of research on Web site designs that are accessible to persons with schizophrenia. The promising nature of these findings suggests the critical importance of additional research in this area.

References

Allen, S., & Mor, V. (1997). The prevalence and consequences of unmet needs: Contrasts between older and younger adults with disability. *Medical Care, 35*(11), 1132–1148.

Allott, P., Loganathan, L., & Fulford, K. W. M. (2002). Discovering hope for recovery. *Canadian Journal of Community Mental Health, 21,* 13–34.

Anderson, C. M., Hogarty, G. E., & Reiss, D. J. (1980). Family treatment of adult schizophrenic patients: A psycho-educational approach. *Schizophrenia Bulletin, 6,* 490–505.

Anderson, C. M., Reiss, D. J., & Hogarty, G. E. (1986). *Schizophrenia and the family.* New York: Guilford Press.

Anderson, R. M., Funnell, M. M., Butler, P. M., Arnold, M. S., Fitzgerald, J. T., & Feste, C. C. (1995). Patient empowerment. Results of a randomized controlled trial. *Diabetes Care, 18,* 943–949.

Architectural and Transportation Barriers Compliance Board. (2000). Proposed access standards for electronic and information technology: An overview. *Federal Register,* 36 CFR Part 1194, [Docket No. 2000-01], RIN 3014-AA25. Retrieved February 3, 2005 from http://www.access-board.gov/sec508/508standards.htm

Arieti, S. (1955). *Interpretation of schizophrenia.* New York: R. Brunner.

Bandura, A. (1981). *Social foundations of thought and action: A social cognitive theory.* Englewood, NJ: Prentice Hall.

Bandura, A. (1994). Self-efficacy. In V. S. Ramachaudran (Ed.), *Encyclopedia of human behavior* (Vol. 4, pp. 71–81). New York: Academic Press.

Barrowclough, C., Tarrier, N., & Johnston, M. (1996). Distress, expressed emotion, and attributions in relatives of schizophrenia patients. *Schizophrenia Bulletin, 22,* 691–702.

Biegel, D. S. E. S. R. (1991). *Family caregiving in chronic illness.* Newbury Park, CA: Sage Publications, Inc.

Bulger, J. R. (2002). *A usability study of mental health websites with an emphasis on homepage design: Performance and preferences of those with anxiety disorders.* Unpublished master's thesis, University of North Carolina at Chapel Hill.

Cohen, S., & Wills, T. A. (1985). Stress, social support, and the buffering hypothesis. *Psychological Bulletin, 98,* 310–357.

D'Zurilla, T., & Nezu, A. M. (1982). Social problem solving in adults. In P. Kendall (Ed.), *Advances in cognitive-behavioral research and therapy* (Vol. 1, pp. 202–274). New York: Academic Press.

Dixon, L. (1999). Providing services to families of persons with schizophrenia: Present and future. *Journal of Mental Health Policy Econ, 2,* 3–8.

Edwards, A. D. N. (1995). Extra ordinary computer interaction: Interfaces for users with disabilities. In A. D. N. Edwards (Ed.), *Computers and people with disabilities* (pp. 19–43). Cambridge & New York: Cambridge University Press.

Eng, T. R., Maxfield, A., Patrick, K., Deering, M. J., Ratzan, S. C., & Gustafson, D. H. (1998). Access to health information and support: A public highway or a private road? *Journal of the American Medical Association, 280,* 1371–1375.

Falloon, I. R., Held, T., Roncone, R., Coverdale, J. H., & Laidlaw, T. M. (1998). Optimal treatment strategies to enhance recovery from schizophrenia. *Australia New Zealand Journal of Psychiatry, 32,* 43–49.

Falloon, I. R. H., & Liberman, R. P. (1983). Behavioral family interventions in the management of chronic schizophrenia. In W. McFarlane (Ed.), *Family therapy in schizophrenia* (pp. 141–172). New York: Guilford.

Flexer, R., & Solomon, P. (1993). Rehabilitation in community support systems. In R. Flexer & P. Solomon (Eds.), *Psychiatric rehabilitation in practice.* (pp. 45–61). Boston: Andover Medical Publishers.

Funnell, M. M., & Anderson, R. M. (2004). Empowerment and self-management of diabetes. *Clinical Diabetes, 22,* 123–127.

Glasgow, R. E., Funnell, M. M., Bonomi, A. E., Davis, C., Beckham, V., & Wagner, E. H. (2002). Self-management aspects of the improving chronic illness care breakthrough series: Implementation with diabetes and heart failure teams. *Annals of Behavioral Medicine, 24,* 80–87.

Green, M. F. (1998). *Schizophrenia from a neurocognitive perspective: Probing the impenetrable darkness.* Needham Heights, MA: Allyn & Bacon.

Greenfield, S., Kaplan, S., & Ware, J. E., Jr. (1985). Expanding patient involvement in care. Effects on patient outcomes. *Annals of Internal Medicine, 102,* 520–528.

IBM (2001): *IBM Web Design Guidelines.* Retrieved March 5, 2004 from http://www-306.ibm.com/ibm/easy/eou_ext.nsf/Publish/572

IOM (Institute of Medicine). (2001). *Institute of Medicine Committee on Quality of Health Care in America. Crossing the quality chasm: A new health system for the 21st century.* Washington, DC: National Academies Press.

Kessler, R. C., Berglund, P. A., Bruce, M. L., Koch, J. R., Laska, E. M., Leaf, P. J., et al. (2001). The prevalence and correlates of untreated serious mental illness. *Health Services Research, 36,* 987–1007.

Koyani, S. J., Bailey, R. W., & Nall, J. R. (2003). *Research-based web design and usability guidelines.* National Cancer Institute, U.S. Department of Health and Human Services. Retrieved March 12, 2005 from http://usability.gov/pdfs/guidelines_book.pdf

Lefley, H. P. (1992). Expressed emotion: Conceptual, clinical, and social policy issues. *Hospital Community and Psychiatry, 43,* 591–598.

Lehman, A. F., et al. (2004). The definition of self-management as developed by the Centre for Advancement of Health (1996). *Schizophrenia Bulletin, 30,* 193–217.

Long, P. W. E. (1989). Families in the treatment of schizophrenia (part II). *The Harvard Medical School Mental Health Letter. 6*(1), 1–3.

Loring, K. R., Ritter, P., Stewart, A. L., et al. (2001). Chronic disease self-management: 2-year health status and health care utilization outcomes. *Medical Care, 39,* 1217–1223.

Loring, K. R., Sobel, D. S., Stewart, A. L., Brown, B. W., Jr., Bandura, A., Ritter, P., et al. (1999). Evidence suggesting that a chronic disease self-management program can improve health status while reducing hospitalization: A randomized trial. *Medical Care, 37,* 5–14.

McCubbin, M. A., Thompson, A. I., & McCubbin, M. A. (1996). *Family assessment: Resiliency, coping and adaptation.* Madison, WI: University of Wisconsin Publishers.

McFarlane, W. (2002). *Multifamily groups in the treatment of severe psychiatric disorders.* New York: Guilford Press.

McFarlane, W. R., Link, B., Dushay, R., Marchal, J., & Crilly, J. (1995). Psychoeducational multiple family groups: Four-year relapse outcome in schizophrenia. *Family Process, 34,* 127–144.

McFarlane, W. R., Lukens, E., Link, B., Dushay, R., Deakins, S. A., Newmark, M., et al. (1995). Multiple-family groups and psychoeducation in the treatment of schizophrenia. *Archives of General Psychiatry, 52,* 679–687.

Microsoft Corporation. (2005). *Microsoft accessibility. Guides by impairment.* Retrieved February 8, 2005 from http://www.microsoft.com/enable/guides/default.aspx

Murray, C., & Lopez, A. (Eds.) (1996). *The global burden of disease.* Cambridge, MA: Harvard University Press.

NIMH (National Institute of Mental Health). (2003). *Translating behavioral science into action: Report of the National Advisory Mental Health Council Behavioral Science Workgroup.* Bethesda, MD: National Institute of Mental Health.

Nuechterlein, K. H., Dawson, M. E., Gitlin, M., Ventura, J., Goldstein, M. J., Snyder, K. S., et al. (1992). Developmental processes in schizophrenic disorders: Longitudinal studies of vulnerability and stress. *Schizophrenia Bulletin, 18,* 387–425.

Onken, S., Dumont, J., Ridgway, P., Dorman, D., & Ralph, R. O. (2002). *Mental health recovery: What helps and what hinders? A national research project for the development of recovery facilitating system performance indicators.* National Technical Assistance Center for State Mental Health Planning.

President's New Freedom Commission on Mental Health. (2003). *Achieving the promise: Transforming mental health care in America: Executive summary.* Rockville, MD: Mental Health Commission.

Rotondi, A. J., Haas, G. L., Anderson, C. M., Newhill, C. E., Spring, M. B., Ganguli, R., et al. (2005). A clinical trial to test the feasibility of a telehealth psychoeducational intervention for persons with schizophrenia and their families: Intervention and three-month findings. *Rehabilitation Psychology, 50,* 325–336.

Rotondi, A. J., Sinkule, J., Haas, G. L., Spring, M. B., Litschge, C. M., Newhill, C. E., et al. (2007). Designing Websites for persons with cognitive deficits: Design and usability of a psychoeducational intervention for persons with severe mental illness. *Psychological Services, 4,* 202–224.

Rotondi, A. J., Sinkule, J., & Spring, M. (2005). An interactive Web-based intervention for persons with TBI and their families: Use and evaluation by female significant others. *Journal of Head Trauma Rehabilitation, 20,* 173–185.

SAMHSA (Substance Abuse and Mental Health Services Administration). (2003). "Results from the 2002 National Survey on Drug Use and Health: National Findings," Substance Abuse and Mental Health Services Administration. Rockville, MD: Office of Applied Studies, NHSDA Series H-22, DHMS, SMA 03-3836.

SAMHSA (Substance Abuse and Mental Health Services Administration). (2004). *Results from the 2003 National Survey on Drug Use and Health: National findings* (Rep. No. SMA 04-3964). Rockville, MD: Substance Abuse and Mental Health Services Administration (Office of Applied Studies, NSDUH Series H-25, U.S. Department of Health and Human Services).

Schulz, R., Tompkins, C. A., & Rau, M. T. (1988). A longitudinal study of the psychosocial impact of stroke on primary support persons. *Psychology of Aging, 3,* 131–141.

Sharma, T., & Harvey, P. (2001). *Cognition in schizophrenia: Impairments, importance and treatment strategies.* Oxford & New York: Oxford University Press.

Steffy, R. A. (1993). Cognitive deficits in schizophrenia. In K. S. Dobson & P. C. Kendall (Eds.), *Psychopathology and cognition* (pp. 429–467). San Diego, CA: Academic Press, Inc.

Stephanidis, C. (2000). *User interfaces for all: Concepts, methods and tools.* Mahwah, NJ: Lawrence Erlbaum Associates, Inc.

Stewart, M., Brown, J., Weston, W., et al. (2003). Patient-centered medicine: Transforming the clinical method. Thousand Oaks, CA, Sage Publications.

Von Korff, M., Gruman, J., Schaefer, J., Curry, S. J., & Wagner, E. H. (1997). Collaborative management of chronic illness. *Annals of Internal Medicine, 127,* 1097–1102.

W3C (World Wide Web Consortium). (1999). *Web content accessibility guidelines 1.0.* Retrieved January 4, 2005 from http://222.23.org/TR/WAI-WEBCONTENT/wai-pageauth.html

Wagner, E. H., Austin, B. T., & Von Korff, M. (1996a). Improving outcomes in chronic illness. *Managed Care Quarterly, 4,* 12–25.

Wagner, E. H., Austin, B. T., & Von Korff, M. (1996b). Organizing care for patients with chronic illness. *Millbank Quarterly, 74,* 511–544.

Wang, P. S., Demler, O., & Kessler, R. C. (2002). Adequacy of treatment for serious mental illness in the United States. *American Journal of Public Health, 92.*

Wortman, C. B. (1984). Social support and the cancer patient: Conceptual and methodologic issues. *Cancer, 53,* 2339–2360.

Zubin, J., & Spring, B. (1977). Vulnerability: A new view of schizophrenia. *Journal of Abnormal Psychology, 86,* 103–126.

chapter five

Substance use disorders

Daniel D. Squires and Monte D. Bryant

Overview

The focus of this chapter will be on technology-assisted (TA) evidence-based treatment practices (EBPs) for substance use disorders. While there is much to report on the topic, our discussion will be limited to TA practices based largely on, or related to, limited-duration behavioral treatment practices. It is critically important for the reader to understand, however, that treatment for substance use disorders (SUDs) often requires more than limited-duration treatment episodes to produce the best outcomes. In addition to traditional talk or TA therapies, medication and self-help involvement are common, often critical components in the process of promoting and achieving long-term recovery from SUDs. While the use of medication to support recovery is not without controversy, there is clear evidence to support the efficacy of medication-assisted treatment for substance use and co-occurring disorders (Miller & Willbourne, 2002; Stuyt, Sajbel, & Allen, 2006). Additionally, the benefits of self-help involvement (e.g., 12-step, Rational Recovery) are significant and undeniable. Put simply, more people have successfully addressed problems related to SUDs through self-help than all formal treatment approaches combined in the United States (Vaillant, 1995).

When it comes to achieving long-term recovery (most commonly defined as abstinence lasting for one year or longer), clinicians should actively assist those seeking change by promoting a combination of treatment options that might include both behavioral and medication therapies, along with self-help involvement. The good news regarding TA programs for behavioral health issues is that such programs are increasingly more available and sophisticated. Another significant benefit of TA programs for SUDs is their ability to promote structured, high-fidelity dissemination of EBPs from research to practice—an area that has long plagued behavioral health and especially the field of SUDs treatment (Squires, 2004).

Background

During the past two decades there has been an explosion in the effectiveness and numbers of evidence-based practices for treating SUDs (Meyers & Miller, 2001; Miller & Willbourne, 2002). Currently, the National Registry of Evidence-Based Programs and Practices (http://www.nrepp.samhsa.gov/index.htm), maintained by the Substance Abuse and Mental Health Services Administration, lists 31 interventions specific to SUDs treatment, along with another 13 focusing on co-occurring disorders. Some of the more well-known interventions on the list include motivational interviewing, multisystemic therapy, 12-step facilitation, contingency management, the matrix model, coping-skills training, the community reinforcement approach, behavioral couples therapy, and dialectical behavior therapy, among others. Increasingly, many of the most effective EBPs available for SUDs today are broad in scope and designed to improve one's overall quality of life beyond the traditional focus on abstinence.

Despite increasing resources, however, issues affecting the dissemination of EBPs have long been a challenge for many areas of health care, and none more so than SUDs treatment. For a variety of reasons, including differing perspectives among researchers and treatment providers on issues surrounding service delivery, structural and financial barriers, education and training, policies that impede treatment options, and stigma, SUDs treatment seems especially resistant to adopting novel treatment approaches (Lamb, Greenlick, & McCarty, 1998). To address these issues there is, and *should be*, more pressure on developers of EBPs for SUDs to focus on packaging interventions in ways that will optimize transportability and utilization. Increasingly, this includes the use of technological aides such as interactive software programs available both for personal computers and on the Internet (Bickel, Marsch, Buchhalter, & Badger, 2008; Hester, Squires, & Delaney, 2005; Ondersma, Chase, Svikis, & Schuster, 2005; Stretcher, Shiffman, & West, 2005), the use of handheld or touch-screen computers for a variety of treatment-oriented tasks (Gwaltney, Shiffman, & Sayette, 2005; Ondersma et al., 2005), and interactive voice response (IVR) systems for telephone-based services (Schneider, Schwartz, & Fast, 1995).

The use of TA programs in the assessment and treatment of SUDs represents an exciting prospect for the field. Furthermore, TA programs are unparalleled with respect to their ability to accelerate the dissemination of EBPs for SUDs. Rather than being viewed as a threat to the interpersonal nature of treatment or dismissed as far-fetched, TA programs have the potential to aid clinicians and their patients in a number of valuable ways. Aside from providing conveniently packaged, high-fidelity interventions that can be used as adjuncts to face-to-face treatment, TA programs can also assist clinicians in engaging clients in a creative, interactive manner that can foster increased interest in a given topic. In fact, research has shown that TA programs can stimulate earlier and more honest exploration of therapeutic topics (Kobak et al., 1997; Servan-Schreiber, 1986), and technology-assisted programs may also be an effective way to offer low-cost, flexible-format treatment options to a greater number of clients, including those who are waiting to receive services following referral. Indeed, there is an impressive body of literature to support the efficacy of TA programs for numerous behavioral health issues, including SUDs (Marks, Cavanagh, & Gega, 2007).

When it comes to SUDs treatment, the range of TA programs currently available varies significantly according to the target class of drug. For example, while several TA programs have been reported in the literature for alcohol (6) and tobacco (7) treatment, only three have been reported for illicit drugs. A significant limitation of TA programs for treatment of SUDs, however, is that only a fraction of those programs reported in the literature are commercially available (6 of 16), and of these, a smaller proportion yet (4 of 6) have produced data to support their efficacy from randomized controlled trials (RCTs). To that end, while our presentation of TA programs will include a discussion of programs across the three principal areas of tobacco, alcohol, and illicit drugs, we will highlight those programs that are both evidence based (via RCTs) and commercially available. We will also touch upon how clinicians might utilize these programs in day-to-day practice. For an exhaustive review of the recent research literature on the topic, see Marks et al. (2007).

Technology-assisted programs for tobacco

Seven distinct TA programs focusing on tobacco (smoking) treatment are reported in the literature, and four of these are commercially available at no charge. However, only two of the four available commercially have produced data from randomized clinical trials to support their efficacy (Etter, 2005; Stretcher et al., 2005). Across the three drug classes discussed in this chapter, smoking is well known to be the most difficult to address both in terms of retaining people in treatment and in terms of long-term abstinence rates. These challenges are reflected clearly in the programs we will review. That said, considerable effort has gone into creating and evaluating TA programs for treating smoking, and as a group, smoking cessation programs spearheaded the TA movement for SUDs.

Almost all TA programs for smoking are Internet based. In some of the earliest work, Schneider (1986) developed an Internet-assisted smoking cessation program (*Go Nosmoke*) and conducted a controlled trial of its efficacy (Schneider, Walter, and O'Donnell, 1990). They found that tailoring a smoking cessation program to individuals significantly improved abstinence rates at 1- and 3-month follow-up compared with a control group. In related work, Schneider et al. (1995) developed an interactive voice response (IVR) program called *Ted*. Ted guides callers through an assessment process that enables the program to personalize preparation and planning for an interactive 10-day action plan to assist callers in quitting smoking. While the program showed early promise, only 14% of users remained abstinent at six months, which is lower than face-to-face rates. While Ted was commercially available on a limited basis in the early 1990s, it is no longer so.

Three additional programs reported in the literature have shown promising results, but fail to meet both criteria of having produced evidence of efficacy via RCTs and being commercially available. The *QuitSmokingNetwork* (Feil, Noell, Lichtenstein, Boles, & McKay, 2003) helps users focus on the benefits of quitting while addressing craving, social support, and self-efficacy, and also includes e-mail resources in the form of expert advice and a peer-support forum. Another program called *QuitNet* (Cobb, Graham, Bock, Papandonatos, & Abrams, 2005) assesses users based on stages of change (Prochaska & DiClemente, 1986) and uses a calendar to guide change efforts. The program also provides support for the use of FDA-approved medications for smoking cessation and provides access to experts via e-mail. Finally, the Internet-based program *StopSmoking* (Lenert et al., 2003) is a brief, one-time usage Web site designed to provide users with educational materials about smoking that include information on readiness for change based on stages of change theory. As with cessation the others mentioned above, it provides a calendar to clarify goals and to set a quit date.

Two TA programs for smoking cessation meet criteria for having produced both evidence of efficacy via RCT and being commercially available. That said, while both programs have produced evidence of efficacy, the magnitude of the effect sizes in both cases is not only marginal, but further limited by response bias given extremely low follow-up rates. The first of these programs is the Internet-based *Committed Quitters™ Stop Smoking Plan* (Strecher et al., 2005). The Committed Quitters program utilizes cognitive-behavior therapy (CBT) techniques for users of a specific brand of nicotine replacement gum (NiQuitin CQ 21). The program tailors responses to a series of registration questions, with a goal of increasing sustained compliance with nicotine replacement. The program was evaluated against a standard face-to-face behavioral smoking-cessation program and produced modest results favoring the Internet-based program. However, only 22% of those who originally enrolled in the program reported at the final 12-week follow-up. While efficacy data regarding the program are far from conclusive, the program is available free of charge (http://www.committedquitters.com) and can be utilized by anyone who has Internet access. While data to support the stand-alone efficacy of the program are inconclusive, clinicians may find Committed Quitters to be a useful CBT adjunct to face-to-face treatment.

The second program is *Stop-Tabac* (Etter, 2005). Like many of the other TA programs for smoking, Stop-Tabac employs an initial assessment that is used to establish personalized feedback. To that end, the program provides users with a comprehensive booklet matched to the user's appropriate stage of change. The program also encourages the use of online discussion forums and chat rooms, and provides e-mail support in the form of advice and encouragement at one- and two-month intervals into the program. Testing of the program via an RCT was based on a unique design whereby the full Stop-Tabac program was compared to an alternate version of the program with fewer components offered. While the full program did produce evidence of efficacy superior to that of the scaled-down (less individually tailored information) version, there was no alternative treatment control group that would allow either group to be compared to standard treatment. In addition, the study suffered from the all-too-common issue of extremely low follow-up rates (35% at 11 weeks), indicating a likely response bias.

While the original Stop-Tabac program was in French, an English version is available free of charge on the Internet (http://www.stoptabac.ch/en/welcome.html). As with the Committed Quitters program, clinicians may find the program to be a useful homework adjunct to face-to-face treatment for smoking cessation. However, caution is warranted for both programs given the tentative, likely biased, evidence of efficacy.

Technology-assisted programs for alcohol

In contrast to smoking, TA programs focusing on alcohol treatment are slightly fewer in number and less commercially available (2 of 6). Overall, technology-assisted programs for alcohol assessment and/or treatment have been studied in health-care settings (Hester, Squires, & Delaney, 2005; Skinner, 1994; Squires and Hester, 2002, 2004), in college student health centers (Kypri et al., 2004; Kypri & McAnally, 2005), and on the Internet (Cunningham, Humphreys, & Koski-James, 2000; Cunningham, Humphreys, Koski-James, & Cordingley, 2005; Hester et al., 2005; Riper, Kramer, Smit, Conijn, Schippers, & Cuijpers, 2008).

Some of the earliest work on TA programs focusing on alcohol was done by Skinner (1994) who developed the *Computerized Lifestyle Assessment* (CLA). The CLA was one of the first programs to address problem drinking by screening drinkers and providing feedback. The CLA assesses behavior across a number of domains and provides feedback to both the individual and his or her health-care provider on lifestyle strengths, areas of concern, and risks. These domains include, but are not limited to, use of alcohol, tobacco, and other drugs. Skinner's work on the CLA was also among the first to report patient and staff acceptance of a computer-based assessment and feedback program in a treatment setting. However, the alcohol component of the CLA is limited to a relatively brief screening procedure, not a comprehensive assessment. As a result, the feedback provided to clients is also limited, and the program does not appear to be commercially available.

In related work, Cunningham and colleagues (2000) developed a brief screen on the Internet for problem drinkers referred to as *Assessment and Personalized Feedback for Problem Drinking* and reported initial data from a survey of its users. Fourteen percent of the first 1,729 users responded to a survey and generally expressed interest in and satisfaction with the screening and feedback they received. A follow-up RCT of the program revealed modest reductions in drinking, but the findings were preliminary (Cunningham et al., 2005).

More recently, Kypri et al. (2004) developed the Internet-based *Screening and Brief Intervention* Program, which provides college students with personalized, normative feedback about drinking. Data from the original RCT of the approach boasted outstanding follow-up rates of 90% and revealed significant reductions in alcohol-related problems at six-month follow-up. However, consumption did not change significantly. In a follow-up study of an extended version of the program, which included focal areas of exercise, diet, and smoking in addition to alcohol, six-week outcomes revealed no differences on the alcohol-related variables, but did demonstrate a positive impact on exercise and diet (Kypri & McAnally, 2005).

Another Internet-based TA program for alcohol reduction is *Drinking Less* (Riper et al., 2008). Drinking Less automates CBT modules focusing on moderate drinking for those experiencing only mild to moderate alcohol-related problems. The program also includes the use of a moderated Web-based discussion board. In an RCT of the program that compared the program to a version of itself that did not include a moderated Web-based discussion board, outcomes indicated that the full version resulted in greater reductions to both drinking and related consequences.

While results from these trials have been favorable in terms of reduced consumption and/or associated alcohol-related problems, most have produced only preliminary or modest evidence of efficacy, indicating that more development and testing is necessary. Of the six TA programs reported in the literature for alcohol assessment and treatment, only two are commercially available (for a fee), and both have produced clear evidence of efficacy based on RCTs.

The first of these is the *Behavioral Self-Control Program for Windows* (BSCPWIN; Hester & Delaney, 1997). The BSCPWIN program was developed to assist motivated, moderate problem (non-severe) drinkers with a goal of moderate drinking. The program utilizes a modular approach with eight topic sections including goal setting, self-motivation, rate control, establishing rewards and penalties, developing alternatives, identifying/anticipating high-risk situations, relapse prevention, and review and feedback. The program allows users to enter self-monitoring data for a total of eight weekly sessions, and can easily be incorporated by clinicians as an adjunct to in-session work to structure homework. Twelve-month outcomes from an RCT of the program showed that stand-alone use of the BSCPWIN program produced reductions in alcohol consumption and associated problems comparable with face-to-face delivery of the same procedures. The BSCPWIN program is available commercially on CD-ROM for use on a personal computer (http://www.behaviortherapy.com). There is no Internet version of the program.

Following the success of the BSCPWIN program, Squires and Hester (2002, 2004) developed and evaluated the *Drinker's Check-up* (DCU; Hester, Squires, & Delaney, 2005; Hester & Squires, in press). The DCU is an automated, brief motivational intervention for problem drinkers, including heavy drinkers. The DCU utilizes three comprehensive, integrated modules that include assessment, feedback, and decision-making exercises. The program assesses users' risk factors, alcohol-related problems, level of dependence, and motivation for change, and then uses this information to craft a personalized feedback report. Once users of the program receive their personalized feedback they can continue to a decision-making module where they can explore ways to address their drinking, or exit the program if they are not interested in considering change options. The DCU also offers a therapist version that allows treatment providers to tailor assessment instruments and intervals for clients, manage client records, and produce detailed reports to track progress.

As with the BSCPWIN program, the DCU is particularly useful as an adjunct to regular treatment services. It can provide a range of helpful services from a brief screen for clients where alcohol may be a suspected issue, to a first step in addressing alcohol dependence. Twelve-month follow-up results from an RCT of the DCU indicated not only that outcomes were comparable to brief motivational interventions delivered in a face-to-face format, but that they actually improved over time. Notably, the DCU is the only evidence-based TA program to be included in the National Registry of Evidence Based Programs and Practices mentioned earlier. The DCU is available for individual consumers in English or Spanish on CD-ROM (http://www.behaviortherapy.com), and on the Internet (http://www.drinkerscheckup.com). The therapist version is available only on CD-ROM.

Technology-assisted programs for illicit drugs

Unlike TA programs for tobacco and alcohol, which have been under development for the past 15 to 20 years, programs focusing on treatment for illicit drugs only recently appear in the literature. While three programs have produced promising initial findings (Bickel et al., 2008; Carroll et al., 2008; Ondersma et al., 2006), no programs are currently available commercially.

In their extensive work with postpartum women, Ondersma and colleagues (2005, 2006) created the *Motivational Enhancement System*, which utilizes a touch-screen computer with headphones to provide assessment, intervention, and goal-setting activities relating to drug use. In an RCT of the program, results indicate some to moderate reduction in drug use, when the intervention component was included. The program is one of only a few that utilize mobile computer technology to deliver adjunctive treatment services in a *bedside* manner.

In recent work, Bickel et al. (2008) created and evaluated the efficacy of the first (unnamed) interactive, computer-based intervention for the Community Reinforcement Approach (CRA), which includes a voucher-based contingency management component. Results indicate that the computer-based CRA intervention produced comparable outcomes to the same protocol delivered on a face-to-face basis. One notable aspect of the program is that it is specifically designed to work in conjunction with face-to-face treatment, and it requires only one-sixth of the intervention to be delivered by the clinician, while the remaining components are delivered by computer. In this respect, it is one of only three TA programs reported in the SUDs treatment literature specifically designed to work in collaboration with a treatment provider.

Finally, Carroll et al. (2008) report data supporting the efficacy of a six-module computer-based CBT skills program called *CBT4CBT*. The program was evaluated with a sample of individuals seeking treatment for a substance use disorder (including alcohol, cocaine, opioids, or marijuana) who received either standard treatment as usual (TAU) or TAU plus the CBT4CBT program over eight weeks in an outpatient community-based clinic. Overall, findings indicate that those who utilized the CBT4CBT program in addition to TAU submitted significantly fewer positive urine specimens, and that treatment involvement including the completion of homework were strongly related to positive outcomes. The CBT4CBT modules include (1) understanding and changing patterns of substance use, (2) coping with craving, (3) refusing offers of drugs and alcohol, (4) problem-solving skills, (5) identifying and changing thoughts about drugs and alcohol, and (6) improving decision-making skills. The program requires approximately 45 minutes per module and was evaluated specifically as an adjunct to TAU in anticipation that this would be the most compatible way to integrate the program into existing treatment settings. While the CBT4CBT program is not currently available commercially, more research is underway to replicate initial findings and to clarify mechanisms.

Section summary

Technology-assisted programs for SUDs have been evolving over the past 15 to 20 years. Programs focusing on smoking cessation were the first to arrive on the scene, followed closely by programs targeting alcohol. Programs for illicit drugs have recently emerged. Despite having been around for two decades, only 4 TA programs of the 13 distinct programs as reported in the literature for the treatment of smoking and alcohol misuse are both evidence based via RCTs and commercially available. Of these, the two available at no cost for smoking cessation have produced only preliminary evidence of efficacy, with very poor follow-up rates. In contrast, the two TA programs currently available for alcohol misuse have produced strong evidence of efficacy, but are available only for a fee. Despite promising preliminary work, no TA programs are currently commercially available for illicit drug use.

Developing Internet Web sites and software applications

While our discussion thus far has been limited to the existing literature on TA programs for SUDs treatment, we understand that some readers may also have an interest in creating resources specific to their own circumstances. Whether the purpose is to create a simple Web site to provide information and resources for patients, collect data, or develop interactive TA programs, there are several things to consider and several resources to choose from. Indeed, creating Web sites today is easier than ever, and there are resources that allow even novice Internet users to build creative sites using easy-to-follow instructions, menus, and customizable templates. We will also touch briefly on resources for those technologically gifted among us who are inclined toward creating more complex, dynamic programs either on their own or with the help of an experienced programmer.

Web site basics

Whether the purpose of building a Web site is to increase the Internet presence of an organization, utilize tools for patient data collection and management, develop clinical applications, or any other purpose, one needs to first determine the desired function and goals of building a Web site. Additional preliminary considerations include available funding for the project, existing technology infrastructure within the organization, and time and resources needed to develop and maintain a site.

For organizations with existing information technology (IT) support and in-house Web servers, the development of clinical applications will usually involve working with existing IT resources to discuss a proposed site. For those lacking IT resources, it is often most efficient to utilize a third-party Web hosting company for a small monthly fee than to purchase a new Web server. New Web servers range in cost from $7,000 to $20,000, not including the cost of a system administrator, programmer, and software. Although numerous Web site hosting companies offer free disk space, we strongly recommend choosing a host based on the available features provided, reliability (fewer service interruptions), response to service needs, and most importantly, security policies, especially if goals include collecting and storing confidential information. For objective, independent ratings of numerous Web hosting companies, go to http://www.webhostinggeeks.com.

There are numerous Web hosting companies to choose from, and most provide a variety of service and pricing plans. Rates typically range from $5 to $15 per month, and include multiple e-mail addresses and the ability to store up to 1,000 Web pages. Links to some highly rated Web hosting sites include

- Network Solutions: http://www.networksolutions.com
- Inmotion: http://www.inmotionhosting.com
- Act Now Domains: http://www.actnowdomains.com

Once goals for a site have been carefully determined, selecting a domain name is usually a first step (e.g., http://www.anyname.com). Although a domain name can contain up to 63 alpha-numeric characters, care should be taken to choose a domain name that is short, captures the essence of the organization or program, and is easy to remember. While many common domain names are already taken, a little creativity will go a long way to finding a suitable name. Searching for and registering an available domain name and renting disk space can be done in several easy steps through a single Web hosting company. Traditionally dot-com names are utilized for-profit businesses and dot-org is utilized for non-profit entities. At Network Solutions (http://www.networksoultions.com), for example, consumers simply type in a domain name they wish to register and the search engine tells them if it is available. While it is not uncommon to have to search several times before finding an available name, the process is generally very quick.

Once a domain name has been decided upon, it can be registered immediately. Network Solutions, like many other hosts, will actually register domain names for free if a Web hosting package is also purchased. At the time the domain is registered, a security certificate can also be purchased, which is especially important if confidential data will be collected. In addition to assuring future Web site users that they are actually accessing the site they think they are, the security certificate also, when invoked, encrypts all data being transferred to and from the site. Although the industry standard is 64-bit encryption, 128-bit certificate is recommended.

With a domain name in hand, building a Web site can be a simple process involving friendly do-it-yourself site builder programs that are generally provided by Web hosting companies, through which Web pages and data-collection forms can be created from templates with relative ease. Alternatively, consumers can opt to work with a programmer from the Web hosting company to create custom pages for a Web site (not from templates), which will average about $500; or can hire an altogether independent programmer/designer to build a site that can later be uploaded to any Web hosting company. The latter option will usually average about $100/hr and, depending on the site and function desired, can require from as few as 10 to hundreds of hours to complete.

Starter Web sites with basic functions are relatively simple, and usually contain five to ten static pages, meaning they contain content only and are not interactive or database driven. These sites allow visitors to provide feedback through posted contact information or e-mail. Since most Web hosting companies offer simple-to-use applications that people familiar with Windows or Macintosh operating systems can use for creating their own Web site, most novices with a little computer know-how can created a basic site.

Beyond the basics: Customized Web design tools and basic advice

For the more ambitious or technically inclined, there are numerous Web-editing and authoring applications available for creating custom Web content, such as Front Page® and Dreamweaver®. Programs like these are often referred to as WYSIWYG (what you see is what you get) editors, because what one sees on the screen during programming is exactly what visitors to the eventual site will see. One of the benefits of WYSIWYG editors is that while creating the Web page, both the "rendered" (actual) Web page and the programming code (typically Hypertext Markup Language, or HTML) used to create it are visible at the same time.

While HTML is the primary programming code for rendering Web pages, there are more advanced applications and programming codes that allow for the enhancement of HTML pages and the development of dynamic database-driven Web applications, through the use of computer languages such as Java®, CGI®, PHP®, and ColdFusion®. The use of more advanced coding languages allows for the development of dynamic applications through the use of interactive dynamic database-driven Web sites. Although these powerful applications are menu driven and generally user friendly, there is a steep learning curve involved in understanding and implementing the numerous features they offer. For a lengthy list on Web authoring applications and tools, as well as a survey that can assist in selecting the best application(s), readers can visit About.com and search for "web authoring tools." As of the date of this writing, the current link for that specific page is http://webdesign.about.com/od/htmleditors/HTML_Editors_Web_Page_Authoring_Tools.htm).

When and how to seek Web design assistance

The more complicated an idea and goals for a site or TA program, the more likely it is that an experienced designer/programmer/consultant will be needed to, at least in part, execute the programming. The benefits of working with a professional Web designer are numerous—not the least of which is assurance that the application will be secure, which is of paramount importance for those bound by the Health Insurance Portability and Accountability Act (HIPAA). Working with a professional Web designer can also help to focus discussion around design and functionality in a way that can rapidly bring goals and vision to fruition. To build more complex interactive sites, most designers will utilize PHP® and Cold Fusion®, which involve intricate logic statements and interdependent table and database relationships.

As with simpler options for designing and building a custom site with template-based programs, it is extremely important that one carefully analyze the desired objectives, and that the structure and content of a site be well thought out before beginning to work with a Web designer. Security needs must also be clarified, along with the type of information (both verbal and nonverbal) the site will convey, data collection needs, product support, and the advanced interactive functionality desired, to name a few.

To that end, it can be very helpful early on in the development of an application to lay out the structure, including main and subdirectories, and the initial content that will be included. This can provide a good overview and visual representation of how many pages the site will include as well as the goals for each page. If a site or program is to include assessment, evaluation, feedback, or other instruments, those should be organized and included within the layout along with variable names, values, and value labels. Guided interviews, Web-based awareness trainings, or surveys that provide intermediate feedback should be carefully planned, including the logic, flow, and all possible response scenarios (e.g., if Yes to Question 1, lead visitor to page 2; if No, then page 3).

When submitting a proposal to a Web designer/programmer, the more planning and information that have been organized ahead of time, the smoother and cheaper the developmental process will be. Costly project overruns can also be avoided if modifications to the original plan are kept to a minimum. It is important to remember that most Web designers will not have expertise on the specific content area of any given site or application (e.g., SUDs treatment). If the logic and flow of a proposed site or clinical application have not been thoughtfully conveyed, programmers will likely have difficulty translating desires and vision into code. To that end, finding a programmer with not only solid technical skills, but good listening and interpretative design skills, will go a long way toward insuring timely and accurate completion of a Web site or TA program. When seeking a web designer, consumers should plan to visit designers' sites and review their list of clients, portfolio, references, technical assistance provided, and pricing structure. To locate a Web designer/programmer, Internet searches using a search engine such as Google (www.google.com) can be done for terms such as *web designer, web programmer, web designer ratings,* or *web programmer ratings*. If a local designer is preferred, consumers can simply include the name of a specific city or region in the queries, or consult local yellow pages.

Ethical considerations of TA programs for substance use treatment

Whether the interest is in identifying and utilizing an existing TA program or in creating a custom application for one's own circumstances, two ethical issues permeate the use of clinical TA programs: the privacy of data and the extent to which TA programs should be utilized in the intimate, interpersonal process of treatment. We will briefly touch upon both.

Privacy

Of foremost importance to most clinicians today is compliance with HIPAA. The Health Insurance Portability and Accountability Act requires that health-care professionals take specific and thorough action to protect the privacy of protected consumer information, with a special focus on information transmitted electronically. Obviously, this is especially relevant for TA programs that store or transmit such information. When considering the use of a TA program or working to design a new application, clinicians should take great care to ensure that the management of all data is in compliance not only with HIPAA, but with industry standards for security and encryption. Questions regarding HIPAA compliance can be addressed online (http://www.hipaa.org) or with local credentialing and funding sources. Questions regarding encryption and data security can be directed to Web hosting companies and/or programmers.

Utilization of TA programs in practice

Despite increasing interest in, and evidence to support the use of, TA programs across a variety of behavioral health issues, there are some who have cautioned against the widespread use of computer technology in behavioral health. In one of the earliest and strongest challenges, Weizenbaum (1976) argued against the analogy of man and computers being essentially of the same "information processing" genus stating, "The public vaguely understands—but is nonetheless firmly convinced—that any effective procedure can, in principle, be carried out by a computer" (p. 157). Indeed, we live in an increasingly computer-driven era where cars "vocally" guide us through city streets using global positioning guidance systems, where information on almost any conceivable topic is available at the touch of a button, and where virtual stores have revolutionized the ways in which consumers purchase goods and services. Given these advances and the many conveniences they offer, one might understandably conclude that the bounds of computer-based technologies are, indeed, limitless. Nonetheless, Weizenbaum asserts, "There are some human functions for which computers *ought* not be substituted" (p. 270).

While cautions regarding the overuse of computer technology in behavioral health are well received by the current authors, there are undeniable benefits as well. First, day-to-day use of computer technology (including the Internet) has exploded since the late 1980s. Today, people from around the world share information, experiences, and data at the touch of a button. Rather than being forced to use this medium, however, most users of computer technology today have adopted it presumably because there is a relative advantage for them associated with doing so (Rogers, 2003). Therefore, rather than being viewed as a threat to the interpersonal nature of the therapeutic relationship, it may be the case that TA programs utilized in conjunction with face-to-face treatment serve to enhance both the therapeutic relationship and the outcomes. Technology-assisted programs, and especially those that are Internet based, may also serve to increase individual perceptions of having greater control over one's experience, along with the choice of when to have that experience.

If creative TA programs can assist in improving the quality of treatment, and hence peoples' lives, the public health benefits could be substantial. With respect to the resolution of substance use disorders where the dissemination of EBPs through traditional means is exceedingly slow, this may be especially true. While there are undoubtedly limitations with respect to the degree to which TA programs can and *should* be utilized in the provision of treatment services, one thing that a free market economy guarantees is that consumers will decide for themselves what is useful and worthy of adoption and utilization, and what is not.

Key Considerations. For providers considering the integration of TA programs into their clinical practice, the following items may provide a starting point:

1. What is the desired scope and function for a given TA program within the context of face-to-face services (e.g., assessment, facilitation of CBT modules and/or homework, feedback)?
2. What type of platform is desired (Internet, office based, or both)?
3. What type of patients will a TA program be appropriate for (e.g., age, educational level, acuity)?
4. What additional time will a TA program require from both clinicians and patients to utilize?
5. What is the degree of, and mechanism for, clinician access to information stored and reported by a TA program?
6. What about billing issues (e.g., parameters regarding third-party payer reimbursement for services supplemented by TA programs)?

Answers to these, and perhaps additional questions, are critically important first steps for establishing a solid foundation for integrating TA programs into clinical practice.

Summary and conclusion

We have covered three distinct areas in this chapter, including a review of existing TA programs for SUDs treatment, general steps and recommendations for creating custom Web site and TA programs, and a brief discussion of ethical and practical issues.

While the literature on TA programs for SUDs treatment has been developing over the past 20 years, it remains in its infancy. Use of many of the programs reported in the literature has been limited to the studies within which they were evaluated, and few are commercially available. As the field progresses beyond the initial phase of feasibility research, it is anticipated that the variety, quality, and availability of TA programs for treating SUDs will increase. However, important questions remain about when, where, for whom, and to what degree such programs should be utilized. Future studies will need to address these important questions, in addition to general efficacy.

We have also provided preliminary technical guidance, which we hope will allow readers to consider creating their own TA resources or programs. While the process can be as simple as setting up a basic Web site to increase visibility or accessibility on the Internet, it can also lead to the creation of dynamic TA programs. Indeed, the possibilities are bound only by the limits of imagination.

Finally, we have briefly touched upon ethical considerations involving privacy and the clinical use of TA programs. One of the true pleasures (and challenges) of working with individuals to resolve behavioral health issues is that each person's needs are unique and always multifaceted. To that end, we encourage clinicians to actively involve patients in the process of considering a variety of treatment and other support options to support immediate and long-term goals. As with TA programs for treatment, the idea is collaboration, and when it is done well, everyone benefits.

References

Bickel, W. K., Marsch, L. A., Buchhalter, A. R., & Badger, G. J. (2008). Computerized behavior therapy for opioid-dependent outpatients: A randomized controlled trial. *Experimental and Clinical Psychopharmacology, 16,* 132–143.

Carroll, K. M., Ball, S. A., Martino, S., Nich, C., Babuscio, T. A., Nuro, K. F., et al. (2008). Computer-assisted delivery of cognitive-behavioral therapy for addiction: A randomized trial of CBT4CBT. *American Journal of Psychiatry, 165,* 881–888.

Cobb, N., Graham, A. L., Bock, B. C., Papandonatos, G., & Abrams, D. B. (2005). Initial evaluation of a real-world Internet smoking cessation system. *Nicotine and Tobacco Research, 7,* 207–216.

Cunningham, J. A., Humphreys, K., & Koski-James, A. (2000). Providing personalized assessment feedback for problem drinking on the Internet: Pilot project. *Journal of Studies on Alcohol, 61,* 794–798.

Cunningham, J. A., Humphreys, K., Koski-James, A., & Cordingley, J. (2005). Internet and paper self-help materials for problem drinking: Is there an additive effect? *Addictive Behaviors, 30,* 1517–1523.

Etter, J. F. (2005). Comparing the efficacy of two Internet-based computer-tailored smoking cessation programs: Randomized trial. *Journal of Medical Internet Research, 7,* 6.

Feil, E. G., Noell, J., Lichtenstein, E., Boles, S. M., & McKay, H. G. (2003). Evaluation of an Internet-based smoking cessation program: Lessons learned from a pilot study. *Nicotine and Tobacco Research, 5,* 189–194.

Gwaltney, C. J., Shiffman, S., & Sayette, M. A. (2005). Situational correlates of abstinence self-efficacy. *Journal of Abnormal Psychology, 114,* 649–660.

Hester, R. K., & Delaney, H. D. (1997). Behavioral self-control program for Windows: Results of a controlled clinical trial. *Journal of Consulting and Clinical Psychology, 65,* 685–693.

Hester, R. K., & Squires, D. D. (in press). Web-based norms for the Drinker Inventory of Consequences from the Drinker's Check-up. *Journal of Substance Abuse Treatment*.

Hester, R. K., Squires, D. D., & Delaney, H. D. (2005). The computer-based Drinker's Check-up: 12 month outcomes of a controlled clinical trial with problem drinkers. *Journal of Substance Abuse Treatment, 28*, 159–169.

Kobak, K. A., Taylor, L. H., Dottl, S. L., Greist, J. H., Jefferson, J. W., Burroughs, D., et al. (1997). A computer-administered telephone interview to identify mental disorders. *Journal of the American Medical Association, 278*, 905–910.

Kypri, K., & McAnally, H. (2005). Randomized control trial of a Web-based primary care intervention for multiple health risk behaviours. *Preventive Medicine, 41*, 761–766.

Kypri, K., Saunders, J. B., Williams, S. M., McGee, R. O., Langley, J. D., Cashnell-Smith, M. L., et al. (2004). Web-based screening and brief intervention for hazardous drinking: A double-blind randomized controlled trial. *Addiction, 99*, 1410–1417.

Lamb, S., Greenlick, M. R., & McCarty, D. (1998). *Bridging the gap between practice and research: Forging partnerships with community-based drug and alcohol treatment*. Washington, DC: National Academy Press.

Lenert, L., Munoz, R. F., Stoddard, J., Delucchi, K., Banson, A., Skoczen, S., et al. (2003). Design a pilot evaluation of an Internet smoking cessation program. *Journal of the American Medical Informatics Association, 10*, 16–20.

Marks, I. M., Cavanagh, K., & Gega, L. (2007). *Hands on help: Computer-aided psychotherapy*. New York: Psychology Press.

Meyers, R. J., & Miller, W. R. (2001). *A community reinforcement approach to addiction treatment*. Cambridge, U.K.: Cambridge University Press.

Miller, W. R., & Willbourne, P. L. (2002). Mesa Grande: A methodological analysis of clinical trials of treatments for alcohol use disorders. *Addiction, 97*, 265–277.

Ondersma, S. J., Chase, S. K., Svikis, D. S., & Schuster, C. R. (2005). Computer-based brief motivational intervention for perinatal drug use. *Journal of Substance Abuse Treatment, 28*, 305–312.

Ondersma, S. J., Svikis, D. S., & Schuster, C. R. (2006). Leveraging technology: Evaluation of a computer-based brief intervention for post-partum drug use and a dynamic predictor of treatment response. Poster presented at the 68th Annual Scientific Meeting of the College on Problems of Drug Dependence, Scottsdale, AZ.

Prochaska, J. O., & DiClemente, C. C. (1986). Toward a comprehensive model of change. In W.R. Miller and N. Heather (Eds.), *Treating addictive behaviors: Processes of change* (pp. 3–27). New York: Plenum Press.

Riper, H., Kramer, J., Smit, P., Conijn, B., Schippers, G., & Cuijpers, P. (2008). Web-based self-help for problem drinkers: A pragmatic randomized trial. *Addiction, 103*, 218–227.

Rogers, E. M. (2003). *Diffusion of innovations* (5th ed.). New York: Free Press.

Schneider, S. J. (1986). Trial of an on-line behavioural smoking cessation program. *Computers in Human Behavior, 2*, 277–286.

Schneider, S. J., Schwartz, M. D., & Fast, J. (1995). Computerized, telephone-based health promotion: 1. Smoking cessation program. *Computer in Human Behavior, 11*, 135–148.

Schneider, S. J., Walter, R., & O'Donnell, R. (1990). Computerized communication as a medium for behavioral smoking cessation treatment: Controlled evaluation. *Computers in Human Behavior, 6*, 141–151.

Servan-Schreiber, D. (1986). Artificial intelligence and psychiatry. *Journal of Nervous and Mental Disease, 174(4)*, 191–202.

Skinner, H. A. (1994). *Computerized lifestyle assessment*. Toronto, Ontario: Multi-Health Systems, Inc.

Squires, D. D. (2004). The research to treatment gap: Disseminating effective methods into practice. *The Brown University Digest of Addiction Theory and Application, 23*, 8.

Squires, D. D., & Hester, R. K. (2002). Development of a computer-based brief intervention for drinkers: The increasing role for computers in the assessment and treatment of addictive behaviors. *The Behavior Therapist, 25*, 59–65.

Squires, D. D., & Hester, R. K. (2004). Utilizing technical innovations in clinical practice: The Drinker's Check-up software program. *Journal of Clinical Psychology/In Session, 60*, 159–169.

Strecher, V. J., Shiffman, S., & West, R. (2005). RCT of a Web-based computer-tailored smoking cessation program as a supplement to nicotine patch therapy. *Addiction, 100*, 682–688.

Stuyt, E. B., Sajbel, T. A., & Allen, M. H. (2006). Differing effects of antipsychotic medications on substance abuse treatment with co-occurring psychotic and substance abuse disorders. *American Journal on Addictions, 15*, 166–173.

Vaillant, G. E. (1995). *The natural history of alcoholism revisited*. Cambridge, MA: Harvard University Press.

Weizenbaum, J. (1976). *Computer power and human reason: From judgment to calculation*. San Francisco, CA: W. H. Freeman & Co.

chapter six

Smoking cessation via the Internet

Yan Leykin, Alinne Z. Barrera, and Ricardo F. Muñoz

Despite smoking being the leading cause of preventable deaths in the United States, smoking cessation is not adequately addressed in health care practice. Approximately 30% of providers address smoking as part of routine health care (Schroeder, 2005). As patient loads increase and time available per individual patient shrinks, clinicians are often forced to address problems that they consider more pressing, which often leads to the exclusion of smoking from the discussion. In part, however, failure to address tobacco use could be affected by training patterns. For example, Prochaska and colleagues report that the majority of psychiatry students and residents are ill prepared to adequately treat smoking in their practices (Prochaska, Fromont, & Hall, 2005; Prochaska, Fromont, Louie, Jacobs, & Hall, 2006). The combination of inadequate training and constraints of clinical practice result in failure to initiate the smoking cessation process, which may be one of the most important interventions for health maintenance.

Smoking cessation interventions delivered over the Web can provide numerous benefits to smokers. Clinicians' ability to refer patients to an evidence-based cessation Web site can enhance their efforts to guide a patient through the cessation process. Patients can have access to efficiently presented information about smoking and its effects, and to a variety of quitting and support tools to optimize the quitting experience.

The World Health Organization (WHO, 2008) warns that the rates of smoking in developing countries are rising. This means that persons wishing to quit will increasingly come from areas of the world where quality health care may not be easily accessible and may not have access to, or be able to afford, medications that are known to aid in smoking cessation, such as nicotine replacement therapy (NRT). The same can be said about many who live in the United States without consistent medical care, such as those on fixed incomes, those without medical insurance, persons who do not speak English, or those residing in rural areas. For these populations, multilingual Internet interventions could provide accessibility by being available around the clock, without needing to wait for difficult-to-produce changes in local health care systems.

Dangers of tobacco use

Worldwide, tobacco is responsible for over five million deaths annually, a number that is estimated to grow to eight million by the year 2030 (WHO, 2008). In the United States, smoking is the leading preventable cause of death, accounting for 440,000, or one in five, deaths per year (http://www.cdc.gov/mmwr/preview/mmwrhtml/mm5114a2.htm), including 38,000 who die from secondhand smoke (Centers for Disease Control and Prevention [CDC], 2005, 2007). Smoking is responsible for over two times as many deaths as AIDS, homicide, suicide, drugs, and alcohol combined. Tobacco shortens the smoker's life span by up to 14 years (CDC, 2005), and increases the time the smoker will live with a disability by 2 years (Nusselder, Looman, Marang-van de Mheen, van de Mheen, & Mackenbachet, 2000).

In 1999, tobacco use cost an estimated $76 billion in health care expenses in the United States, which amounted to 6% of all health care expenses (Mackay & Eriksen, 2002). This figure has risen by about $20 billion in the decade since that estimate (CDC, 2007). Exposure to secondhand smoke is associated with $10 billion in health care costs annually (Behan, Eriksen, & Lin, 2003). Smokers were found to use up to 40% more sick days than nonsmokers (Mackay & Eriksen, 2002), costing over $97 billion annually in lost productivity (CDC, 2007).

Worldwide tobacco use

According to the World Health Organization (2007), 29% of the world population smokes tobacco. Almost half of all men worldwide smoke, as do about 10% of women. In the United States, over 23% of adults smoke currently, with men smoking slightly more than women; the number may be even higher for youth, with over 25% smoking currently.

The WHO reports that the tobacco industry has been increasingly focusing on youth in the developing countries in its marketing efforts. In fact, around 80% of the world's smokers are now found in low- and middle-income countries. The WHO predicts that tobacco use is on track to becoming the largest cause of preventable death or disability in the world, with up to one billion deaths attributable to tobacco in the 21st century. Given the current demographic trends, 80% of those deaths will take place in the developing world (WHO, 2007). Given that tobacco kills up to half of its regular users (WHO, 2008) and likely affects many others with a variety of health complications, an easily accessible health support system becomes of paramount importance. However, developing countries are more likely to lack a robust health care system capable of handling an influx of patients dealing with complications related to tobacco use. They also are more likely to lack the capacity to make preventive health care widely and commonly available.

Current emphasis on evidence-based treatments

The health field has recently begun to demand that its practitioners provide evidence-based, or empirically supported, treatment for their patients, in an effort to improve the value of health care (Bernal & Scharron-del-Rio, 2001; Chambless & Ollendick, 2001; Porter & Teisberg, 2007). Interventions other than traditional face-to-face interventions should be subjected to the same criteria. We recommend that the Internet intervention field carry out studies to document the level of evidence for specific interventions for specific health problems, such as smoking.

One of the most common criteria for evaluating a treatment for consideration as an evidence-based intervention was developed by the American Psychiatric Association (APA) Division 12 Task Force. As described in Chambless and Ollendick (2001), a "well-established treatment" must have had two or more between-group design trials that demonstrated either superiority to a placebo or another treatment, or equivalence to an established treatment; used sound experimental strategies; clear treatment manuals, and been tested by two different research teams. To be considered a "probably efficacious treatment," an intervention must have been demonstrated to be superior to wait-list control in two randomized trials or to meet most of the criteria for "well-established treatment," with an exception of the requirements for two separate research teams.

Clearly, some aspects of these criteria are more applicable for Internet-based interventions than others. For instance, it is unlikely that two different teams of investigators would develop exactly the same Web site to test its efficacy. Nonetheless, in order to help clinicians make determinations regarding referrals to quitting Web sites, evaluations of Internet interventions should strive for a high degree of methodological rigor and sophistication.

Evidence-based practices for smoking cessation

Numerous treatments exist to assist smokers in their attempts to quit smoking and remain abstinent. Clinicians and health care providers are charged with the responsibility of delivering empirically supported methods and, therefore, hold a certain level of responsibility for assisting smokers in choosing cessation methods. The Clinical Practice Guideline for

"Treating Tobacco Use and Dependence: 2008 Update," published by the U.S. Department of Health and Human Services (Fiore et al., 2008), encourages the use of brief interventions (e.g., physician advice), given their cost effectiveness and good results. Clinicians are advised to identify and document tobacco use at every health care visit and, at minimum, clinicians should (1) ask about tobacco use, (2) advise all users to quit, (3) assess whether the user is ready to quit at the moment, (4) assist users with a detailed quit plan, and (5) arrange for follow-up visits to review quit progress.

Traditional face-to-face therapy is an effective treatment option when available. A key role of a clinician working with an individual who is ready to quit smoking is helping him or her prepare for quitting by (1) setting a quit date (preferably within two weeks), (2) encouraging the patient to share his or her plan to quit with family and friends and to seek their support, (3) brainstorm potential obstacles that may interfere with quitting, including nicotine withdrawal symptoms, (4) recommend the patient remove all tobacco products from his or her environment, and (5) reduce smoking in common places. Improved outcomes result when the therapy (individual, group, telephone) becomes more intense and when it is combined with medications. Counseling should reinforce complete abstinence from smoking and reduction of or abstinence from alcohol; focus on problem-solving triggers; identify other smokers in the patient's environment, failed past quit attempts, etc.; and provide additional support either through clinical contact or via additional support materials or outlets (e.g., national quit line: 1-800-QUIT-NOW). Finally, follow-up sessions are critical and should begin the week of the established quit date, with the goal of assisting patients with relapse prevention and further problem solving. Likewise, a variety of cessation medications are available. Buproprion, varenicline, and NRTs (gum, patch, nasal spray, inhaler, and lozenges) are all approved by the U.S. Food and Drug Administration for the treatment of tobacco use and dependence and are considered first-line treatment options (alone and in combination). Clonidine and nortriptyline are recommended as second-line medications. All the recommended medications demonstrate similar efficacy and few adverse effects (Fiore et al., 2008).

Traditional smoking cessation methods yield 6-month cessation rates in the range of 14% to 22% for the nicotine patch (Fiore, Smith, Jorenby, & Baker, 1994) and 24% to 27% with smoking cessation groups (Lando, McGovern, Barrios & Etringer, 1990). Nicotine replacement therapies consistently produce higher quit rates when compared to a placebo control (Schroeder, 2005). Newer medications, such as varenicline (Gonzales et al., 2006) can increase and maintain abstinence rates to 30% or even higher levels (see http://www.fda.gov/CDER/Drug/infopage/varenicline/default.htm for warnings related to rare but severe side effects).

Internet interventions

For individuals logging into the smoking cessation Web sites, Internet interventions will provide a new type of quitting experience. Unlike bibliotherapy or booklets and handouts that might be available in physicians' offices, the medium of the Internet allows for an interactive and personalizable cessation aid. By being engaging and appealing to a user, Web interventions offer the possibility of actively involving smokers in their own recovery, including monitoring their own progress and gaining a feeling of empowerment, and may potentially increase the chances of quitting successfully.

Automated self-help Web-based interventions are time and cost effective. Unlike most other treatments, including bibliotherapy, physician and mental health encounters, or support groups, they are nonconsumable and can be used by many people at the same time. A visit to a clinician by one person serves only that person, during that period of time. Upon the conclusion of the visit, that time with the clinician is irretrievably spent, and no other person is able to make use of it. An individual visiting a Web site does not use up that resource and does not prevent another person from using it during the same time span. Furthermore, fidelity of information delivery and treatment implementation, which is often not consistent when administered by live clinicians, remains constant on the Web. The flexibility of Web sites allows for rapid updates and extensions with the introduction of new knowledge, interventions, and information, as well as the expansion of the Web site's contents to a variety of languages.

The unparalleled reach of interventions available on the Internet is especially important. An intervention that offers small improvement in the health of large numbers of people can be considered, in terms of public health, more valuable than one that offers greater improvement to few people (see RE-AIM model, e.g., Glasgow, McKay, Piette, & Reynolds, 2001). Fortunately, as will be discussed in the following text, quality Web interventions are comparable to other traditional cessation tools in terms of quit rates, yet can reach a great deal more people at little to no cost to the individual consumer.

Overview of past investigations of smoking cessation Internet interventions

Public interest. Smokers appear to be interested in Web interventions. McClure and colleagues (2006) invited patients from two major health care organizations to quit smoking. They reported that the level of interest, as measured by visiting the Web site and enrolling in the program was comparable to those in non-Web-based programs. Of those eligible, almost two thirds agreed to enroll in the program.

The question of interest was also addressed by Muñoz and colleagues (2006), who reported on a series of trials conducted in English and Spanish. Each trial demonstrated the feasibility of attracting large geographically diverse samples in both languages, as well as evaluated the benefits of two of the components of their cessation Web site, educational e-mails, and mood management. The first goal, the passive recruitment strategy, which consisted of links from search engines and press releases about the Web site, was highly successful, recruiting over 4,000 users from 74 countries. This clearly illustrates the global reach of Internet-based interventions, as well as potential interest in and demand for such interventions. The individually timed educational messages delivered via e-mail proved to be effective, increasing quit rates in the Spanish-speaking samples (26% at 6 months, using the convention of missing = smoking). Interestingly, the mood management component appeared to be associated with lower abstinence rates, though still in the range provided by the nicotine patch, that is, 20% at 6 months for Spanish speakers (Muñoz et al., 2006, Table 4, p. 82). This finding was surprising, as it does not agree with some of the previous research on mood management methods increasing quit rates (e.g., Muñoz, Marín, Posner, & Pérez-Stable, 1997). A more recent randomized trial yielded 20% quit rates for Spanish speakers and 21% quit rates for English speakers at 12 months (missing = smoking). Although in that randomized controlled trial (RCT) the mood management component did not yield lower quit rates, it also failed to improve them, suggesting that the standard quitting guide is likely a powerful stand-alone intervention (Muñoz et al., in press).

Attrition. A common concern for clinical trials conducted on the Internet is the prevalence of Internet users who are one-time browsers and are unlikely to return to smoking cessation sites (Eysenbach, 2005). While tens of thousands may click onto a Web site, a smaller proportion of them choose to sign up, fewer will visit the site repeatedly, and fewer still will respond to requests for follow-up data. A concern is that Internet samples may represent a small and likely unrepresentative proportion of the original visitors. Of course, randomized samples in traditional face-to-face trials are also composed of small and unrepresentative proportions of potential participants, in addition to being highly constrained geographically. It would be misleading to assume that visiting a Web site is similar to showing up at a research clinic. A more proper analogy would be comparing site visitors to people reading an advertisement for a face-to-face trial, and comparing those who fill out the initial eligibility questionnaire on the Web to those making the initial phone call to learn more about a clinical trial. The response cost of initial contact is lower for Web trials as compared to face-to-face trials—not as much motivation is needed to click on a Web site as opposed to going into a clinic, and once committed, the social obligation of continuing with the Web site is presumably lower than continuing to go to scheduled appointments. Nonetheless, some studies have achieved an admirable follow-up rate, comparable to a traditional RCT. For instance, Saul et al. (2007) attained a respectable 78% 6-month follow-up rate, reaching those who did not respond to the initial online solicitation (roughly half) by telephone. This example opens the possibility of conducting a high-quality RCT on the Internet, with interpretable and meaningful results.

Brendryen and Kraft (2008) report on an automated cessation program conducted in Norway. This program is unique for two reasons. First, the user interacted with the program by a variety of media, including cell phone, Web site, and e-mail and SMS (text messages). Second, the amount of contact that was expected from the user was considerable. Every day for the first 6 weeks of the 54-week program, the user would be expected to make a phone call to listen to a recording, log on to the Web site, receive a text message, receive an e-mail, and receive an automated phone call. Somewhat remarkably, only 12% of subjects withdrew consent in the first 6 weeks, and the program attained a repeated abstinence rate of 22% (abstinent at 1, 3, 6, and 12 months), as compared to 13% of information-only controls (a majority of both groups were also on NRT). This suggests that for at least some individuals, an intensive program delivered electronically can be well tolerated and produce superior results.

Tailoring and personalization. The importance of tailoring the intervention to the individual user was revealed in a study conducted by Stretcher, Shiffman, and West (2005). The authors found that, when used in addition to a nicotine replacement patch, an Internet intervention tailored to the individual user produced significantly better outcomes than a standard, nontailored intervention. Not only were the odds of participants quitting and staying quit higher for the tailored intervention (at 6 weeks, 29% vs. 23.9%, missing = smoking, over 1.30 odds ratio), but they also reported being more satisfied with it. This finding was also echoed by a different Web cessation program, *1-2-3 Smokefree* (Swartz, Noell, Schroeder, & Ary, 2006). This program is video based and provides tailored smoking cessation tools and motivational materials to smokers wishing to quit within the next month. By 90 days, those receiving program materials were three times more likely to quit as compared to wait-list controls, 24% to 8% (12% to 5% with the intent-to-treat sample, missing = smoking).

Strecher and colleagues (2008) conducted a type of dismantling study to understand the most potent communication components in intervention delivery. They focused on the following attributes: expected outcomes (highly vs. minimally personalized), expected efficacy (highly vs.

minimally personalized), hypothetical success stories (highly vs. minimally personalized), personalization of source (organizational vs. personable), and exposure to information schedule (massed vs. distributed over time). Though all highly personalized variations attained slightly better outcomes, for only two variables—success story (34.3% vs. 26.8% quit at 6 months, missing = smoking) and message source (33.6% vs. 27.4% quit at 6 months)—did the difference between high and low personalization reach significance. This further suggests the advantages of tailoring and "humanizing" the interventions to the extent that it is possible.

The studies described above represent a new frontier in clinical research, with the Internet serving both as a vehicle and an active agent of change. In the majority of studies and for the majority of users, the Web site was used as a complement to medication or nicotine replacement therapy, often increasing quit rates found when using these products alone. In their review of computerized smoking cessation tools, Walters, Wright, and Shegog (2006) observed, perhaps not too surprisingly, that interventions were most effective for those individuals already wishing to quit. This, of course, is also the case with other interventions: those not ready to quit generally do not buy nicotine gum or patches or attend smoking cessation groups. Web sites have the advantage of being available at any time, at the user's convenience, instead of requiring a trip to a pharmacy or to a smoking cessation clinic. This again suggests a key role for health care professionals, who can initiate the process of change by instilling the initial idea of quitting, actively encouraging their patients to quit smoking, and referring them to quit lines or evidence-based Internet interventions. An e-mail from a physician with a link to an evidence-based site may be a powerful (and convenient) incentive for a patient to start the smoking cessation process.

Toward evidence-based Internet smoking cessation interventions

Looking back at the available literature on Web-based smoking cessation, several conclusions can be reached. Internet cessation programs seem to attract plenty of interest from the smoking public, at least on par with more traditional nonpharmacological cessation interventions. It is likely that the demand for Web interventions will continue to grow as Internet penetration increases, and as people become more comfortable with and accustomed to looking for health management on the Web.

Web cessation programs are moderately successful in helping individuals to quit smoking. When used in conjunction with NRT or other biological agents, they are generally shown to improve outcomes over the contribution of NRT alone, and they have also been shown to work as a stand-alone treatment, on par with other behavioral treatments. This is of particular interest for individuals to whom pharmacological or behavioral treatments may be either unavailable or prohibitively expensive. While the quit rates for all interventions, Web based or otherwise, are relatively modest, given the scope of the problem, any decrease in the number of tobacco users worldwide can have a considerable impact on the health of entire populations.

Published reports on Internet-based interventions suggest several themes when it comes to improving quit rates.

Breadth. In their review of the available smoking cessation Web sites, Bock et al. (2004) noted that a large majority fail to mention one or more techniques or tools that constitute best practices for quitting. There is an inherent dilemma: in order to maximize the person's chances of quitting, the person must be given many options to do so, and that means being diligent and thorough in providing comprehensive treatment that includes a variety of components. However, too many choices may overwhelm the user. Providing a limited set of options might be more efficient and perhaps more effective. There will always be a conflict between breadth of information and its value. As the size of the Web site and the information contained therein increases, the value of each piece of information may be perceived as reduced. Tailoring information to the individual user can help in counteracting this perception, as discussed below.

Tailored and personalized treatment. Tailoring of interventions to an individual user can be important. No Web site may fully emulate the responsiveness of a live clinician (at least not with current technology). However, Web sites may be able to provide users with the most effective interventions tailored to their characteristics, in a manner that is more consistent than that of a live clinician. Strecher, Shiffman, and West (2006) reported on two versions of their Web smoking cessation intervention (CQ PLAN): a version automatically tailored to each user and a generic nontailored version. They found that greater differences between the two versions of the intervention occurred when users reported having tobacco-induced illness, nonsmoking children in the house, and higher alcohol usage. For each of these, the tailored intervention delivered customized information to users (e.g., a person reporting having diabetes would receive additional information describing the influence of tobacco use on development of diabetes).

This level of customization is beneficial in two important ways. First, by focusing on the unique circumstances of the individual, the user is likely to feel more understood; and believe that the program is relevant and beneficial to him- or herself; in fact, perceived relevance was found to be an important mediator of quit rates in the above study. Second, it reduces the burden of unnecessary information. This is similar to a live clinician, who notes the patient's concerns and produces necessary information, rather than overwhelming the patient with all the knowledge at his or her disposal.

There is controversy regarding whether tailored interventions yield higher quit rates than do nontailored interventions. One study found that a full, tailored cessation program was superior to a modified version at 11-week follow-up, with abstinence rates of 10.9% versus 8.9% for current smokers (Etter, 2005). Another study comparing a tailored Web-based program and a nontailored Web-based program found higher abstinence rates for the tailored program at 6 weeks (25.6% vs. 21.0%) and 12 weeks (20.1% vs. 15.9%) for concurrent users of a nicotine patch (Strecher et al., 2005). A major published Internet trial that addressed this point is the American Cancer Society's QUITLINK (Pike, Rabius, McAlister, & Geiger, 2007; Rabius, Wiatrek, Pike, Hunter, & McAlister, 2006). This study compared six interactive sites to a static site that provided a brochure online. Prolonged intent-to-treat abstinence rates were not significantly different, and were, on average, 11.0% and 9.9% for the interactive sites at 4 and 7 months compared to 10.9% and 9.5% for the static site. This study suggests that tailored Web interventions or an online brochure are both moderately efficacious.

Interactivity. Unlike the fairly passive patient population of yesteryear, more individuals are actively researching information pertaining to their medical conditions, arming themselves with necessary knowledge, coming to doctor's visits with printouts from the Internet, and asking pointed questions. The degree to which this development is helpful or harmful is debatable. However, it seems clear that the mood and the attitudes of the public seeking medical services have changed. A modern Internet-based intervention can place the user in control of his or her quitting progress by providing meaningful, interactive quit tools, such as charts, graphs, logs, and many others. Several studies have pointed out that success rate is associated with site utilization. By making a Web intervention more interactive, the user is more likely to be engaged and interested in visiting it and using its tools, rather than consider it a necessary burden.

Usability. The best intervention is likely to fail if it is presented in a way that is confusing or bothersome to the intended audience. Because designers of the interventions are likely to be experts in the field of smoking cessation, their idea of the most clear and cohesive presentation of material may differ from the way it is perceived by a recipient unfamiliar with smoking cessation methods. Modern Web building tools allow for eye-pleasing, catchy graphical presentation of information. However, many cessation Web sites presume reading ability at above 8th grade level, which would systematically exclude those with less education or those who are struggling with language barriers—all of which can be addressed by conducting extensive and diligent prelaunch usability testing (Stoddard, Augustson, & Mabry, 2006).

Support E-mails. The Web sites that use some version of support e-mails tend to attain better quit rates than those without this component. For instance, in the van Osch et al. (2009) study, inclusion of supportive e-mails predicted both short-term and long-term abstinence. The same was found in the Lenert, Muñoz, Perez, & Banson (2004), Lenert (2003), and Muñoz et al. (2006) studies, as well as in several others. The information, support, or reminders delivered by e-mails are useful for several reasons. They may remind the person trying to quit of his or her goals, values, and priorities. They may reassure him or her and point to a helpful resource if the person is struggling. They can also provide useful information that would be helpful in the person's quit attempt.

Forums. Most modern Web interventions include as part of the program some form of a forum or a discussion board where users can ask questions, post their own messages, get support from peers, and engage in a discussion on topics related to smoking. Some of these are moderated (i.e., a live person is responsible for monitoring the board's content to prevent offensive or dangerous postings), and others are unmoderated. There is limited data regarding the usefulness of forums in Web-based intervention. For instance, in their QuitNet intervention, Cobb and colleagues (2005) reported that, in addition to number of site logins and minutes spent on the Web site, a strong predictor of quitting was usage of the social support system (i.e., the forum). The authors report that those individuals who have made at least one post, or sent or received an e-mail to another user, or selected "buddies" on the Web site were considerably more likely to quit, which suggests a potential benefit of forum participation. Additional research is needed to examine the contributions of online support groups for smoking cessation.

Internet cessation intervention as a clinician's resource

For individuals who do not have access to a clinician for financial, situational, or personal reasons, Web interventions can provide a convenient and effective cessation tool. For those who have the benefit of a clinician's involvement in their health maintenance, Internet-based interventions can serve as an adjunct to maximize the chances of quitting successfully. In at least some of the studies listed above, a sizable proportion of participants were treated concurrently with an NRT (e.g., Brendryen & Kraft, 2008; Stretcher et al., 2008). A cessation Web site thus increased quit rates over and above what was possible with the standard pharmaceutical regimen.

Web sites can save clinicians' time while facilitating cessation efforts. We can envision three ways in which Internet interventions can be particularly helpful to clinicians. First, they can function as an information repository, to educate the patient about nicotine and its effects on health. Rather than spending valuable clinical time familiarizing the patient with the effects of nicotine, clinicians can refer him or her to a Web site that contains this information. A clinician may then clarify the points that were misunderstood or address any follow-up questions.

Second, clinicians can keep the patient on track by suggesting that he or she make use of the interactive tools available on Web interventions. The same tools can be used to share information with the clinician. The better cessation Web sites contain interactive progress trackers, which track the patient's cigarette use, along with other related factors, such as mood or smoking-related thoughts or activities, which are summarized in graph form. A printout of such a graph brought by a patient to an appointment can give clinicians a quick yet thorough picture of the patient's smoking and quitting patterns, along with potential problems or pitfalls. The ability to print information found on the Internet can also enable those without consistent Internet connectivity to have ready access to those materials they found most helpful, as well as to have an environmental cue (i.e., a printed page to post on their refrigerator or keep at their bedside) to remember their quit plan and related useful tools.

Finally, cessation Web sites can promote continued recovery by being available to the user long after the quit date or long after therapy has ended, to help with relapse prevention. A clinician may encourage patients to use their preferred Web resource to help in combating cravings, staying away from smoking-promoting situations, or learning additional skills. Most smokers have to quit several times before quitting for good. A Web site that is available at any time of the day or night can be a useful resource to help the smoker go back to abstinence after a slip or a relapse. With most Web cessation tools having a forum or a discussion board feature, a recovering smoker can also use his experience to help other smokers online.

There has been a steady trend among consumers of health services to take the proverbial reins into their own hands, with the Internet playing an increasingly larger role in the lay public information seeking on health and wellness. Whether or not this development is helpful may be a matter of opinion. However, to the extent that patients will continue to seek health-related information on the Internet, it may be especially important for clinicians to channel their patients to trusted Web sites and to keep systematic records of which Web sites work best for which of their patients. The latter would provide an empirical basis for recommending specific sites for the types of patients they see. By enabling patients to take more control over their health and their recovery, a clinician may indirectly increase the patient's motivation—a strong common factor for success in quitting smoking, as well as in many other areas of personal change and growth.

Existing Internet resources

There is no shortage of Web sites related to smoking cessation. The degree to which these sites are helpful is unclear. Bock and colleagues (2004) conducted a thorough survey of 46 Web sites claiming to either provide information about or to help individuals with smoking cessation. They found that in general, sites were not easy to read and did not take advantage of the possibilities offered by the interactive medium of the Internet, and more importantly, over four fifths of the sites surveyed failed to include at least one of the treatment components known to be effective. This sentiment was echoed by Etter (2006), who found that the majority of cessation Web sites was perceived by the users to be less than helpful.

We propose to distinguish between *information* Web sites and *intervention* Web sites. We call the latter "Internet interventions," following the terminology of the International Society for Research on Internet Interventions (ISRII; http://www.isrii.org). The vast majority of smoking cessation Web sites are information sites; that is, they are not interactive and do not offer a formal cessation program. In a sense, they are not different from information one can obtain from a booklet at a clinician's office or by looking through a cessation book in the library. In Table 6.1, we list several Web sites in English, Spanish, French, and Russian that appear to conform to accepted best practices and provide easy-to-follow, helpful advice. Other Web sites offer limited information, but offer products reputed to help quit smoking, from herbal concoctions to electronic cigarettes. Because these Web sites' purpose is commercial rather than public health–related (a user must pay to gain access to the information), they will not be discussed further.

There are only a handful of bona fide public Internet intervention sites, that is, Web sites where a user is expected to sign up for the intervention, and that are either minimally personalizable or delivered over the course of several days in a manner similar to an in-person program. Information about each of the sites can be found in Table 6.2. While we focused on programs for adults in English, as it is by far the most spoken language on the Internet (http://www.internetworldstats.com/stats7.htm, 2008), programs in Spanish, French, and Russian were also reviewed. We will not discuss each program at length here; however, we will point out noteworthy features of some for the benefit of designers of future programs.

Stopsmoking.ucsf.edu, available in English and Spanish, was created by the senior author of this chapter (Ricardo F. Muñoz) and his

Table 6.1 Information-Only Cessation Web Sites in English, Spanish, French, and Russian

English

http://www.anti-smoking.org
http://www.cancer.org/docroot/PED/content/PED_10_13X_Guide_for_Quitting_Smoking.asp
http://www.cdc.gov/tobacco/
http://www.mayoclinic.com/health/quit-smoking/QS99999
http://www.quitsmoking.com/
http://Quitsmoking.About.com
http://www.smokefree.gov/quit-smoking/index.html
http://thescooponsmoking.org/xhtml/quittingHome.php
http://www.webmd.com/smoking-cessation/
http://whyquit.com/

Spanish

http://www.americanheart.org/downloadable/heart/110186768682334-Quit%20Smoking.pdf
http://www.ahrq.gov/consumer/tobacco/helpsmokerssp.htm
http://www.cancer.gov/espanol/cancer/tabaco
http://www.help-eu.com/
http://www.saludymedicinas.com.mx/nota.asp?id=1799

French

http://www.hc-sc.gc.ca/hl-vs/tobac-tabac/quit-cesser/index-fra.php
http://www.je-veux-arreter-de-fumer.com
http://tabac.stop.free.fr/
http://www.tabac-cigarette.com

Russian

http://antismoke.org.ua/
http://www.lyzhnik.al.ru/
http://www.nekurim.ru/
http://nonsmoke.chat.ru
http://www.nosmokes.ru
http://www.nosmoking.ru

Chapter six: Smoking cessation via the Internet 123

Table 6.2 Public Smoking Cessation Internet Intervention Sites

Site name	Site URL	Personalizable	Reasons for quitting	Allows smoker to set own quit date	Stimulus control/ triggers	Link to biological aids	Email reminders	Related skills (e.g., mood or anxiety manag.)	Relapse prevention	Forum	Notes
Become an Ex	http://www.becomeanex.org/	Yes	Yes	Yes, contingent on several factors	Yes, strongly encouraged	Not ready to set quit date without it. Can be bypassed.	No	How to get support from people	Yes	Yes	English and Spanish
Jarrete	http://www.jarrete.qc.ca/	Yes	List, plus own	Yes	Yes	Encouraged, not required	No	Yes, various	Yes	Yes	French only. Adolescent and Adult version (reviewed). Very engaging.
Курение - нет (*Kureniyu – nyet*)	http://www.kureniyu.net	No. Some interactive elements	List. Encourages to write own	No	Yes, advice	Mentioned as a possibility	No	Limited	Not obvious	Yes	Russian only
Madrid + Salud Programa Dejar de Fumar	http://www.munimadrid.es	No	No	Yes, within 10–20 days after enroll	Yes	No, recommended/ instructions on use	No	Willpower	No	No	Spanish only. Primarily focused on how to use the patch.
Never Smoke Again	www.neversmokeagain.com	No	Listed, not personalized	Encouraged, not set online	Discussed	Actively discouraged	No	Willpower building	Some	No	English only
Quit 4 Life	http://www.quit4life.com	Yes	Yes, via Goals and Values	Not apparent	Yes	No	No	Some	Yes	No	English (reviewed) and French. Somewhat confusing, some functions (e.g., search) don't work
QuitNet	quitnet.com	Limited in the free version	Listed, not personalized	Yes	Buried in Q/A	Must select. Hard to bypass	Yes, quit tips	A bit in Q/A	Q/A only	Yes	Actively encouraging to pay. Cluttered, confusing layout

Continued

Table 6.2 Public Smoking Cessation Internet Intervention Sites *(Continued)*

Site name	Site URL	Personalizable	Reasons for quitting	Allows smoker to set own quit date	Stimulus control/ triggers	Link to biological aids	Email reminders	Related skills (e.g., mood or anxiety manag.)	Relapse prevention	Forum	Notes
Quit Smoking eCourse	http://www.quitsmokingaids.org	No	Listed, not personalized	Yes, plus recommendation	Yes, advice	No	Yes—entire course is on emails	Create support group, self-motivation, rewards	Some discussion	No	5-day course, daily emails
San Francisco Stop Smoking Site	http://www.stopsmoking.ucsf.edu; http://www.dejardefumar.ucsf.edu	Yes	Listed, not personalized	Yes	Discussed	Yes, thorough discussion, not required	Yes, if user chooses to get them	Yes, various, incl. mood and anxiety management	Yes	Yes, if user chooses	English and Spanish (both reviewed)
Stop Smoking Center	http://www.stopsmokingcenter.net/	Yes	Asked about costs/benefits	Yes	Yes, select from list	Available, not required or encouraged	Not reminders, but support emails	Reward self	Some	Yes	English only
Stop Tabac Coach	http://www.stop-tabac.ch	Yes	Questionnare	Yes	Not explicit	Yes, part of plan	Yes	Not explicit	List of strategies	No	Several languages (French and Spanish reviewed)
Vida Sin Tabaco	http://www.vidasintabaco.com/public/esp	Yes	Asked about costs	Yes	Yes	No	Yes, strongly encouraged/ key component of quit plan	Mostly willpower, request support from those around you	No	No	Spanish only. Heavy encouragement of NRTs; little advice for other resources.

Note: These sites were found by the authors in September 2008. We would appreciate being informed of additional public smoking cessation Internet interventions via e-mail to ricardo.munoz@ucsf.edu.

colleagues (Lenert et al., 2003, 2004; Muñoz et al., 2006; Stoddard et al., 2005), and goes beyond the standard cessation methods. It provides a smoking cessation guide with basic information about self-help methods to stop smoking as well as information about nicotine replacement and other pharmacological agents, e-mail reminders to come back to the site timed to the quit date set by the user, mood management skills related to smoking cessation, and a "virtual group" composed of an asynchronous bulletin board that allows users to provide information and support to each other. The mood management program borrows heavily from the version of cognitive-behavioral therapy based on *Control Your Depression* (Lewinsohn, Muñoz, Youngren, & Zeiss, 1992). As such, this program teaches mood management, basic cognitive restructuring, behavioral activation, and anxiety management. It also provides graphic tools for keeping track of each of the above, allowing users to monitor their progress. These methods speak to the social learning goals of increasing self-efficacy or agency (Bandura, 1977, 2001), that is, of individuals holding the reins of their own recovery, which adds to user involvement and motivation. The smoking cessation guide and the mood management elements stem from an earlier study (Muñoz et al., 1997) conducted using printed material delivered via surface mail to Spanish-speaking smokers. The materials from that study have been designated as a Research-Tested Intervention Program (RTIP) by the National Cancer Institute. Clinicians working with Spanish-speaking smokers with no access to the Web can download the printed materials or order them on a free CD from http://cancercontrol.cancer.gov/rtips (click on Tobacco Control and then on #15, *Programa Latino Para Dejar de Fumar*).

Designwise, the Canadian Web site *J'Arrête* (I Quit) represents a welcomed departure from the design template of most other cessation sites. *J'Arrête* is a French-language-only Web site, with separate versions for adolescents and adults, created via a collaborative effort by several Quebecoise research and public service organizations. The site appears to rely heavily on Flash animation technology for navigation and presentation of information, uses animated characters, and generally makes the user feel engaged, as if he or she is playing a game. This in no way makes it simple or superficial, as it is a comprehensive cessation Web site with many useful tools and thorough information.

A similar English (and Spanish) language Web site is Become an Ex, created by the National Academy of Smoking Cessation. This program likewise relies on Flash animation for presentation of information and quitting tools, for a clear, professional appearance (though not as playful as *J'Arrête*). This site seems to position itself as an addition to pharmacological interventions—users attempting to set a quit date without specifying their chosen pharmacological agent are strongly encouraged to do so and are precluded from going forward (though this can be bypassed). An excellent feature of this Web site is its strong emphasis on relapse prevention, which is one of the chief problems in smoking cessation.

The Web site Stop Tabac, created by a group at the University of Geneva (Etter, 2005), is noteworthy for being one of the few sites with truly international aspirations. The Web site is currently available in at least six languages (French, German, Italian, English, Spanish, and Portuguese), and appears to continue to expand its global reach. The benefit of this approach is that testing, developing, and dissemination of information is easier and faster from a centralized location, as compared with a situation where each independent Web site attempts to conduct separate testing and integrate new developments.

While Russia has a very high percentage of smokers (63% men, 10% women, as well as 41% male and 30% female adolescents) and a fair degree of technological sophistication, both international and Russian search engines failed to turn up a comprehensive Web-based cessation program. The only possible exception is Курению – Нет ("No to Smoking"), which aside from some interactive elements and a virtually unused blog, presents only text information.

Canada-based WATI (Web-Assisted Tobacco Interventions) maintains an inventory of Web sites devoted to smoking cessation on the Internet (http://www.wati.net/resources.html). While not all links listed in their document are operational, they nonetheless list a number of helpful sites, including Web sites specifically designed to help younger smokers.

Further considerations of Internet-based cessation interventions

Clinical considerations. Internet-based interventions show a great deal of promise in their ability to aid clinicians, including those in medical and mental health professions. Web sites are able to present information consistently and thoroughly, involve the patient in his or her recovery, and give the user a greater sense of control over his or her progress. Web sites also have the potential to reach underserved populations, providing information, advice, and personalized intervention to anyone accessing the Internet, thus serving as a helpful adjunct to clinicians in settings where they serve smokers who do not speak their language and where there are a limited number of providers fluent in alternative languages.

While Web interventions may be flexible, interactive, and offer an impressive breadth of information, they cannot offer material outside of their programming or ask questions based on the client's unique presentation or experience. More importantly, unlike skilled clinicians, Web sites cannot go beyond the health concern they are programmed to address; that is, they cannot note or manage related or comorbid issues. This ability may be especially crucial for smoking given the vast array of medical problems that can arise as a result of smoking, including a variety of cancers (e.g., lung, mouth and throat, kidney, stomach), heart conditions, impotence (in men) or infertility (in women), osteoporosis, and diabetes, among many others.

A virtual health clinic, affiliated with a trusted institution or institutions, would remedy this problem. Much like a traditional clinic, referrals to experts on a variety of topics (i.e., other Web sites) may be made to those who screen positively for a particular health concern, and they may be redirected to other "experts" as the clinical picture becomes more coherent. The potential danger here is that such a system might become overly involved and cumbersome for the user, effectively negating the possible benefits Web interventions are likely to provide.

Ethical Considerations. In many ways, automated self-help interventions are akin to bibliotherapy. A self-help book does not promise individualized personal attention from the author, or even contact with the author. The buyer of the book understands that whatever benefit he or she will gain will come from and be limited to the book and the information contained therein. Because Web interventions are relatively new and a consumer may not be similarly aware of their constraints, the creator of a Web site is ethically bound to make a concerted effort to inform the site visitor about the limitations of the intervention. Automatic Web sites need to clarify that they do not offer a therapeutic contract with a provider. At the same time, a consumer selecting a Web site created by health professionals to help him or her quit smoking needs to be assured that the information he or she is about to receive is consistent with current best practices and comes from a reputable source. An affiliation with a respectable medical or educational institution may address this concern, as would a certification of a well-regarded health organization, for example, the Health on the Net Foundation (HON; http://www.hon.ch/).

The limitations as to the scope and breadth of the Web site need to be expressly acknowledged in order for the user to have reasonable and appropriate expectations. They should entirely avoid making unsubstantiated assurances and promises (e.g., "You are guaranteed to quit in 10 days"), as no ethical clinician would consider giving such a guarantee. Further, Internet intervention sites should routinely gather outcome data regarding the effectiveness of the intervention. Ideally, these data would eventually yield expected success rates for specific subgroups of smokers. Additionally, these data could provide information about potential iatrogenic effects. In such a case, a decision to modify the Web site should be made or a warning provided to the participants whose characteristics are associated with the negative effects.

Considering the sensitivity of data obtained via Web interventions, data safety and security is a major concern. In order to personalize the intervention, a considerable amount of data needs to be gathered, processed, and stored in order to optimize the individual's future visit(s). These data are likely to contain private, sensitive information, similar to that shared with providers of medical or mental health services. Data security and confidentiality therefore needs to be of utmost concern, steps taken to ensure these need to be clearly presented as part of the Web site, and every effort should be made to collect only the required minimum of identifying information.

Accessing and navigating the Internet presupposes some level of technological sophistication. This causes some to be concerned that Web interventions may systematically exclude those from the lowest tiers of the socioeconomic status, including those individuals who are illiterate. However, world Internet penetration is growing rapidly. In the United States, around 70% of households are connected to the Internet, and many others are able to get access through libraries or schools. Internet penetration throughout the world is also rapidly increasing. For instance, since 2000, Latin America has experienced almost a sevenfold increase in the number of online households; tenfold and higher increases have been observed in the Middle East and Africa (http://www.internetworldstats.com, 2007). We propose to begin preparing these evidence-based Web interventions now rather than later, so that they will be readily available when more of the world is online. In addition, work is currently underway to develop interventions delivered via cell phones, by means of text messages (Bramley et al., 2005; Rodgers et al., 2005). It has been shown that some underserved and minority populations use text messages more frequently than the Web (Fox & Livingston, 2007), leaving open the possibility of using this medium to deliver interventions.

Future directions

As the health care field differentiates between health information Web sites and health intervention Web sites, it will begin to conceptualize and appreciate Internet interventions as bona fide health interventions. It will begin to require similar levels of empirical information about sites that are intended to yield health outcomes, such as smoking cessation, as are required of traditional prevention and treatment interventions. Gradually, a set of Internet interventions will acquire the status of *evidence-based smoking cessation interventions*.

Our experience with working toward evidence-based Internet interventions in the smoking cessation area has encouraged us to call on global health agencies and professions to contribute to a systematic effort to create Web interventions targeting a variety of conditions, in multiple languages. We also call for ways to make these Internet interventions accessible to those who currently have little access to the Web. For example, we call for the creation of Internet Resource Rooms at public sector health clinics worldwide so that individuals using such clinics can be screened for common health problems and given access to evidence-based Internet interventions addressing these problems. Moreover, Internet interventions should provide versions that can be used by people who cannot read or write, using video, sound, and graphics to convey information.

There are over one billion smokers in the world as of 2008 (http://www.who.int/tobacco/mpower/mpower_report_tobacco_crisis_2008.pdf). We will never be able to train enough clinicians to help them all to quit. Developing public, self-help, automated, Internet interventions that have been empirically tested and found to be effective in helping individuals to stop smoking will provide an additional cost-effective tool in the quest to reduce the number of deaths and disabilities caused by smoking.

Acknowledgment

This research was supported by grant 13RT-0050 from the Tobacco-Related Disease Research Program (Muñoz, P.I.), which helped establish the University of California, San Francisco/San Francisco General Hospital Internet World Health Research Center (http://www.health.ucsf.edu). The authors thank the Center for Health and Community (Nancy Adler, Director) for providing office space and additional resources to target Latino smokers.

References

Bandura, A. (1977). *Social learning theory*. Englewood Cliffs, NJ: Prentice Hall.
Bandura, A. (2001). Social cognitive theory: An agentic perspective. *Annual review of psychology* (Vol. 52, pp. 1–26). Palo Alto: Annual Reviews.
Behan, D. F., Eriksen, M. P., & Lin, Y. (2003) Economic effects of environmental tobacco smoke report. Society of Actuaries. Retrieved August 10, 2008 from http://www.soa.org/ccm/content/areas-of-practice/life-insurance/research/economic-effects-of-environmental-tobacco-smoke-SOA/
Bernal, G., & Scharron-del-Rio, M. R. (2001). Are empirically supported treatments valid for ethnic minorities? Toward an alternative approach for treatment research. *Cultural Diversity and Ethnic Minority Psychology, 7*, 328–342.

Bock, B., Graham, A., Sciamanna, C., Krishnamoorthy, J., Whiteley, J., Carmona-Barros, R., et al. (2004). Smoking cessation treatment on the Internet: Content, quality, and usability. *Nicotine and Tobacco Research, 6*, 207–219.

Bramley, D., Riddell, T., Whittaker, R., Corbett, T., Lin, R., Wills, M., et al. (2005). Smoking cessation using mobile phone text messaging is as effective in Maori as non-Maori. *New Zealand Medical Journal, 118*, 1–10.

Brendryen, H. & Kraft, P. (2008). Happy ending: A randomized controlled trial of a digital multi-media smoking cessation intervention. *Addiction, 103*, 478–484.

Centers for Disease Control and Prevention. (2003). Cigarette smoking-attributable morbidity—United States, 2000. *Morbidity and Mortality Weekly Report, 52*.

Centers for Disease Control and Prevention. (2005). Annual smoking-attributable mortality, years of potential life lost, and productivity losses—United States, 1997–2001. *Morbidity and Mortality Weekly Report, 54*.

Centers for Disease Control and Prevention. (2007). *Best practices for comprehensive tobacco control programs—2007*. Atlanta, GA: U.S. Department of Health and Human Services, Centers for Disease Control and Prevention, National Center for Chronic Disease Prevention and Health Promotion, Office on Smoking and Health.

Chambless, D. L., & Ollendick, T. H. (2001). Empirically supported psychological interventions: Controversies and evidence. *Annual Review of Psychology, 52*, 685–716.

Cobb, N. K., Graham, A. L., Bock, B. C., Papandonatos, G., & Abrams, D. B. (2005). Initial evaluation of a real-world Internet smoking cessation system. *Nicotine & Tobacco Research, 7*, 207–216.

Etter, J. (2005). Comparing the efficacy of two Internet-based computer-tailored smoking cessation programs: A randomized trial. *Journal of Medical Internet Research, 7*, e2.

Etter, J., F. (2006). A list of the most popular smoking cessation Web sites and a comparison of their quality. *Nicotine and Tobacco Research, 8S1*, S27–S34.

Eysenbach, G. (2005). The law of attrition. *Journal of Medical Internet Research, 7*, e11.

Fiore, M. C., Jaén, C. R., Baker, T. B., Bailey, W. C., Benowitz, N. L., Curry, S. J., et al. (2008). *Treating tobacco use and dependence: 2008 update*. Rockville, MD: U.S. Department of Health and Human Services.

Fiore, M. C., Smith, S. S., Jorenby, D. E., & Baker, T. B. (1994). The effectiveness of the nicotine patch for smoking cessation: A meta-analysis. *Journal of the American Medical Association, 271*, 1940–1947.

Fox, S., & Livingston, G. (2007). Latinos online: Hispanics with lower levels of education and English proficiency remain largely disconnected from the Internet. Pew Internet & American Life Project. Retrieved September 30, 2008 from http://www.pewinternet.org/PPF/r/204/report_display.asp

Glasgow, R. E., McKay, H. G., Piette, J. D., & Reynolds, K. D. (2001) The RE-AIM framework for evaluating interventions: What can it tell us about approaches to chronic illness management? *Patient Education and Counseling, 44*, 119–127.

Gonzales, D., Rennard, S. I., Nides, M., Oncken, C., Azoulay, S., Billing, C. B., et al. (2006). Varenicline, an alpha4beta2 nicotinic acetylcholine receptor partial agonist, vs. sustained-release bupropion and placebo for smoking cessation: A randomized controlled trial. *Journal of the American Medical Association, 296*, 47–55.

Lando, H. A., McGovern, P. G., Barrios, F. X., & Etringer, B. D. (1990). Comparative evaluation of American Cancer Society and American Lung Association smoking cessation clinics. *American Journal of Public Health, 80*, 554–559.

Lenert, L., Muñoz, R. F., Perez, J., & Banson, A. (2004). Automated e-mail messaging as a tool for improving quit rates in an Internet smoking cessation intervention. *Journal of the American Medical Informatics Association, 11*, 235–240.

Lenert, L., Muñoz, R. F., Stoddard, J., Delucchi, K., Bansod, A., Skoczen, S., & Pérez-Stable, E. J. (2003). Design and pilot evaluation of an Internet smoking cessation program. *Journal of the American Medical Informatics Association, 10*, 16–20.

Lewinsohn, P. M., Muñoz, R. F., Youngren, M. A. & Zeiss, A. (1992). *Control your depression*. New York: Fireside Books.

Mackay, J. & Eriksen, M. (2002). *The tobacco atlas*. Geneva: World Health Organization

McClure, J. B., Greene, S. M., Wiese, C., Johnson, K. E., Alexander, G., & Strecher, V. (2006). Interest in an online smoking cessation program and effective recruitment strategies: Results from Project Quit. *Journal of Medical Internet Research, 22*, e14.

Muñoz, R. F., Barrera, A. Z., Delucchi, K., Penilla, C., Torres, L. D., & Pérez-Stable, E. J. (In press). Worldwide Spanish/English Internet smoking cessation trial yields 20% abstinence rates at one year. Nicotine and Tobacco Research.

Muñoz, R. F., Lenert, L. L., Delucchi, K, Stoddard, J., Perez, J. E., Penilla, C., & Pérez-Stable, E. J. (2006). Toward evidence-based Internet interventions: A Spanish/English Web site for international smoking cessation trials. *Nicotine & Tobacco Research, 8*, 77–87.

Muñoz, R. F., Marín, B., Posner, S. F., & Pérez-Stable, E. J. (1997). Mood management mail intervention increases abstinence rates for Spanish-speaking Latino smokers. *American Journal of Community Psychology, 25*, 325–343.

Nusselder, W. J., Looman, C. W. N., Marang-van de Mheen, P. J., van de Mheen, H., & Mackenbachet, J. P. (2000). Smoking and the compression of morbidity. *Journal of Epidemiology and Community Health, 54*, 566–574.

Pike, K. J., Rabius, V., McAlister, A., & Geiger, A. (2007). American Cancer Society's QuitLink: Randomized trial of Internet assistance. *Nicotine & Tobacco Research, 9*, 415–420.

Porter, M. E., & Teisberg, E. O. (2007). How physicians can change the future of health care. *Journal of the American Medical Association, 297*, 1103–1111.

Prochaska, J. J., Fromont, S. C., & Hall, S. M. (2005). How prepared are psychiatry residents for treating nicotine dependence? *Academic Psychiatry, 29*, 256–261.

Prochaska, J. J., Fromont, S. C., Louie, A. K., Jacobs, M. H., & Hall, S. M. (2006). Training in tobacco treatments in psychiatry: A national survey of psychiatry residency training directors. *Academic Psychiatry, 30*, 372–378.

Rabius, V., Wiatrek, D., Pike, K. J., Hunter, J., & McAlister, A. (2006, October). *American Cancer Society's QUITLINK: A randomized trial of Internet assistance for smoking cessation*. Paper presented at the 11th World Conference on Internet in Medicine, Toronto, Canada.

Rodgers, A., Corbett, T., Bramley, D., Riddell, T., Wills, M., Lin, R., & Jones, M. (2005). Do u smoke after txt? Results of a randomised trial of smoking cessation using mobile phone text messaging. *Tobacco Control, 14*, 255–261.

Saul, J. E., Schillo, B. A., Evered, S., Luxenberg, M. G., Kavanaugh, A., Cobb, N., & An, L. C. (2007). Impact of a statewide Internet-based tobacco cessation intervention. *Journal of Medical Internet Research, 30*, e28.

Schroeder, S. A. (2005). What to do with a patient who smokes. *Journal of the American Medical Association, 294*, 482–487.

Stoddard, J. L., Augustson, E. M., & Mabry, P. L. (2006). The importance of usability testing in the development of an Internet-based smoking cessation treatment resource. *Nicotine and Tobacco Research, 8S1*, S87–S93.

Stoddard, J. L., Delucchi, K. L., Muñoz, R. F., Collins, N. M., Pérez-Stable, E. J., Augustson, E., & Lenert, L. L. (2005). Smoking cessation research via the Internet: A feasibility study. *Journal of Health Communications, 10*, 27–41.

Strecher, V. J., McClure, J. B., Alexander, G. L., Chakraborty, B., Nair, V. N., Konkel, J. M., Greene, S. M., Collins, L. M., Carlier, C. C., Wiese, C. J., Little, R. J., Pomerleau, C. S., & Pomerleau, O. F. (2008). Web-based smoking-cessation programs: Results of a randomized trial. *American Journal of Preventive Medicine. 34*, 373–381.

Strecher, V. J., Shiffman, S., & West, R. (2005). Randomized controlled trial of a Web-based computer-tailored smoking cessation program as a supplement to nicotine patch therapy. *Addiction, 100*, 682–688.

Strecher, V. J., Shiffman, S., & West, R. (2006). Moderators and mediators of a Web-based computer-tailored smoking cessation program among nicotine patch users. *Nicotine and Tobacco Research, 8S1*, S95–101.

Swartz, L. H. G., Noell, J. W., Schroeder, S. W., & Ary, D. V. (2006). A randomised control study of a fully automated internet based smoking cessation programme. *Tobacco Control, 15*, 7–12.

van Osch, L., Lechner, L., Reubsaet, A., Steenstra, M., Wigger, S., & de Vries, H. (2009). Optimizing the efficacy of smoking cessation contests: An exploration of determinants of successful quitting. *Health Education Research, 24*, 54–63.

Walters, S. T., Wright, J., & Shegog, R. (2006). A review of computer and Internet-based interventions for smoking behavior. *Addictive Behaviors, 31*, 264–277.

World Health Organization (WHO) (2007). 10 facts about tobacco and second-hand smoke. Retrieved August 10, 2008, from http://www.who.int/features/factfiles/tobacco/en/index1.html

World Health Organization (WHO) (2008). *WHO report on the global tobacco epidemic.* Retrieved August 10, 2008, from http://www.who.int/tobacco/mpower/en/

chapter seven

Pain management

*Jeffrey J. Borckardt, Alok Madan,
Arthur R. Smith, and Stephen Gibert*

Chapter aims and scope

Pain is a major physical and mental health care problem in the United States. Chronic pain is defined as prolonged and persistent pain lasting at least three months (Gatchel, Peng, Peters, Fuchs, & Turk, 2007). Recurrent pain affects 10 to 20% of adults in the general population (Blyth et al., 2001; Gurje, Von Korff, Simon, & Gater, 1998) and costs more than $70 billion annually in health care costs and lost productivity (Gatchel 2004a, 2004b). Advances in the area of pain management are needed to help providers address the problem of pain, and several emerging technologies may be of considerable use to practitioners. This chapter is not intended to be a comprehensive guide to the use of technology in providing behavioral pain management services, but rather is a broad overview of some of the predominant technology-supported interventions for chronic pain. We first provide a brief overview of psychological and behavioral aspects of chronic pain management. Next, we will review some of the many Internet technology resources for clinicians and patients. We also provide a brief review of applied physiology techniques, systems and computer programs available to clinicians, as well as some technology-based techniques for behavioral tracking in chronic pain populations. We will touch on the use of computers to evaluate and triage patients with chronic pain to the appropriate level of care and provide a brief review of multimedia resources for supporting relaxation training interventions. We will provide a brief overview of virtual reality technologies for pain management as well as some preliminary work in the area of computer-assisted hypnotic intervention and assessment. Lastly, we touch on possible future directions for technology-based (more specifically brain stimulation technology) interventions for chronic pain management in behavioral health care.

Background of behavioral and psychological aspects of chronic pain

Historically, pain has been viewed as a symptom secondary to tissue or nerve pathology, and thus has been viewed as of secondary importance to the underlying pathology. With greater understanding of the complexity of pain comes the appreciation that it is not a single syndrome. Acute or nociceptive pain resulting from real or potential tissue damage is different from chronic pain, in that the latter can persist long after any injury has resolved. In such a case, the plasticity of the pain projection system can result in irreversible changes in neurological anatomy and physiology, so that pain itself is the disease process. For some time, the roles of emotional and behavioral factors in pain experience were ignored; however, variability in subjective perception of similar painful stimuli has led to the development of more comprehensive evaluation of pain phenomena. Pain is now widely accepted as a composite of sensory-discriminative, affective/motivational, and cognitive-evaluative dimensions (Melzack and Casey, 1968). Further, as defined by the International Association for the Study of Pain, pain is "an unpleasant sensory and emotional experience" (Mersky, 1986). Additionally, negative emotion may (1) predispose people to experience pain, (2) be a precipitant of symptoms, (3) amplify pain, or (4) perpetuate it (Gatchel et al., 2007; Fernandez & Turk, 1992; Turk & Monarch, 2002). It is not surprising then that chronic pain is highly comorbid with mood and anxiety disorders (Banks & Kerns, 1996; Dersh, Gatchel, Mayer, Polatin, & Temple, 2006; Wolfe at al., 1990).

There is good evidence to support the effectiveness of several psychological interventions in the management of a variety of chronic pain conditions. Behavioral and cognitive-behavioral therapies have been shown to be effective in the management of spine-related pain disorders (McCracken & Turk, 2002). Meta-analytic studies (see Morley, Eccleston, & Williams, 1999) have concluded that behavioral and cognitive-behavioral therapies are effective in improving the pain experience, coping, and pain behaviors of adults with non-headache chronic pain compared to control groups. Applied physiological interventions and relaxation training techniques have been shown to help patients with various types of headache, especially migraine and tension-type headaches (Nestoriuc & Martin, 2007; Nestoriuc, Rief, & Martin, 2008). There is some evidence that short-term psychodynamic interventions may be effective adjunctively for chronic pain management (Basler, Grzesiak, & Dworkin, 2002). Hypnosis and imagery are frequently used and are effective psychological approaches for chronic pain management (Syrjala & Abrams, 2002).

Fortunately, there are many technology-based resources for clinicians and patients to aid with treatment.

Internet technology as a resource for clinicians and patients

Numerous good Internet resources are available to clinicians and patients to support educational and behavioral approaches to pain management. Painedu.org is a free online organization that offers articles on pain management, downloadable pain measurement scales, links to other pain management–related resources, and educational content. Further, clinicians can get continuing education credits by participating in their online training programs. Another good, free resource for clinicians and patients is painknowledge.org. Educational materials are available (along with continuing education credit opportunities) and Webcasts as well as podcasts are hosted by the site. Additionally, there is a slide library available so that clinical educators can download educational materials to enhance lectures and provide educational material to patients. Further, patients can learn from reputable Web sites about available treatment options and common behavioral and cognitive pitfalls associated with chronic pain.

Patients with chronic pain often have limited mobility and may have difficulty participating in community-based treatment programs and groups. Therefore, one viable option is the use of online support groups and message boards. See http://www.chronicpainsupport.org. These virtual communities allow such patients to connect with others who are suffering from similar ailments and conditions. This may be useful for patients as they learn adaptive behavioral strategies for coping with chronic pain from others. Further, as patients with pain may find themselves socially isolated (either as a consequence of physical limitations due to pain or as a consequence of comorbid depression), online communities with a social support network can be therapeutic.

Telemedicine

Another emerging technology is telemedicine (or telehealth). Broadly speaking, *telemedicine* refers to the use of telecommunications or computing devices to provide health care services (educational, diagnostic, or interventional) from a distance. Its popularity has grown considerably over the past two decades, and this new technology has been used to advance established psychological interventions for chronic pain. For example, Naylor and colleagues (2008) recently developed a telephone-based tool for maintenance therapy for patients completing group cognitive-behavioral therapy (CBT) for chronic pain. Based on the observation that many patients' initial gains from CBT decline within weeks of completion of treatment, they developed Therapeutic Interactive Voice Response (TIVR), which allows patients to interact with a computer via a touch-tone telephone. Patients call a toll-free number daily to (1) complete a self-monitoring questionnaire, (2) access a verbal review of previously presented pain management skill sets, (3) access guided behavioral rehearsals of coping skills taught during the initial group sessions, and (4) retrieve a once-monthly personalized therapist message. Though in its infancy, recent studies have shown improved outcomes at four and eight months following treatment among treatment participants compared to controls. Similar work has been done in the area of diabetes management (Piette, 2002, 2007).

Traditional biofeedback instruments and paradigms have been modified to be delivered over distances. Using basic audio–video equipment, low-bandwidth communication networks (i.e., standard telephone lines), computers with remote control software, and specialized biofeedback equipment, Folen et al. (2001) have demonstrated the success of remote application of biofeedback for the treatment of headache. Initially developed at the Tripler Army Medical Center in Hawaii for use around the islands, this system has been extended to distant stations in Seoul, South Korea; Yokosuka, Japan; and Guam. The potential for cost savings is obvious, especially when such long distances of travel are easily avoided. One example of a virtual telemedicine pain clinic (Telepain) for soldiers, family members, and retirees estimated a savings of 54,400 miles of patient and clinician travel time (Burton & Boedeker, 2000). Additionally, with the promotion of higher-bandwidth fiber-optic infrastructures (e.g., cable Internet, T3 lines), the ease of implementing real-time telemedicine applications of biofeedback has further improved in recent years.

Though telemedicine technologies have been available for decades and have been successfully implemented with psychological approaches to pain management, research and clinical applications are still in their infancy. Most published accounts are based on small sample sizes with limited control groups and selected patient populations. Generalization of efficacy is not possible at this point. Research suggests that there is some utility to virtual psychological intervention, either alone or as an adjunct to direct patient contact. Of the limited research in the area, patients appear to adopt the use of new technologies well (Elliot, Chapman, & Clark, 2007), and some reports suggest that they may prefer them to in-clinic visits (e.g., Cottrell, Drew, Gibson, Holroyd, & O'Donnell, 2007). However, the best approach may be to integrate telemedicine technologies with traditional psychotherapeutic arrangements so that patients can benefit from face-to-face interactions as well as the convenience of video follow-up appointments.

There is reason to believe that the application of telemedicine technologies in general, and more specifically with psychological approaches to pain management, is likely to increase in coming years. Specific advantages of this technology include cost savings, increased efficiency and continuity of care with centralization of services, and improved access to underserved areas (Bynum, Irwin, Cranford, & Denny, 2003). At this point, most of the current efforts assessing the utility and efficacy of telemedicine technologies have significant military-related funding and/or expertise, but the same advantages apply to using these services for civilian populations.

Computers and applied physiological interventions: Biofeedback

Computer technology has evolved greatly in the past 10 years with respect to the availability of clinician- and patient-friendly applied physiology resources. There is a vast literature on the effectiveness of relaxation training interventions and biofeedback in the management of diverse chronic pain conditions. In particular, biofeedback techniques have been shown to help patients with various types of headache, especially migraine and tension-type headaches (Nestoriuc & Martin, 2007; Nestoriuc et al., 2008). Systems for displaying physiological indices of stress and elevated sympathetic nervous system tone such as skin temperature, muscle tension, heart rate variability, and skin conductance are becoming more reliable, affordable, and engaging. For around $200, clinicians can purchase basic and simple biofeedback systems with software geared toward engaging patients in the process of learning to regulate physiological responses to pain. While traditional formats for displaying physiological indices are typically available in these systems (i.e., simple displays of skin temperature or heart-rate variability over time in a line-graph format), more interactive game formats are available as well. For example, patients can learn to make a hot-air balloon fly on the computer screen by increasing their heart-rate variability coherence, or open colorful worlds and three-dimensional (3D) scenery by reducing their skin conductance. These approaches (whether traditional feedback or more immersive game formats are used) provide a platform for patients to learn physiological regulation by providing immediate feedback regarding the effectiveness of mental or behavioral strategies in changing key physiological indices. If emerging data continue to support biofeedback as a viable and effective treatment for various chronic pain conditions, then this realm of technology is likely to continue to expand over the coming years with respect to increased availability, improved accuracy, and design flexibility.

Symptom tracking and behavioral logs

Symptom monitoring is a common and effective tool of many behavioral and cognitive-behavioral therapies for pain management. Traditional data collection tools have relied on paper-and-pencil approaches with their inherent and well-documented limitations. New technologies, such as personal digital assistants (PDAs), wireless-enabled pain drawings (Serif & Ghinea, 2005), pain dairies using digital pens (Lind, Karlsson, & Fridlund, 2007), and home-based health monitors (Dobscha, Corson, Pruitt, Crutchfield, & Gerrity, 2006) have all been used to facilitate symptom monitoring, reporting, and storage of data.

With the increasing availability of affordable handheld PDAs and assessment software, practitioners now have the ability to track variables of interest electronically, and each entry into the system can be time-stamped so that clinicians can verify when the measures were completed. Free software for these applications is available at http://www.experience-sampling.org. In addition to tracking and timing data input, this software allows clinicians to program devices to sound an alarm at various points during each day to prompt patients to complete electronic ratings. Alternatively, patients can initiate ratings whenever some specified event occurs (e.g., collect mood, anxiety, and activity ratings whenever pain reaches a certain level during the day). Such devices vary in cost, but some can be purchased for as little as $80.

Recently, it has been suggested that cell phone text messaging services might be used as reminders for patients to engage in various behaviors or activities in conjunction with outpatient psychotherapy protocols. Messages can be sent to patients at various times between therapy sessions to prompt them to record behaviors or pain experience, to engage in health behaviors or health-related activity, or to remind them about completing CBT homework. Alternatively, patients are able to use text messaging services to report experiences or activities in real time to clinicians as well.

Patient screening and outcomes assessment

Clinical outcome assessments should be an integral component of behavioral health care in the United States. Ideally, they can be used for surveillance of the quality of clinical services as well as the identification of potentially problematic areas before crises occur. Recently, Madan and Borckardt (2008) described a computerized system designed to collect relevant psychiatric symptom ratings and pain experience among patients enrolled in outpatient behavioral health care in a manner that allows for real-time feedback at the patient and clinician levels regarding clinical improvement (also see Chinman, Young, Schell, Hassell, & Mintz, 2004). In practice, comprehensive use of outcomes assessments is rarely the norm, often limited to areas that are mandated by an accreditation agency or required by an insurance provider. While clinical trials and other research endeavors require formal assessment (e.g., symptomatology, functional status) before and after treatment, the use of pre- and postintervention assessments is often excluded from clinical practice. As a likely consequence of limited resources and reimbursement available, the development, implementation, and reporting of outcomes assessments is often ignored and by default has become optional in behavioral health practice.

From a strictly methodological standpoint, outcomes assessments refer to any means of evaluating whether clinical change has occurred in patients. The greatest initial obstacle to the systematic use of outcomes assessment is identifying what to measure. Disparate target audiences vary in the nature of outcome assessments that they consider important. In clinical practice, symptom reductions are used as an indication of treatment response, and clinicians may find these data more meaningful than reports of adverse events. Unlike the other medical specialties, clinical outcomes in behavioral health are rarely based on laboratory values. Rather, self-reported and behavioral observations tend to be the basis for assessing clinical improvement in behavioral health practice. Informal outcomes assessments in behavioral health care have typically employed questionnaires or behavioral observation as the primary means of collecting pertinent data, both of which have considerable costs associated with them. While questionnaires are relatively easy to administer, appropriate measures are not always readily available. Developing specialized questionnaires is possible, but conducting the requisite psychometric analysis of a newly developed measure can increase the burden of data collection as well (Borckardt & Nash, 2001). Behavioral observations are even less appealing in clinical practice, as considerable time and energy must be devoted to training observers, while still having to address the psychometric concerns of the collected data (e.g., reliability). Regardless of method chosen, inherent in the practice of outcomes assessments is the necessity for data to be collected at multiple points over an extended period of time. This becomes another limitation that increases the cost and burden of conducting systematic outcomes assessments in behavioral health clinical settings. The effectiveness of laboratory-tested, evidence-based practices in the real world clinical trenches is not enough. Without evidence supporting that findings from randomized, placebo-controlled clinical trials actually generalize to real-world clinical practice, we risk undermining the fundamental tenet of health care—reducing the burden of illness in served patient populations.

In response to the need for a clinical outcomes measurement system within an expanding quality management program, the authors have developed the Web-based Clinical Quality Management System (ClinQMS). This tool allows the clinician to assess patients' endorsements of symptomatology at each outpatient clinic visit. Rather than create an infrastructure limited to providing only evaluation of overall clinic performance, the program was developed to be capable of providing additional evaluation of clinical functioning at the individual patient and clinician levels. While providing the necessary outcomes data related to clinical efficacy, we developed the ClinQMS to be brief, simple, valuable, relevant, acceptable, and available, thereby adhering to recommendations in the scientific literature regarding outcomes assessment protocols (Slade, Thornicraft, & Glover, 1999).

One of the goals in the development of ClinQMS was to keep the focus on practicality and utility rather than conventionality. As such, a *brief* but comprehensive 12-item original symptom checklist was developed and designed to capture a full array of psychiatric problems. The rating system was designed to mimic numeric rating scales (*simple*) that are often used by clinicians in clinical practice to quantify clinical functioning (particularly in pain management, where clinicians routinely ask patients to rate their level of pain using a 1 to 10 scale). We modified this approach to capture a variety of pain and psychiatric issues (*valuable and relevant*) rather than focus on disease-specific symptom clusters (Slade et al., 1999). The checklist has not yet been psychometrically evaluated (with respect to reliability and validity). Our goal was to get a quick snapshot of patients' symptom distress across an array of clinical problems including pain. Brevity was important in order to maintain both the practicality and utility of this evaluation for use at every outpatient visit for every patient.

We use our intranet system at the Medical University of South Carolina (MUSC) to manage this project. Standard computer programming languages (Hypertext Markup Language [HTML] and Hypertext Preprocessor [PHP]) were used to develop a Web interface that allowed for online administration of the checklist to patients, automatic logging of data into a centralized database, and real-time clinical progress reports at the patient and clinician levels. Additionally, higher-level clinic performance dashboards can be made available online for administrative review of clinic performance over time. For security, the intranet firewall prevents access from outside the network, and limited patient information is stored in the database (e.g., only encrypted medical record numbers, age, and sex) to further minimize Internet security risks and eliminate the possibility that unauthorized persons could identify patients by breaking into the database.

Full reports can be generated automatically in real time, which provide current symptom ratings as well as summary/trend tables and graphs that show patient ratings as a function of visits to the clinic. Additionally, the online reports include a brief narrative summary of patient progress across all relevant clinical symptoms, which clinicians can copy and paste into progress notes or treatment summaries.

As discussed previously, depression, anxiety, and other psychiatric disorders are frequently comorbid with chronic pain. While the importance of identifying psychosocial concerns among patients with chronic pain is widely recognized, there are few resources available to clinicians to help identify patients who would benefit from specialized mental health services. Recently, Madan and Borckardt (2008) developed a computer-based psychosocial screening and triage system (Outpatient Psychosocial Triage System; OPTS). OPTS allows treatment teams to administer standardized psychosocial screening measures easily (either by paper and pencil or via a secure computer interface) as part of the standard patient registration procedures. The system automatically scores the psychosocial measures and applies empirically derived clinical cutoffs to the patients' scores. The system then provides a narrative summary complete with recommendations regarding the level of mental health treatment that is warranted. These narratives can be copied to the medical record. Further, all of the psychosocial screening data is amassed in a secure central database, which can be used as a research tool for current and future investigations into psychosocial concomitants of cancer and cancer treatment. OPTS is currently in use with the heart and liver transplant programs at MUSC. A different version of the system designed to track patient functioning over time (using only a patient–computer interface; ClinQMS), is currently in use in all of the MUSC psychiatry outpatient clinics and in the Digestive Diseases Center. Preliminary data on patient and staff burden suggest that the OPTS and ClinQMS programs are well tolerated and found to be useful in enhancing quality of patient care. The OPTS system is immensely flexible and any questionnaires can be implemented including standardized pain questionnaires such as the McGill Pain Questionnaire, Brief Pain Inventory, Pain Anxiety Symptoms scale, and Pain Outcomes Questionnaire once rights have been attained to implement them digitally in a practice or institution.

Relaxation CDs, DVDs, and MP3s

Relaxation training has been shown to be effective in the psychological management of chronic pain. This can be accomplished in session as the clinician guides the patient through a variety of widely used relaxation protocols (including diaphragmatic breathing training, progressive muscle relaxation, or relaxation imagery scripts). These guided relaxation scripts can be recorded live during the session and cassettes or CDs can be given to patients to take with them for home practice. It is not unusual for pain psychologists to prerecord relaxation and/or hypnosis scripts for pain management and provide these more generic recordings to patients at office visits. Further, it is possible to plug patients' portable audio devices (e.g., iPods or other MP3 players) directly into a clinician's computer and download relaxation or hypnosis scripts onto the player for patients to play at home. Recently, in the Behavioral Medicine and Pain Management Clinic at the Medical University of South Carolina, we developed relaxation DVDs that are distributed to appropriate patients to further enhance the degree of patient engagement in relaxation protocols. The audio represents traditional relaxation imagery training scripts imposed over relaxing music. The video represents relaxing images and landscapes designed to correspond with the imagery/relaxation scripts. Numerous commercial products such as relaxation DVDs and CDs are available for purchase and can be used in session, distributed to patients by therapists, or purchased by patients directly.

Computerized hypnosis

There is ample evidence to support hypnosis as a viable intervention for managing certain types of chronic pain in certain patients (Montgomery, Duhamel, & Redd, 2000). While certain hypnotic interventions can be recorded in session and given to participants, there are other options available to pain psychologists. The effectiveness of hypnosis for pain management appears to be mediated by patients' level of hypnotizability, which makes formal assessment of hypnotizability preferable in clinical practice. This type of assessment can be quite time consuming. Fortunately, there is evidence that suggests hypnotizability can be reliably assessed via computer-delivered hypnotizability measurement systems (Grant & Nash, 1995). Participants can sit in a quiet room in a clinician's office and undergo a comprehensive hypnotizability assessment on the computer, thereby minimizing clinician time and minimizing inter-rater variability and error.

While assessment of hypnotizability is important, computer-delivered hypnotic interventions for pain may soon be more prominent. Borckardt et al. (2002) found that a computer-delivered hypnotic intervention for laboratory-induced pain was superior to a computer-delivered control condition. The Computer-Assisted Cognitive Imagery System may be an effective adjunct to more traditional behavioral pain-management treatment, although it has yet to be tested in clinical populations. This program is designed to allow patients to draw their pain onto a human figure on the computer screen. Next, a hypnotic induction is provided by the program. Suggestions are then given for alterations in pain experience and the visual representation of the pain drawn on the computerized human figure begins to change in appearance in a manner that corresponds with the suggestions. Finally, a waking procedure is provided by the program along with posthypnotic suggestions for long-lasting alterations in pain experience. Further development of such interventional systems may provide opportunities for patients to receive hypnotic pain management interventions both in the clinic and potentially at home on their own personal computers.

Virtual reality

It is established that psychological factors (especially attention) can influence the subjective experience of pain (Andrasik, Flor, & Turk, 2005; Melzack, 1999; Melzack & Wall, 1965). The relationship between attention and pain is a complicated one that is influenced by a number of factors (Buck & Morely, 2006; Eccleston, 1995; Fernandez & Turk, 1989; Tan, 1982). Recent research suggests that immersive virtual reality (VR) can be a useful pain control technique (Hoffman et al., 2001; and see Hoffman et al., 2008 for a recent review). Pain is an attention-demanding experience (Chapman & Nakamura, 1999; Eccleston, 1995), and VR is thought to be effective at drawing attention away from painful experiences. Researchers predict that patients who experience a stronger illusion of going into the virtual world are more distracted by VR and will thus report less pain than those who experience a less compelling illusion of "presence" in the virtual world (Hoffman & Sharar et al., 2004).

In a preliminary study, VR distraction reduced pain during staple removal from burn skin grafts more effectively than a two-dimensional video game (Hoffman et al., 2000). Patients reported feeling more present in the computer-generated world during the VR condition than during the video-game condition. Compared with standard of care (no distraction), burn patients consistently report clinically meaningful (i.e., >30%) reductions in pain during wound care and physical therapy sessions while in VR (Hoffman et al., 2008). Although more research is needed, findings to date indicate that VR does not decline in analgesic effectiveness when used repeatedly on multiple occasions (Hoffman & Patterson, 2001).

SnowWorld (http://www.vrpain.com) was the first immersive VR software designed for treating pain. SnowWorld was specifically designed to reduce pain experienced by burn patients during medical procedures. In SnowWorld, patients "go into" an icy, cool, three-dimensional virtual environment. Immersive VR blocks the user's view of the real world and presents patients with a view of a computer-generated world instead. A helmet and headphones exclude sights and sounds from the hospital environment, providing converging evidence from the virtual world to multiple senses (both sight and sound). Patients are able to interact with the virtual world by moving their joystick to look around, aim, and pull the trigger to shoot snowballs. Snowballs serve as a very simple human–computer interface for patients to interact with the virtual world with minimal motion of their bodies.

Functional magnetic resonance imaging research has corroborated participants' subjective pain reports with objective neural correlates of VR (Hoffman et al., 2004; Hoffman & Richards, 2006). Healthy volunteers received brief thermal pain stimuli every 30 seconds for 6 to 7 minutes. Participants received three 30-second pain stimuli with no VR (control condition) and three 30-second pain stimuli while playing SnowWorld (treatment order randomized). Significant reductions in subjective pain ratings during VR were accompanied by significant (>50%) reductions in pain-related brain activity in all 5 regions of interest in the neuroanatomic "pain matrix," consisting of the insula, thalamus, anterior cingulate cortex, and primary and secondary somatosensory cortices 36 (see also Hoffman et al., 2007).

Other preliminary studies have found that VR reduces pain during dental/periodontal procedures (Hoffman et al., 2001); during endoscopic urologic procedures, such as transurethral microwave thermotherapy for ablation of the prostate (Wright, Hoffman, & Sweet, 2005); associated with passive range of motion exercises during physical therapy for burn patients (Hoffman, Patterson, et al., 2001); during physical therapy exercises for cerebral palsy patients; and during painful physical therapy rehabilitation after single event multilevel surgery (Steele et al., 2003). These VR pain distraction physical therapy studies involving limb motion may lead to future studies exploring the use of VR for indirectly treating chronic pain via motion therapy.

Although the analgesic effects of VR are typically thought to disappear when the helmet is removed, there exists the possibility that VR may have longer-term benefits for chronic pain patients because providing periodic temporary relief from pain may provide an obstacle to the development of central sensitization to pain. Also, the skills of distraction and manipulation of attention to manage one's pain may generalize to real-world settings. More research is needed in this area.

Minimally invasive brain stimulation

While not fitting squarely in the realm of psychological pain management (but also not exactly in the realm of traditional neurological or psychiatric practice), minimally invasive brain stimulation technologies may soon have a place in pain management protocols. Transcranial magnetic stimulation (TMS) is a noninvasive (and relatively painless) brain stimulation technology that can focally stimulate the brain of an awake individual (Barker, Jalinous, & Freeston, 1985; George et al., 2003). A localized pulsed magnetic field transmitted through a figure 8 coil (lasting only microseconds) is able to focally stimulate the cortex by depolarizing superficial neurons (George & Belmaker, 2000; George, Lisanby, & Sackeim, 1999), which induces electrical currents in the brain. If TMS pulses are delivered repetitively and rhythmically, the process is called *repetitive TMS (rTMS)*. rTMS with a frequency greater than 1 Hz (*fast rTMS*) is limited to brief trains of stimulation at 25 to 30 Hz or less, as frequencies higher than this are associated with increased seizure risk.

TMS can induce varying brain effects depending on (1) the cortical region stimulated, (2) the activity that the brain is engaged in, and (3) the TMS device parameters (particularly frequency and intensity). TMS has been shown to produce immediate effects (e.g., thumb movement, phosphenes, temporary aphasia; Epstein et al., 1996) that are thought to result from direct excitation of inhibitory or excitatory neurons. TMS at different intensities, frequencies, and coil angles excites different elements (e.g., cell bodies, axons) of different neuronal groups (e.g., interneurons, neurons projecting into other cortical areas; Amassian, Eberle, Maccabee, & Cracco, 1992). Unfortunately, with current TMS technology (and understanding of the neurobiology of behavior), researchers have been unable to produce immediate effects along the lines of complex behaviors or fluid movement.

Intermediate effects of TMS (seconds to minutes) likely arise from transient changes in local pharmacology (e.g., gamma-aminobuteric acid, glutamate) and much research has been focused on whether different TMS frequencies might have different intermediate biological effects. Repeated low-frequency stimulation of a single neuron in culture produces long-lasting inhibition of cell–cell communication (Bear, 1999), while high-frequency stimulation can improve communication (Malenka & Nicoll, 1999). It has been hypothesized that TMS can produce sustained inhibitory or excitatory effects in a way analogous to single-cell electrical stimulation. Several studies have shown that chronic stimulation of the motor cortex can produce inhibitory or excitatory intermediate effects (lasting several minutes) following stimulation (Chen et al., 1997; Wu, Sommer, Tergau, & Paulus, 2000). Investigations of the intermediate effects of TMS have been used to develop a better understanding of brain functioning with respect to movement, vision, memory, attention, speech, neuroendocrine hormones, mood, and pain. Longer-term effects of TMS (days to weeks) are not well understood at a neurobiological level, but there is some evidence to support longer-term effects on mood, seizure activity, and pain.

With respect to mood, it is hypothesized that chronic repetitive stimulation of the prefrontal cortex initiates a cascade of events in the prefrontal cortex and in connected limbic regions. Prefrontal TMS sends information to important mood- and pain-regulating regions including the cingulate gyrus, orbitofrontal cortex, insula, and hippocampus, and there is positron emission tomography (PET) evidence that prefrontal TMS causes dopamine release in the caudate nucleus (and reciprocal activity with the anterior cingulate gyrus).

There is accumulating evidence supporting the effectiveness of repetitive TMS for managing neuropathic pain, headache, fibromyalgia, and idiopathic pain (Avery et al., 2007; Borckardt & Reeves, in press; Borckardt, Smith, et al., in press). TMS was recently approved by the U.S. Food and Drug Administration (FDA) as a treatment option for treatment-resistant depression, and it may not be long before it is available as an approved treatment for certain types of chronic pain.

Summary and future directions

Pain is a major physical and mental health care problem in the United States. As health-related technology continues to develop, clinicians will find more and more resources and treatment options at their fingertips. Many of the approaches and technologies described in this chapter are promising with respect to clinical effectiveness and enhancing clinical efficiency in practice; however, more work is needed to solidify the place for emerging technology in real-world pain management practice. Meanwhile, practitioners have a wide array of technological options available to them to enhance their behavioral pain management practice. One only needs to initiate a simple Internet search on any of the technologies discussed in this chapter to find a wealth of resources, available products, and training opportunities.

References

Amassian, V. E., Eberle, L., Maccabee, P. J., & Cracco, R. Q. (1992). Modelling magnetic coil excitation of human cerebral cortex with a peripheral nerve immersed in a brain-shaped volume conductor: The significance of fiber bending in excitation. *Electroencephalographic Clinical Neurophysiology, 85*, 291–301.

Andrasik, F., Flor, H., & Turk, D. C. (2005). An expanded view of psychological aspects in head pain: The biopsychosocial model. *Neurological Science, 26*, (suppl 2), S87–S91.

Avery, D. H., Holtzheimer, P. E., Fawaz, W., Russo, J., Neumaier, J., Dunner, D. L., et al. (2007). Transcranial magnetic stimulation reduces pain in patients with major depression: A sham-controlled study. *The Journal of Nervous and Mental Disease, 195*(5), 378–381.

Banks, S. M., & Kerns, R. D. (1996). Explaining high rates of depression in chronic pain: A diathesis-stress framework. *Psychological Bulletin, 119,* 95–110.

Barker, A. T., Jalinous, R., & Freeston, I. L. (1985). Non-invasive magnetic stimulation of the human motor cortex. *Lancet, 1,* 1106–1107.

Basler, S., Grzcsiak, R., Dworkin, R. (2002). Integrating Psychodynamic and Action-Oriented Psychotherapies: Treating Pain and Suffering. In: *Psychological Approaches to Pain Management,* 2nd ed. Gatchel, R. and Turk, D., Eds. New York: Guilford Press.

Bear, M. F. (1999). Homosynaptic long-term depression: a mechanism for memory? *Proceedings of the National Acadamy of Science USA, 96,* 9457–9458.

Blyth, F. M., March, L. M., Brnabic, A. J., Jorm, L. R., Williamson, M., & Cousins, M. J. (2001). Chronic pain in Australia: A prevalence study. *Pain, 89,* 127–134.

Borckardt, J. J., & Nash, M. R. (2001). How practitioners (and others) can make scientifically viable contributions to clinical-outcome research using the single-case time-series design. *International Journal of Clinical and Experimental Hypnosis, 50*(2), 114–148.

Borckardt, J. J., Reeves, S. T., George, M. S. (In press). The potential role of brain stimulation in the management of postoperative pain. *Journal of Pain Management.*

Borckardt, J. J., Smith, A. R., Reeves, S. T., Madan, A. M., Shelley, N., Branham, R., et al. (In press). A pilot study investigating the effects of fast left prefrontal rTMS on chronic neuropathic pain. *Pain Medicine.*

Buck, R., & Morley S. (2006). A daily process design study of attentional pain control strategies in the self-management of cancer pain. *European Journal of Pain, 10,* 385–398.

Burton, R., & Boedeker, B. (2000). Application of telemedicine in a pain clinic: The changing face of medical practice. *Pain Medicine, 1,* 351–357.

Bynum, A. B., Irwin, C. A., Cranford, C. O., & Denny, G. S. (2003). The impact of telemedicine on patients' cost savings: Some preliminary findings. *Telemedicine Journal and e-Health, 9*(4), 361–367.

Chapman, C. R., & Nakamura, Y. (1998). A passion of the soul: An introduction to pain for consciousness researchers. *Conscious Cognition, 8,* 391–422.

Chen, R., Classen, J., Gerloff, C., et al. (1997). Depression of motor cortex excitability by low-frequency transcranial magnetic stimulation. *Neurology, 48,* 1398–1403.

Chinman, M., Young, A. S., Schell, T., Hassell, J., & Mintz, J. (2004). Computer-assisted self-assessment in persons with severe mental illness. *Journal of Clinical Psychiatry, 65*(10), 1343–1351.

Cottrell, C., Drew, J., Gibson, J., Holroyd, K., & O'Donnell, F. (2007). Feasibility assessment of telephone-administered behavioral treatment for adolescent migraine. *Headache, 47,* 1293–1302.

Dersh, J., Gatchel, R. J., Mayer, T. G., Polatin, P. B., & Temple, O. W. (2006). Prevalence of psychiatric disorders in patients with chronic disabling occupational spinal disorders. *Spine, 31,* 1156–1162.

Dobscha, S. K., Corson, K., Pruitt, S., Crutchfield, M., & Gerrity, M. S. (2006). Measuring depression and pain with home health monitors. *Telemedicine Journal of e-Health, 12,* 702–706.

Eccleston, C. (1995). The attentional control of pain: Methodological and theoretical concerns. *Pain, 63,* 3–10.

Elliott, J., Chapman, J., & Clark, D. J. (2007). Videoconferencing for a veteran's pain management follow-up clinic. *Pain Management and Nursing, 8,* 35–46.

Epstein, C. M., Lah, J. J., Meador, K., Weissman, J. D., Gaitan, L. E., & Dihenia, B. (1996). Optimum stimulus parameters for lateralized suppression of speech with magnetic brain stimulation. *Neurology, 47,* 1590–1593.

Fernandez, E., & Turk, D. C. (1989). The utility of cognitive coping strategies for altering pain perception: A meta-analysis. *Pain, 38,* 123–135.

Fernandez, E., & Turk, D. C. (1992). Sensory and affective components of pain: Separation and synthesis. *Psychological Bulletin, 112,* 205–217.

Folen, R. A., James, L. C., Earles, J. E., & Andrasik, F. (2001). Biofeedback via telehealth: A new frontier for applied physiology. *Applied Psychophysiological Biofeedback, 26,* 195–204.

Gatchel, R. J. (2004a). Award for distinguished professional contributions to applied research. *American Psychologist, 59,* 794–805.

Gatchel, R. J. (2004b). Comorbidity of chronic pain and mental health: The biopsychosocial perspective. *American Psychologist, 59,* 792–794.

Gatchel, R. J., Peng, Y. B., Peters, M. L., Fuchs, P. N., & Turk, D. C. (2007). The biopsychosocial approach to chronic pain: Scientific advances and future directions. *Psychological Bulletin, 133*(4), 581–624.

George, M. S., & Belmaker, R. H. (2000). Transcranial magnetic stimulation. In George, M. S., & Belmaker, R. H. (Eds.), *Neuropsychiatry.* Washington, DC: American Psychiatric Press.

George, M. S., Lisanby, S. H., & Sackeim, H. A. (1999). Transcranial magnetic stimulation: Applications in neuropsychiatry. *Archives of General Psychiatry, 56,* 300–311.

George, M. S., Nahas, Z., Kozel, F. A., et al. (2003). Mechanisms and the current state of transcranial magnetic stimulation. *CNS Spectrums, 8*(7), 496–514.

Grant, C. D., & Nash, M. R. (1995). The Computer-Assisted Hypnosis Scale: Standardization and norming of a computer-administered measure of hypnotic ability. *Psychological Assessment, 7*(1), 49–58.

Gureje, O., Von Korff, M., Simon, G. E., & Gater, R. (1998). Persistent pain and well being: A World Health Organization study in primary care. *Journal of the American Medical Association, 280,* 145–151.

Hoffman, H. G., Doctor, J. N., Patterson, D. R., et al. (2000). Virtual reality as an adjunctive pain control during burn wound care in adolescent patients. *Pain, 85,* 305–309.

Hoffman, H. G., Garcia-Palacios, A., Patterson, D. R., et al. (2001). The effectiveness of virtual reality for dental pain control: A case study. *Cyberpsychological Behavior, 4,* 527–535.

Hoffman, H. G., Patterson, D. R., Carrougher, G. J., et al. (2001). Effectiveness of virtual reality-based pain control with multiple treatments. *Clinical Journal of Pain, 17,* 229–235.

Hoffman, H. G., Richards, T. L., Coda, B., et al. (2004). Modulation of thermal pain-related brain activity with virtual reality: Evidence from fMRI. *Neuroreport, 15,* 1245–1248.

Hoffman, H. G., Richards, T. L., Van Oostrom, T., et al. (2007). Analgesic effects of opioids and immersive virtual reality distraction: Evidence from subjective and functional brain imaging assessment. *Anesthesia Analgesia, 105,* 1776–1783.

Hoffman, H. G., Sharar, S. R., Coda, B., et al. (2004). Manipulating presence influences the magnitude of virtual reality analgesia. *Pain, 111,* 162–168.

Lind, L., Karlsson, D., & Fridlund, B. (2007). Digital pens and pain diaries in palliative home health care: Professional and caregivers' experiences. *Med Inform Internet Med, 32,* 287–296.

Madan, A., Borckardt, J., Weinstein, B., Wagner, M., Dominick, C., Cooney, H., et al., (2008). Clinical outcomes assessment in behavioral healthcare: Searching for practical solutions. *J. Healthc. Qual. 30*(4), 30–37.

Malenka, R. C., & Nicoll, R. A. (1999). Long-term potentiation: A decade of progress? *Science, 285,* 1870–1974.

McCracken, L. M., & Turk, D. C., (2002). Behavioral and cognitive-behavioral treatment for chronic pain: Outcome, predictors of outcome, and treatment process. *Spine, 27*(22), 2564–2573.

Melzack, R. (1999). From the gate to the neuromatrix. *Pain, 82*(Suppl. 6), S121–S126.

Melzack, R., & Casey, K. L. (1968). Sensory, motivational and central control determinants of pain: A new conceptual model. In D. Kenshalo (Ed.), *The skin senses* (pp. 423–443). Springfield, IL: Charles C. Thomas.

Melzack, R., & Wall, P. D. (1965). Pain mechanisms: A new theory. *Science, 150,* 971–979.

Mersky, H. (1986). International association for the study of pain: Classification of chronic pain. Descriptions of chronic pain syndromes and definitions of pain terms. *Pain, 3*(Suppl.), 1–226.

Montgomery, G. H., Duhamel, K. N., & Redd, W. H. (2000). A meta-analysis of hypnotically induced analgesia: How effective is hypnosis? *International Journal of Clinical and Experimental Hypnosis, 48*(2), 138–153.

Morley, S., Eccleston, C., & Williams, A. (1999). Systematic review and meta-analysis of randomized controlled trials of cognitive behavior therapy and behavior therapy for chronic pain in adults, excluding headache. *Pain, 80,* 1–13.

Naylor, M. R., Keefe, F. J., Brigidi, B., Naud, S., & Helzer, J. E. (2008). Therapeutic interactive voice response for chronic pain reduction and relapse prevention. *Pain, 335*–345.

Nestoriuc, Y., & Martin, A. (2007). Efficacy of biofeedback for migraine: A meta-analysis. *Pain, 128*(1–2), 111–127.

Nestoriuc, Y., Rief, W., & Martin, A. (2008). Meta-analysis of biofeedback for tension-type headache: Efficacy, specificity, and treatment moderators. *Journal of Consult Clinical Psychology, 76*(3), 379–396.

Piette, J. D. (2002). Enhancing support via interactive technologies. *Current Diabetes Reports, 2*(2), 160–165.

Piette, J. D. (2007). Interactive behavior change technology to support diabetes self-management: Where do we stand? *Diabetes Care, 30*(10), 2425–2432.

Serif, T., & Ghinea, G. (2005). Recording of time-varying back-pain data: A wireless solution. *IEEE Trans Inf Technol Biomed, 9,* 447–458.

Short, B., Borckardt, J. J., George, M., Beam, W., Reeves, S. T. (In press). Non-invasive brain stimulation approaches to fibromyalgia pain. *Journal of Pain Management.*

Slade, M., Thornicroft, G., & Glover, G. (1999). The feasibility of routine outcome measures in mental health. *Social Psychiatry and Psychiatric Epidemiology, 34,* 243–249.

Steele, E., Grimmer, K., Thomas, B., et al. (2003). Virtual reality as a pediatric pain modulation technique: A case study. *Cyberpsychology and Behavior, 6,* 633–638.

Syrjala, K. L., & Abrams, J. R. (2002). Hypnosis and imagery in the treatment of pain. In: *Psychological approaches to pain management: A practitioner's handbook,* 2nd ed. Turk, D. C., Ed. New York: Guilford Press.

Tan, S. Y. (1982). Cognitive and cognitive-behavioral methods for pain control: A selective review. *Pain, 12,* 201–228.

Turk, D. C., & Monarch, E. S. (2002). Biopsychosocial perspective on chronic pain. In D. C. Turk & R. J. Gatchel (Eds.), *Psychological approaches to pain management: A practitioner's handbook* (2nd ed., pp. 3–30). New York: Guilford Press.

Wolfe, F., Smythe, H. A., Yunus, M. B., Bennett, R. M., Bombardier, C., Goldenberg, D. L., et al. (1990). The American College of Rheumatology 1990 criteria for the classification of fibromyalgia: Report of the Multicenter Criteria Committee. *Arthritis and Rheumatism, 33,* 160–172.

Wright, J. L., Hoffman, M. G., Sweet, R. M. (2005). Virtual reality as an adjunctive pain control during transurethral microwave thermotherapy. *Urology, 66(6),* 1320.

Wu, T., Sommer, M., Tergau, F., & Paulus, W. (2000). Lasting influence of repetitive transcranial magnetic stimulation on intracortical excitability in human subjects. *Neuroscience Letters, 287,* 37–40.

chapter eight

Body image and eating disorders

Susan J. Paxton and Debra L. Franko

Introduction

Interventions using modern technologies in the context of body image problems and eating disorders have been widely evaluated and their usefulness is increasingly being recognized. Body image and disordered eating lie on a continuum from healthy body image and eating behaviors, to moderate body dissatisfaction and disordered eating behaviors, to subclinical body dissatisfaction and eating disorder symptoms, to clinical eating disorders. Body image refers to cognitions, emotions, and perceptions of one's body (Thompson, Heinberg, Altabe, & Tantleff Dunn, 1999). Body image attitudes and emotions may be positive, especially in girls and boys in primary and elementary school years (Holt & Ricciardelli, 2008), during which time the prevention of future development of negative body attitudes and emotions is the principal focus of intervention. Uses of modern communication technologies have been explored in the prevention context.

However, negative cognitions (e.g., I have an unattractive body) and emotions (e.g., I hate my body) about one's body, often summarized as body dissatisfaction, frequently occur in adolescence and adulthood (Ricciardelli & McCabe, 2001). Body dissatisfaction, especially weight dissatisfaction, is a risk factor for the development of low self-esteem and depressive symptoms (Paxton, Eisenberg, & Neumark-Sztainer, 2006) and use of unhealthy and extreme weight loss strategies such as skipping meals, crash dieting, fasting, laxative abuse, and self-induced vomiting (Neumark-Sztainer, Paxton, Hannan, Haines, & Story, 2006). In one study, serious purging behaviors were reported by over 9% of adolescent girls and boys (Ackard, Fulkerson, & Neumark-Sztainer, 2007). Use of unhealthy weight control techniques such as dieting and purging behaviors in turn are risk factors for binge eating with loss of control (Neumark-Sztainer, Wall, et al., 2006). At a subclinical level, body dissatisfaction and disordered eating warrant intervention because of the distress they cause, but in addition, they are risk factors for the development of clinical eating disorders (Franko & Striegel-Moore, 2007; Stice 2002). Consequently, modern technologies have been used to facilitate the implementation of targeted prevention or early intervention designed to reduce body dissatisfaction and disordered eating behaviors.

A number of clinical eating disorders have been identified. Anorexia nervosa is a disorder in which there is a relentless pursuit of thinness, resulting in weight loss or failure to gain weight during growth, extreme fear of gaining weight or becoming fat, and overvaluation of appearance and body image disturbances (American Psychiatric Association [APA]), 2000). Although not a common disorder, having a point prevalence of up to 0.5% (Aalto-Setälä, Marttunen, Tuulio-Henriksson, Poikolainen, & Lönnqvist, 2001) and a lifetime prevalence of about 2% (Keski-Rahkonen et al., 2007; Wade, Crosby, & Martin, 2006), it is a severely debilitating disorder associated with high morbidity and mortality (Keel et al., 2003).

Bulimia nervosa is a disorder in which there are regular binge eating episodes, use of compensatory behaviors, and an overvaluation of appearance (APA, 2000). The community point prevalence of bulimia nervosa is around 1% of women (Hay, Mond, Buttner, & Darby, 2008) and lifetime prevalence has been estimated at approximately 3% in women (Wade et al., 2006). Eating disorder not otherwise specified (EDNOS) is a diagnosis given to other clinically relevant eating disorders that are not encompassed by anorexia or bulimia nervosa, and includes binge eating disorder (BED) in which frequent binge eating, but not compensatory behaviors, is observed (APA, 2000). In fact, EDNOS is the most frequently occurring diagnosis and has a point prevalence of up to 5% in women, of whom approximately 2 to 3% have BED (Hay et al., 2008). EDNOS has lifetime prevalence in women of greater than 8% (Wade et al., 2006). In light of the prevalence and burden of disease associated with eating disorders, a wide range of treatment approaches have been used, and recently the use of modern technologies to enhance and extend these treatment options has been explored.

Due to the complex combination of genetic, neurochemical, psychological, and sociological factors involved in the etiology of eating disorders, a multidisciplinary team approach to treatment has been found to be most effective, and communication technologies have been used largely to assist in providing additional support and monitoring (Yager, 2003). For the treatment of bulimia nervosa and BED, however, there is a growing body of scientific evidence to support the use of cognitive-behavioral therapy (CBT; Shapiro, Berkman et al., 2007), which has been readily adapted for delivery using computer-based communication technologies.

In this chapter we explore the use of modern technologies across the spectrum of body image and eating disorders. In the following section, we selectively review research describing use of these technologies in different contexts including psychoeducation and prevention; delivery of intervention; use of palm pilots, e-mail, and text messaging to enhance traditional treatments; and online support groups. In the final section we explore clinical challenges in this area, with the aim of alerting the reader to practical issues and pitfalls to avoid.

Review of the evidence for use of modern technologies for body image and eating disorders

Psychoeducation and prevention

A recent innovation in health education, prevention, and mental health treatment is the use of a wide range of multimedia technologies that include computer-assisted learning tools, CD-ROMs, and Web sites (Budman, 2000). The use of such technology is appealing because it is accessible to large groups, cost effective, and much less labor intensive than face-to-face interventions. Moreover, multimedia technology has several distinct benefits over traditional formats. Computer-assisted tools provide a multisensory experience, which conveys information more vividly and memorably than do single-medium presentations. People are more naturally "literate" in learning from visual images and have to do less decoding than when presented with text (Seeck, Schomer, Mainwaring, & Ives, 1995). The multimedia approach encourages active participation in the experience by allowing the learner to physically interact with the content. Such approaches offer the possibility of reaching a wide audience, are not bound by geography, and are substantially less costly than in-person contact. Finally, multimedia programs provide flexible delivery and can be tailored to suit a wide range of learning styles and needs.

Psychoeducation, as well as both universal and selective prevention interventions, is central to public health approaches for addressing body image issues and disordered eating attitudes and behaviors. Psychoeducation offers information about psychological topics and can be provided in both classroom and clinical settings. Prevention from a public health perspective attempts to eliminate risk factors in the community at large, thereby reducing the risk of individuals of developing an eating disorder (Austin, 2001). In females, research suggests risk factors for body dissatisfaction include low self-esteem, peers' dieting, internalization of the idealized thin body ideal, weight and shape teasing, and perfectionism (Paxton et al., 2006; Presnell, Bearman & Stice, 2004; Wade & Lowes, 2002; Wertheim, Paxton, & Blaney, 2008). Risk factors for disordered eating include body dissatisfaction and extreme dieting behaviors (Franko & Striegel-Moore, 2007; Wertheim, Paxton, & Blaney, 2004). If the goal of preventing eating disorders is to be attained, affordable means of reaching whole communities (particularly, but not exclusively, communities of girls and young women) and reducing the presence of these risk factors are essential. Modern communication technologies, including CD-ROMs, DVDs, and the unassisted use of the Internet, offer this possibility.

The use of computer-assisted psychoeducation for eating disorders is relatively new, though one recent program (Franko & Cousineau, 2006) has shown promise. The Internet-based program, *Trouble on the Tightrope: In Search of Skateboard Sam*, provides information about puberty, body image, self-esteem, and sociocultural attitudes toward thinness using an interactive format aimed at young adolescents. In a recently conducted randomized controlled trial (RCT), 190 participants (mean age 11.6 years) were randomized to either the intervention or an attention placebo control condition and were assessed at baseline, after 3 Internet-based sessions and 3 months later, on a variety of psychosocial measures. Increases in knowledge about puberty were documented in participants who completed *Skateboard Sam* relative to controls. Pubertal status moderated the effects on appearance- and weight-related body esteem and several domains of self-esteem, resulting in positive effects for participants in the intervention group who had begun puberty. The study was conducted in health classes in middle schools, suggesting the possibility of disseminating such a program in school settings.

Two computer-based eating disorder prevention programs have been published. *Food, Mood, and Attitude (FMA)* is a 2-hour CD-ROM developed to reduce risk factors for eating disorders in college women (Franko et al., 2005). In a controlled trial, 240 first-year college women were randomly assigned to the intervention or a control condition. Half in each condition were at risk and half were low risk to develop an eating disorder. Participants in the *FMA* condition improved on several risk factors (internalization of thin ideal, shape concerns, weight concerns) and eating behavior measures relative to controls. Additionally, at-risk participants in the intervention group showed greater improvement on measures of risk and disordered eating behaviors than did low-risk participants (effect sizes .24–.33). Further, low-risk women in the *FMA* group increased their knowledge of risk factors relative to controls. Exploratory analyses also suggested that *FMA* had a protective effect *at follow-up* for overeating, excessive exercise, and purging behaviors in women at risk to develop eating disorders.

A well-researched Internet-based program, *Student Bodies*, was developed to decrease weight and shape concerns (risk factors for eating disorders) as well as unhealthy weight regulation behaviors in students at high risk for development of eating disorders (Low et al., 2006; Zabinski, Celio, Jacobs, Manwaring, & Wilfley, 2003; Zabinski, Wilfley, Calfas, Winzelberg,

& Taylor, 2004). The program is 8 to 10 weeks in length, and uses a self-help cognitive-behavioral approach. Each week, participants engage in (a) psychoeducational readings and reflection (e.g., media influences, nutrition, physical activity, general eating disorder information), (b) a cognitive-behavioral exercise, and (c) a Web-based body image journal to monitor events that trigger body dissatisfaction. The program also includes discussion groups, which have varied in terms of being moderated or unmoderated and either synchronous (i.e., real time) or asynchronous.

Taylor et al. (2006) evaluated the asynchronous, moderated chat version of Student Bodies in a sample of 480 at-risk college women. At baseline and for each year of a three-year follow-up, structured clinical interviews were used to determine if *Student Bodies* could prevent the onset of eating disorders. Participation in the intervention was associated with a decrease in the onset of eating disorders in two subgroups: those who had baseline compensatory behaviors (self-induced vomiting, laxative use, diuretic use, driven exercise) and those who had elevated baseline weights. In a secondary analysis examining the impact of adherence on the participants in the Taylor et al. study, Manwaring et al. (2008) found that the total number of weeks participating in the study, as well as the frequency with which participants utilized the online Web pages and journal entries, predicted changes on the restraint subscale of the Eating Disorder Examination Questionnaire (Fairburn & Beglin, 1994).

In summary, a small but growing body of literature supports the efficacy of psychoeducational and preventive approaches delivered using modern communication technologies. As these programs allow for greater reach of programs and appear to facilitate healthy changes in body image attitudes and eating behaviors, we recommend the use of such programs in settings where eating disorder risk is high.

Interventions for body image and eating disorders

Modern computer-based technologies have been used as the principal means of delivering therapy. The major rationale for using technologies in this manner is as a means of extending access to those in need of such an intervention. Technological approaches attempt to overcome practical difficulties to receiving in-person intervention, such as high geographic centralization of specialist services, and can reduce costs to the participants (Heinicke, Paxton, McLean, & Wertheim, 2007; Myers, Swan-Kremeier, Wonderlich, Lancaster, & Mitchell, 2004). Online intervention delivery can also assist in overcoming obstacles unique to body image and eating disorders. In particular, high levels of shame associated with disordered eating symptomatology have been found to be associated with reluctance to seek help (Hepworth & Paxton, 2007). Skarderud (2003) suggests that the anonymity associated with an Internet-delivered intervention can assist in overcoming this hurdle to seeking treatment.

Research evidence suggests that CD-ROM programs can provide an effective and acceptable form of delivery for self-help treatment for bulimia nervosa and binge eating disorder. Using a CD-ROM-delivered CBT self-help program for bulimia nervosa, Bara-Carril et al. (2004) found significant reductions in bingeing and compensatory behaviors in this treatment-seeking sample. Shapiro, Reba-Harrelson et al. (2007) compared CD-ROM-delivered CBT for binge eating disorder with therapist-led group CBT and a control group. The CD-ROM and group interventions were equally effective in reducing the frequency of bingeing. Pretreatment, more participants chose the CD-ROM than the group treatment, and attrition in this condition was lower compared to the group condition, indicating that this delivery approach was acceptable to patients. However, the authors suggested that combining the CD-ROM intervention with an Internet-based therapist and group interaction could meet some individuals' needs for contact and support. Interestingly, the addition of three brief face-to-face support sessions with a therapist to a CD-ROM-delivered CBT self-help program did not result in significant additional benefits (Murray et al., 2007).

In a recent study to reduce binge eating and overweight in adolescents, Jones et al. (2008) compared an Internet-facilitated intervention, *StudentsBodies2-BED*, with a wait-list control, and reported that body mass index (BMI), binge episodes, and weight and shape concerns decreased in the intervention group. Overall, these findings add to the growing literature that indicates CD-ROM-based and Internet programs have potential as a viable first step in a stepped-care approach to the treatment of eating disorders.

Access to early intervention for body dissatisfaction and disordered eating has also been enhanced by using synchronous online chat-room technology. The most notable evaluated interventions of this kind are group interventions delivered using chat-room technology, in which a small group of participants and a therapist log in to a secure synchronous (i.e., real-time) chat room site simultaneously and work through a treatment program (Heinicke et al., 2007; Zabinski et al., 2003). Heinicke et al. evaluated an Internet-based intervention that consisted of six 90-minute weekly small group, synchronous online sessions that were facilitated by a therapist and manual. Seventy-three girls (mean age = 14.4 years, SD = 1.48), who self-identified as having body image or eating problems, were randomly assigned to an intervention group (n = 36) or a delayed treatment control group (n = 37). Clinically significant improvements in body dissatisfaction, disordered eating, and depression were observed at postintervention and maintained at both the 2- and 6-month follow-up. The program offers a promising approach to improving body image and eating problems that also addresses geographic access problems.

An important question to address when using computer-based technologies in treatment is the extent to which they are comparable to face-to-face interventions (Gollings & Paxton, 2006). A study comparing synchronous online and face-to-face group delivery of a CBT intervention for adult women with body dissatisfaction and disordered eating indicated that the Internet approach may not have as immediate an impact as face-to-face delivery (Paxton, McLean, Gollings, Faulkner, & Wertheim, 2007). Specifically, *Set Your Body Free* is an 8-session program in which participants work through a manual on their own and meet weekly in a small group with a therapist either online or face-to-face. In a randomized controlled trial (Paxton et al., 2007), although both face-to-face and Internet groups showed improvements in measures of body image and eating behaviors compared to a delayed treatment control, larger effect sizes were found in the face-to-face group relative to the Internet group. Interestingly, however, the Internet group did continue to make significant improvements in body image and eating behaviors, and by the end of the 6-month follow-up period there were no clear differences on most measures. It seems that for adult women, it is desirable to provide treatment in a face-to-face setting where possible, but where this is not feasible, a synchronous online approach is a viable and effective alternative.

Technologies as adjuncts to therapy: Palm pilots, e-mail, and text messaging in treatment

Modern communication technologies have been widely used as adjuncts to traditional face-to-face interventions for a range of clinical eating disorders. In this context, the immediacy of monitoring may be enhanced by use of palm pilot technology, or the frequency and intensity of interactions with a therapist can be increased through e-mail or text messaging (SMS or short message service) contact between regular sessions (Bauer, Percevic, Okon, Meermann, & Kordy, 2003; Yager, 2003).

Palmtop computers may prove to be a feasible and effective means of extending eating disorder treatment, although little research has been reported to date. Norton, Wonderlich, Myers, Mitchell, and Crosby (2003) described the use of palmtop computers in integrative cognitive therapy for bulimia nervosa, which allows patients to self-monitor, plan, and store psychoeducational and personalized data. Ongoing research is needed to explore the potential use of this modality in clinical settings and to evaluate the efficacy of palmtop computers in treatment. Palm pilots can be used to extend therapy and have the additional capacity for storage and management of psychoeducational and therapeutic data (Norton et al.). Palmtops have been shown to be an effective and valid research tool for measuring eating disorder behaviors using ecological momentary assessment (Stein & Corte, 2003). Ecological momentary assessment provides a means for rating behaviors and moods as they occur over the course of the day, and has been used successfully to examine antecedents and consequences of episodes of binge eating and purging (Smyth et al., 2007; Stein et al., 2007; Wonderlich et al., 2007) and predictors of eating behavior across treatment (Wild et al., 2006).

CBT is the treatment of choice for bulimia nervosa and binge eating disorder. A hallmark technique in CBT is self-monitoring, in which patients are asked to record thoughts and feelings before and after eating behaviors. As self-monitoring can be a cumbersome (and sometimes embarrassing) process that patients often avoid, the use of computer-assisted technology might greatly enhance compliance. To date, such self-monitoring has been used in research studies (see Wonderlich et al., 2007), but certainly could be adapted to the clinical setting for patients who currently carry PDAs and even cell phones, some of which now have sophisticated data recording capabilities. Clinicians might work with patients to set up simple self-monitoring schemes and then review the data at the start of each session. Alternatively, patients could send information to the therapist prior to the next session so that the therapist could review and identify patterns to discuss with the patient.

Yager (2003) describes the potential advantages of using e-mail with patients with anorexia nervosa to increase the frequency, amount, and time flexibility of contact between patient and clinician, to promote written reflection from patients, to remind patients about therapeutic tasks, and to transmit information such as food diaries and symptom logs. Studies by Robinson and Serfaty (2001, 2003) examining the efficacy and acceptability of e-mail therapy found that individual therapy in e-mail form produced significant reductions in bulimic symptoms. They reported that e-mail therapy successfully engaged individuals in treatment who may not have presented to face-to-face treatment. Participants' qualitative responses to e-mail therapy revealed mixed opinions that included both positive aspects (anonymity, accessibility, effectiveness) and negative aspects (lack of personal interaction, easier avoidance) of this form of treatment.

Recently, e-mail has also been used successfully in a guided self-help therapy for patients with bulimia and binge eating disorder (Ljotsson et al., 2007). In this research, patients worked through a self-help manual (Fairburn, 1995) over 12 weeks and received feedback via e-mail about homework tasks and guidance about the program once or twice weekly. Ljotsson et al. found a 64% reduction in binge eating and that at the end of treatment 37% of patients reported no episodes of binge eating or purging. These favorable outcomes were maintained at 6-months follow-up. These findings are most encouraging in light of the capacity to deliver guidance in this manner over a wide geographic area; however, comparison to a group who had no e-mail contact will be an important next step in this research.

E-mail may also be employed in the context of consultation. Grunwald and Wesemann (2007) documented e-mail requests from users of an online eating disorder consultation service in Germany and identified three groups who made contact: those who described themselves as having an eating disorder, those who were related to someone with an eating disorder, and people who were interested in eating disorders. The service was predominantly used by individuals with bulimia nervosa or their families and friends. One third (33.3%) of the posted e-mails dealt with behavioral patterns of the illness and the affected person, and 18.7% of the e-mails were inquiries for information or requests for help in finding specialized clinics or therapists. The authors suggested that online consulting may have a role in the complementary care of those with eating disorders.

Communication technologies may also be useful in extending the final stages of treatment. Text messaging (SMS) has been examined as a means of reducing cost and improving time efficiency in aftercare for bulimia nervosa patients. Persistent symptoms and relapse are common after treatment for bulimia, suggesting the usefulness of a step-down approach when concluding treatment (Robinson et al., 2006). Bauer et al. (2003) described the use of SMS for patients' weekly symptom reports and the opportunity to receive therapeutic feedback. They found that the intervention was well accepted and potentially useful for patients, but subsequent studies have reported mixed outcomes in relation to acceptability, attrition rates, and efficacy (Bauer, Hagel, Okon, Meermann & Kordy 2006; Robinson et al., 2006). The content and delivery of this particular intervention may require modification, but further research exploring the potential of SMS as a component of treatment seems indicated.

Online support groups

Finally, computer-based technologies are being used as a means of extending the reach of support groups for individuals with eating disorders. Analogous to face-to-face support groups, chat-room support groups have been established in which registered participants may log in to a chat room and discuss issues in an unstructured way in the presence of a facilitator (Brooke, Pethick, & Greenwood, 2007; Darcy & Dooley, 2007). Wesemann and Grunwald (2008) analyzed threads from over 14,000 postings on a "pro-recovery" Web site for individuals with bulimia nervosa and found that most of the online discussion centered on problem-oriented threads (nearly 80%), communication-oriented threads (15%), or meta-communication threads (2.6%). In problem-oriented threads, communication occurs between users based on a question or problem introduced at the start. Communication-oriented threads are focused on private or everyday topics rather than a problem. Meta-communication threads were those that used the forum itself as a topic to encourage users to fight the disorder or to address problematic (e.g., pro-disorder) messages. Active management of the eating disorder was the primary discussion point on this Web site and the authors concluded that their findings indicated that "the people who are affected visit the forum with different objectives and the content of their communication varies accordingly" (p. 7).

Issues in computer delivered interventions
Participant safety

The fact that interventions can be delivered at a distance from the therapist using computer-mediated approaches is an important advantage but also potentially a worrisome disadvantage that may be somewhat unique to eating disorders. The negative aspect relates to the fact that the therapist cannot immediately see how well the client is progressing. Of course, this may also be the case in face-to-face intervention delivery should a client choose not to attend an appointment. Thus, this problem is not unique to computer delivery, but in this instance, distance is the norm rather than the exception and needs to be planned for and arrangements made (e.g., report from primary care physician) to anticipate potential difficulties that might arise (Heinicke et al., 2007; Robinson & Serfarty, 2003; Zabinski et al., 2004).

The most important issue that needs to be addressed is the possibility that, unbeknownst to the therapist, the intervention participant may be deteriorating in a dangerous way. In the eating disorder area this could mean losing a large amount of weight or becoming depressed with suicidal intent. Risk management strategies need to be in place to counter this possibility. It is essential to conduct a thorough assessment of each participant to ensure that she or he is suitable for the particular program, rather than someone who is keen to be involved in the program because it can be delivered across distance but not one that specifically suits his or her needs.

It is also essential to assess the likelihood that a person may be at risk of deteriorating. With this in mind, prior to Internet-delivered therapy groups, patients should be screened thoroughly for self-harm and suicidal ideation, which are common in patients seeking help for body image and eating disorders (Heinicke et al., 2007). These behaviors occur at varying levels of severity and their presence at a mild level may not necessarily be a reason for exclusion. However, in the context of a program for body image and eating disorders, if present, these concerns need to be discussed with the patient, and it needs to be clarified that the program does not specifically address these issues. Should the behaviors persist during the program, the therapist needs to be alerted.

An additional means to reduce the likelihood of unexpected deterioration is to ensure that the therapist has contact or the means of contact with others in the immediate physical location of the client. Having patient permission prior to commencement of a program to contact a significant other (e.g., parent or spouse) if the therapist has unresolved concerns is very helpful. It is also valuable to know where the patient is geographically so that the therapist can make a referral in the area if necessary. Although in treatment trials this may not be the case, in many situations the patients will also be in contact with a health professional in their region. Ensuring the patient is having regular check-ups with that health professional and getting confirmation of these visits is a helpful way to overcome the problems of distance. Because of the possible complications of geographical distance, it is especially important for the patient and therapist to be clear about the limits of confidentiality and to know exactly the circumstances under which the therapist will contact others for reassurance or assistance (Heinicke et al., 2007).

When conducting online synchronous group programs, participants are requested to complete a very brief online questionnaire to assess current mood before logging in to the session. This can alert the therapist to any immediate concerns of the participant (Heinicke et al., 2007; Paxton et al., 2007). If particular distress is expressed by a patient during a session, an immediate follow-up phone call is recommended. However, as many things are often taking place at one time in a session, the therapist may not be able to respond immediately if a concern is raised. When conducting online synchronous support groups, Brooke et al. (2007) have used two therapists as a means of assisting with this difficulty, one conducting the session and the other being available to contact a participant if there is a need for immediate help. An advantage of online delivery is that typically a complete session transcript is available to the therapist. A review of the transcript is recommended to ensure nothing of immediate concern was missed during the session.

Finally, there are potential risks to privacy when using computer-mediated technologies. In particular, there is the possibility that a chat-room session could be hacked into or that a person with destructive intent may join an open online meeting. There is no reported incident of this kind. However, it is important that patients are aware of this possibility and they may wish to consider the use of a pseudonym because of it.

The therapeutic interaction

Although the scope of communication technologies is expanding rapidly, at present, the major means of communicating is with text. This introduces a range of issues related to the loss of visual and spoken communication. Of central importance is the loss of affective information conveyed by facial expressions, gestures, tone of voice, and conversational pauses, the lack of which reduce clear communication and reception of information by both therapist and participant. It is beneficial to suggest to participants in chat room sessions that they be as explicit about emotions as possible and accompany comments with an emotive word or description, e.g., "angry" or "lol" (laugh out loud), an emoticon (e.g., ☺ ☹), or different types of fonts (Heinicke et al., 2007; Paxton et al., 2007; Ragusea & VandeCreek, 2003; Skaderud, 2003). It is also helpful to encourage explorations of feeling by both therapist and participants with questions such as "How did you feel about that?"

It is valuable for a therapist to have specific training in leading an online group prior to beginning such a program, as the range of therapeutic tools normally available is much reduced. In face-to-face interactions, a great deal about the communication of emotions is assumed. However, when using text alone, far less can be assumed, and the capacity for misinterpretation is substantially increased; that is, something written in jest, without emotional tone conveyed by voice, may be interpreted as sarcasm. In addition, the usual ways of expressing empathy with an expression or gesture are not available. The therapist also needs to be more alert to possible difficulties the participants may be having and check in with participants if they have not been very actively involved in the online conversation.

Practical issues in computer-mediated communication may also play a part in the therapeutic interaction. Home computers frequently do not have up-to-date programs, and guidance about downloading appropriate software may be needed prior to commencement of a program. Internet speeds also frequently differ, slowing down the rate that material is received and transmitted. This can interrupt the flow of text communication, which can be very frustrating to participants. Similarly, Internet connections can be lost altogether. Having a phone contact is often helpful for technical troubleshooting in addition to the risk management situations described above.

Although the challenges of computer-mediated programs described in this section at first seem somewhat daunting, with some thought they are largely surmountable. In addition, participants do find this means of intervention delivery acceptable (Heinicke et al., 2007; Zabinski et al., 2004). It is also important to note that the lack of face-to-face interaction frequently conveys an advantage in the treatment of body image and eating problems. In particular, the absence of face-to-face contact reduces the shame associated with discussion of body image problems and disordered eating symptoms (Skaderud, 2003), and self-disclosures and the sharing of pain and suffering may be facilitated by the relative anonymity of the text environment.

Therapist availability and burden

Although the use of e-mail and text messaging allows for greater between-session contact and can be very helpful (e.g., Yager, 2003), use of these technologies also increases therapist availability and, potentially, burden. Providing patients with clear guidelines as to what they can expect from the therapist regarding returning e-mail and text messages should be done at the beginning of treatment. Similar to the way in which many clinicians outline their response to emergency situations, clearly delineating turnaround time is important. Therapists may want to set limits on the number, frequency, and length of such messages. With eating disorder patients who may require more than weekly contact, providing a clear schedule of when online contact will occur will keep boundaries clear. Therapists should also decide at the outset of treatment whether extended online between-session contact is a billed service. Although generally not reimbursable by third-party payers, therapists who spend a lot of online time with patients may want to develop a billing policy so as not to create resentment or counter-transference issues if such practices take more time than the therapist is able or willing to spend in such contact.

Conclusion

In the body image and eating disorder field there has been rapid uptake of modern technologies as a means to deliver prevention programs, treatment interventions, and online support groups. These will be of interest to both clinicians and educators and will serve to broaden the reach to enable more adolescents and young people who struggle with these issues to be helped. The empirical literature documents the success of such programs, which though relatively few in number, are appearing increasingly in the research literature. At this time, there are few commercial entities that supply such programs; however, most are obtainable from the authors, who are in academic institutions and conduct the efficacy research needed to test them. At this time, many of the challenges of these delivery approaches may be minimized by careful consideration of the issues highlighted throughout this chapter. In addition, as the capacity of communication technologies is changing so rapidly, many of these challenges may disappear. There is little doubt that the ways in which technologies may be used to alleviate body image and eating problems will continue to expand dramatically over coming years.

References

Aalto-Setälä, T., Marttunen, M., Tuulio-Henriksson, A., Poikolainen, K., & Lönnqvist, J. (2001). One-month prevalence of depression and other DSM-IV disorders among young adults. *Psychological Medicine, 31*, 791–801.

Ackard, D. M., Fulkerson, J. A., & Neumark-Sztainer, D. (2007). Prevalence and utility of DSM-IV eating disorder diagnostic criteria among youth. *International Journal of Eating Disorders, 40*, 409–417.

American Psychiatric Association (APA). (2000). *Diagnostic and statistical manual of mental disorders* (4th ed., text rev.). Washington, DC: American Psychiatric Association.

Austin, S. (2001). Population-based prevention of eating disorders: An application of the Rose Prevention Model. *Preventive Medicine: An International Journal Devoted to Practice and Theory, 32*, 268–283.

Bara-Carril, N., Williams, C. J., Pombo-Carril, M. G., Reid, Y., Murray, K., Aubin, S., et al. (2004). A preliminary investigation into the feasibility and efficacy of a CD-ROM-based cognitive-behavioral self-help intervention for bulimia nervosa. *International Journal of Eating Disorders, 35*, 538–548.

Bauer, S., Hagel, J., Okon, E., Meermann, R., & Kordy, H. (2006). Experiences with the short message service (SMS) in the aftercare of patients with bulimia nervosa. *PDP Psychodynamische Psychotherapie: Forum der tiefenpsychologisch fundierten Psychotherapie, 5*, 127–136.

Bauer, S., Percevic, R., Okon, E., Meermann, R., & Kordy, H. (2003). Use of text messaging in the aftercare of patients with bulimia nervosa. *European Eating Disorders Review, 11*, 279–290.

Brooke, L., Pethick, L., & Greenwood, K. (2007, August). *Internet-based support for people with an eating disorder.* Paper presented at the Australia and New Zealand Academy of Eating Disorders Conference, Melbourne, Australia.

Budman, S. H. (2000). Behavioral health care dot-com and beyond: Computer-mediated communications in mental health and substance abuse treatment. *American Psychologist, 55*, 1290–1300.

Darcy, A. M., & Dooley, B. (2007). A clinical profile of participants in an online support group. *European Eating Disorders Review, 15*, 185–195.

Fairburn, C. G. (1995). *Overcoming binge eating.* New York: Guilford.

Fairburn, C. G., & Beglin, S. J. (1994). Assessment of eating disorders: Interview or self-report questionnaire? *International Journal of Eating Disorders, 16*, 363–370.

Franko, D. L., & Cousineau, T. (2006, June). *Decreasing risk for eating disorders across the developmental spectrum: Multimedia tools for children, adolescents, and college students.* Paper presented at the International Conference on Eating Disorders, Barcelona, Spain.

Franko, D. L., Mintz, L. B., Villapiano, M., Green, T. C., Mainelli, D., Folensbee, L., et al. (2005). Food, mood, and attitude: Reducing risk for eating disorders in college women. *Health Psychology, 24*, 567–578.

Franko, D. L., & Striegel-Moore, R. H. (2007). Psychosocial risk for eating disorders: What's new? *Annual Review of Eating Disorders, Part 1*, 51–62.

Gollings, E. K., & Paxton, S. J. (2006). Comparison of Internet and face-to-face delivery of a group body image and disordered eating intervention for women: A pilot study. *Eating Disorders: The Journal of Treatment & Prevention, 14*, 1–15.

Grunwald, M., & Wesemann, D. (2007). Special online consulting for patients with eating disorders and their relatives: Analysis of user characteristics and e-mail content. *Cyberpsychology and Behavior, 10*, 57–63.

Hay, P. J., Mond, J., Buttner, P., & Darby, A. (2008). Eating disorder behaviors are increasing: Findings from two sequential community surveys in South Australia. *PLoS ONE, 3*(2), e1541.

Heinicke, B. E., Paxton, S. J., McLean, S. A., & Wertheim, E. H. (2007). Internet-delivered targeted group intervention for body dissatisfaction and disordered eating in adolescent girls: A randomized controlled trial. *Journal of Abnormal Child Psychology, 35*, 379–391.

Hepworth, N., & Paxton, S. J. (2007). Pathways to help-seeking in bulimia nervosa and binge eating problems: A concept mapping approach. *International Journal of Eating Disorders, 40*, 493–504.

Holt, K. E., & Ricciardelli, L. A. (2008). Weight concerns among elementary school children: A review of prevention programs. *Body Image, 5*, 233–43.

Jones, M., Luce, K. H., Osborne, M. I., Taylor, K., Cunning, D., Doyle, A. C., et al. (2008). Randomized, controlled trial of an internet-facilitated intervention for reducing binge eating and overweight in adolescents. *Pediatrics, 121*, 453–462.

Keel, P. K., Dorer, D. J., Eddy, K. T., Franko, D. L., Charatan, D. L., & Herzog, D. B. (2003). Predictors of mortality in eating disorders. *Archives of General Psychiatry, 60*, 179–183.

Keski-Rahkonen, A., Hoek, H. W., Susser, E. S., Linna, M. S., Sihvola, E., Raevuori, A., et al. (2007). Epidemiology and course of anorexia nervosa in the community. *American Journal of Psychiatry, 164*, 1259–1265.

Ljotsson B., Lundin C., Mitsell K., Carlbring P., Ramklint M., & Ghaderi A. (2007). Remote treatment of bulimia nervosa and binge eating disorder: A randomised trial of Internet-assisted cognitive behavioural therapy. *Behaviour Research and Therapy, 45*, 649–661.

Low, K. G., Charanasomboon, S., Lesser, J., Reinhalter, K., Martin, R., Jones, H., et al. (2006). Effectiveness of a computer-based interactive eating disorders prevention program at long-term follow-up. *Eating Disorders: The Journal of Treatment & Prevention, 14*, 17–30.

Manwaring, J. L., Bryson, S. W., Goldschmidt, A. B., Winzelberg, A. J., Luce, K. H., Cunning, D., et al. (2008). Do adherence variables predict outcome in an online program for the prevention of eating disorders? *Journal of Consulting and Clinical Psychology, 76*, 341–346.

Murray, K., Schmidt, U., Pombo-Carril, M.-G., Grover, M., Alenya, J., Treasure, J., et al. (2007). Does therapist guidance improve uptake, adherence and outcome from a CD-ROM based cognitive-behavioral intervention for the treatment of bulimia nervosa? *Computers in Human Behavior, 23*, 850–859.

Myers, T. C., Swan-Kremeier, L., Wonderlich, S., Lancaster, K., & Mitchell, J. E. (2004). The use of alternative delivery systems and new technologies in the treatment of patients with eating disorders. *International Journal of Eating Disorders, 36,* 123–143.

Neumark-Sztainer, D., Paxton, S. J., Hannan, P. J., Haines, J., & Story, M. (2006). Does body satisfaction matter? Five-year longitudinal associations between body satisfaction and health behaviors in adolescent females and males. *Journal of Adolescent Health, 39,* 244–251.

Neumark-Sztainer, D., Wall, M., Guo, J., Story, M., Haines, J., & Eisenberg, M. (2006). Obesity, disordered eating, and eating disorders in a longitudinal study of adolescents: How do dieters fare 5 years later? *Journal of the American Dietetic Association, 106,* 59–568.

Norton, M., Wonderlich, S. A., Myers, T., Mitchell, J. E., & Crosby, R. D. (2003). The use of palmtop computers in the treatment of bulimia nervosa. *European Eating Disorders Review, 11,* 231–242.

Paxton, S. J., Eisenberg, M. E., & Neumark-Sztainer, D. (2006). Prospective predictors of body dissatisfaction in adolescent girls and boys: A five-year longitudinal study. *Developmental Psychology, 42,* 888–899.

Paxton, S. J., McLean, S. A., Gollings, E. K., Faulkner, C., & Wertheim, E. H. (2007). Comparison of face-to-face and internet interventions for body image and eating problems in adult women: An RCT. *International Journal of Eating Disorders, 40,* 692–704.

Presnell, K., Bearman, S. K., & Stice, E. (2004). Risk factors for body dissatisfaction in adolescent boys and girls: A prospective study. *International Journal of Eating Disorders, 36,* 389–401.

Ragusea, A. S., & VandeCreek, L. (2003). Suggestions for the ethical practice of online psychotherapy. *Psychotherapy: Theory, Research, Practice, Training, 40,* 94–102.

Ricciardelli, L. A., & McCabe, M. P. (2001). Dietary restraint and negative affect as mediators of body dissatisfaction and bulimic behavior in adolescent girls and boys. *Behaviour Research and Therapy, 39,* 1317–1328.

Robinson, P., & Serfaty, M. A. (2003). Computers, e-mail and therapy in eating disorders. *European Eating Disorders Review, 11,* 210–221.

Robinson, P. H., & Serfaty, M. A. (2001). The use of email in the identification of bulimia nervosa and its treatment. *European Eating Disorders Review, 9,* 182–193.

Robinson, S., Perkins, S., Bauer, S., Hammond, N., Treasure, J., & Schmidt, U. (2006). Aftercare intervention through text messaging in the treatment of bulimia nervosa: Feasibility pilot. *International Journal of Eating Disorders, 39,* 633–638.

Seeck, M., Schomer, D., Mainwaring, N., & Ives, J. (1995). Selectively distributed processing of visual object recognition in the temporal and frontal lobes of the human brain. *Annals of Neurology, 37,* 538–545.

Shapiro, J. R., Berkman, N. D., Brownley, K. A., Sedway, J. A., Lohr, K. N., & Bulik, C. M. (2007). Bulimia nervosa treatment: A systematic review of randomized controlled trials. *International Journal of Eating Disorders, 40,* 321–336.

Shapiro, J. R., Reba-Harrelson, L., Dymek-Valentine, M., Woolson, S. L., Hamer, R. M., & Bulik, C. M. (2007). Feasibility and acceptability of CD-ROM-based cognitive-behavioural treatment for binge-eating disorder. *European Eating Disorders Review, 15,* 175–184.

Skarderud, F. (2003). Sh@me in cyberspace. Relationships without faces: The e-media and eating disorders. *European Eating Disorders Review, 11*, 155–169.

Smyth, J. M., Wonderlich, S. A., Heron, K. E., Sliwinski, M. J., Crosby, R. D., Mitchell, J. E., et al. (2007). Daily and momentary mood and stress are associated with binge eating and vomiting in bulimia nervosa patients in the natural environment. *Journal of Consulting and Clinical Psychology, 75*, 629–638.

Stein, K. F., & Corte, C. M. (2003). Ecologic momentary assessment of eating-disordered behaviors. *International Journal of Eating Disorders, 34*, 349–360.

Stein, R. I., Kenardy, J., Wiseman, C. V., Dounchis, J. Z., Arnow, B. A., & Wilfley, D. E. (2007). What's driving the binge in binge eating disorder? A prospective examination of precursors and consequences. *International Journal of Eating Disorders, 40*, 195–203.

Stice, E. (2002). Risk and maintenance factors for eating pathology: A meta-analytic review. *Psychological Bulletin, 128*, 825–848.

Taylor, C., Bryson, S., Luce, K. H., Cunning, D., Doyle, A. C., Abascal, L. B., et al. (2006). Prevention of eating disorders in at-risk college-age women. *Archives of General Psychiatry, 63*, 881–888.

Thompson, J. K., Heinberg, L. J., Altabe, M., & Tantleff Dunn, S. (1999). *Exacting beauty: Theory, assessment, and treatment of body image disturbance.* Washington, DC: American Psychological Association.

Wade, T. D., Crosby, R. D., & Martin, N. G. (2006). Use of latent profile analysis to identify eating disorder phenotypes in an adult Australian twin cohort. *Archives of General Psychiatry, 63*, 1377–1384.

Wade, T. D., & Lowes, J. (2002). Variables associated with disturbed eating habits and overvalued ideas about the personal implications of body shape and weight in a female adolescent population. *International Journal of Eating Disorders, 32*, 39–45.

Wertheim, E. H., Paxton, S. J., & Blaney, S. (2004). Risk factors for the development of body image disturbances. In J. K. Thompson (Ed.), *Handbook of eating disorders and obesity* (pp. 463–494). Hoboken, NJ: John Wiley & Sons, Inc.

Wertheim, E. H., Paxton, S. J., & Blaney, S. (2008). Body image in girls. In L. Smolak & J. K. Thompson (Eds.) *Body image, eating disorders and obesity in youth* (2nd ed.). Washington, DC: American Psychological Association.

Wesemann, D., & Grunwald, M. (2008). Online discussion groups for bulimia nervosa: An inductive approach to Internet-based communication between patients. *International Journal of Eating Disorders, 41*, 527–534.

Wild, B., Quenter, A., Friederich, H.-C., Schild, S., Herzog, W., & Zipfel, S. (2006). A course of treatment of binge eating disorder: A time series approach. *European Eating Disorders Review, 14*, 79–87.

Wonderlich, S. A., Rosenfeldt, S., Crosby, R. D., Mitchell, J. E., Engel, S. G., Smyth, J., et al. (2007). The effects of childhood trauma on daily mood lability and comorbid psychopathology in bulimia nervosa. *Journal of Traumatic Stress, 20*, 77–87.

Yager, J. (2003). E-mail therapy for anorexia nervosa: Prospects and limitations. *European Eating Disorders Review, 11*, 198–209.

Zabinski, M. F., Celio, A. A., Jacobs, M., Manwaring, J., & Wilfley, D. E. (2003). Internet-based prevention of eating disorders. *European Eating Disorders Review, 11*, 183–197.

Zabinski, M. F., Wilfley, D. E., Calfas, K. J., Winzelberg, A. J., & Taylor, C. (2004). An interactive psychoeducational intervention for women at risk of developing an eating disorder. *Journal of Consulting and Clinical Psychology, 72,* 914–919.

chapter nine

Obesity

Rebecca A. Krukowski, Jean Harvey-Berino, and Delia Smith West

Using Internet technology to treat obesity

Obesity is an epidemic in the United States, among both adults and children (Ogden et al., 2006). The majority of the U.S. population is overweight or obese, and the cost of caring for those with obesity-related illnesses is estimated to rise to $861 to $957 billion by 2030 (Wang, Beydoun, Liang, Caballero, & Kumanyika, 2008). In addition, obesity is associated with numerous chronic health risk factors, including hypertension, dyslipidemia, and impaired glucose tolerance in adults (Bray, 2004) and children (Weiss et al., 2004). However, effective in-person obesity interventions have been developed, and programs that combine dietary restriction, physical activity, and behavior therapy are considered the most successful in promoting weight loss and maintenance for both adults and children (National Heart, Lung, and Blood Institute, 1998; Spear et al., 2007).

In adults, evidence-based behavioral weight loss programs have been shown to result in clinically significant weight losses of 7 to 10% (Diabetes Prevention Program Research Group, 2002; Look AHEAD Research Group, 2007). Fortunately, this degree of weight loss can ameliorate many of the comorbidities associated with obesity, including diabetes and cardiovascular risk factors (Look AHEAD Research Group, 2007). Furthermore, behavioral weight loss programs have demonstrated greater effectiveness than medication for reducing the incidence of type 2 diabetes and cardiovascular risk factors (Diabetes Prevention Program Research Group, 2002). These results indicate that clinically significantly weight losses can be achieved among adults in evidence-based behavioral weight loss programs, and these weight losses are associated with significant improvements in health risk factors.

Similar beneficial outcomes of behavioral weight loss programs have also been demonstrated among children. A recent meta-analysis illustrated that, in randomized controlled trials of behavioral weight control programs with children, the average decrease in percentage overweight was 8.2%, in comparison to an average increase in percentage overweight of 2.1% for children in the control groups (Wilfley et al., 2007). Improvements in BMI among children have been shown to be associated with improvements in health risk factors, including cholesterol levels (Nemet et al., 2005; Reinehr & Andler, 2004) and insulin sensitivity (Reinehr, Kiess, Kapellen, & Andler, 2004; Savoye et al., 2007). There is also some research indicating that family-based behavioral weight control programs for children may be more successful for long-term maintenance of weight loss (Epstein, Valoski, Wing, & McCurley, 1994) than behavioral weight control programs among adults. A focus on behavioral weight control programs for children may be critical in order to raise a new generation of healthier children.

Unfortunately, access and adherence to obesity treatment programs may be limited for both adults and children due to numerous treatment barriers. Participation in evidence-based weight management programs may be restricted by the availability of such programs in certain areas, proximity to the intervention site, transportation concerns, and time constraints. Appropriate weight management within the primary care setting by physicians is also hampered because delivery of efficacious behavioral treatment within the limited time available during office visits is challenging (Cooper, Valleley, Polaha, & Evans, 2006; Gottschalk & Flocke, 2005) and reimbursement for obesity-related behavioral treatment is restricted (Tsai, Asch, & Wadden, 2006). New weight control technologies that utilize the Internet have the potential to reduce some of these treatment barriers, particularly among minority, lower income, and rural populations. Furthermore, Internet-based weight control programs may also represent an effective supplemental treatment modality for health professionals caring for their overweight patients.

The Internet: An accessible and acceptable modality for obesity interventions

Recent evidence suggests that the Internet may be a promising method for disseminating obesity interventions across many segments of the population. In 2007, most American adults (75%) reported using the Internet, including 56% of African Americans, 61% of adults in the lowest income category, and 64% of adults living in rural areas (Pew Internet & American Life Project, 2008). United States Census Bureau data from 2001 reveals that 31% of 5- to 7-year-olds, 54% of 8- to 10-year-olds, and 68% of 11- to 14-year-olds reported currently using the Internet (DeBell & Chapman, 2003). In addition, in a recent survey, 87% of adolescents (12 to 17 years of age) reported using the Internet, including 77% of African-American teens, 73% of teens from the lowest income families, and 83% of rural teens (Pew Internet & American Life Project, 2006). Of adult Internet users, the majority reported accessing health (Pew Internet & American Life Project, 2005a), diet, exercise, or fitness information online (Pew Internet & American Life Project, 2005b). Of adolescent Internet users, 31% reported accessing health, dieting, or fitness information online (Pew Internet & American Life Project, 2006). Moreover, 93% of adult Internet users reported that it is important for them to be able to access health information when it is convenient for them, at any hour of the day (Pew Internet & American Life Project, 2000).

Based on the high rates of Internet utilization for adults, children, and adolescents, and the growing use of the Internet to obtain health information and information about diet, exercise, and fitness, the Internet appears to be an accessible and acceptable modality for health behavior interventions, which may include obesity treatment programs. Increasing rates of obesity, access to the Internet, and expediency of information on the Internet may be significant factors related to increasing interest in, and feasibility of, Internet-based weight loss programs.

Using Internet technology for weight loss in adults

Research indicates that Internet technology may facilitate successful weight loss (Gold, Burke, Pintauro, Buzzell, & Harvey-Berino, 2007; Micco et al., 2007; Tate, Jackvony, & Wing, 2006) among adults. However, weight losses in Internet-based obesity research programs have ranged from approximately 1 kg (Womble et al., 2004) to 8 kg at one year (Gold et al., 2007; Table 9.1). The variability in weight loss success among different programs demonstrates the importance of determining the most effective components in Internet-based treatment and the pressing need to develop and disseminate strong evidence-based programs. Several elements may explain the variability in weight loss success, including the programmatic structure, inclusion of behavior therapy components, partial meal replacements, synchronous (or real-time) group meetings, and interactive, dynamic program Web sites. Each of these elements appears to enhance outcomes.

First, several studies have illustrated the importance of maintaining the structure of in-person weight loss programs when implementing Internet-based treatments, specifically including such elements as ongoing scheduled contact, specific lesson materials, and homework expectations (Gold et al., 2007; Tate, Wing, & Winett, 2001; Womble et al., 2004). The omission of some or all of these components may be a critical limitation of some commercially available Internet-based weight loss programs, which often fail to incorporate them. There are other aspects of the programmatic structure of successful in-person programs that may

Table 9.1 Examples of Internet-Based Weight Control Programs for Adults

Study	Sample	Program Description	Design	Intervention Components	Weight Change	
Gold et al., 2007	n = 124; 98% Caucasian, 2% race/ethnicity not specified; 80% Women	Weight loss: 6 months; Weight maintenance: 6 months	RCT; 2 conditions	1. Vtrim: Behavior therapy components;* programmatic structure; interactive Web site; synchronous communication; bulletin board; progress feedback by a facilitator; self-monitoring	Weight loss (0–6 months): 1. 8.3 ± 7.9 kg; 2. 4.1 ± 6.2 kg	Between groups: 1>2; p<0.01
				2. eDiets.com: interactive Web site; synchronous communication; bulletin board; automatic feedback; self-monitoring; tailored meal plan	Weight loss (0–12 months): 1. 5.1 ± 7.1 kg; 2. 2.6 ± 5.3 kg	Between groups: 1>2; p<0.01
McConnon et al., 2007	n = 221; 95% Caucasian, 5% race/ethnicity not specified; 77% Women	Weight loss: 12 months	RCT; 2 conditions	1. Internet: behavior therapy components;* automatic feedback; email reminders	Weight loss (0–12 months): 1. 1.3 kg; 2. 1.9 kg	n.s.
				2. Usual physician care		

Chapter nine: Obesity

Micco et al., 2007; Krukowski, Harvey-Berino, Ashikaga, Thomas, & Micco, 2008	n = 123; 99% Caucasian, 1% race/ethnicity not specified; 83% Women	Weight loss: 12 months	RCT; 2 conditions	1. Internet: behavior therapy components;* programmatic structure; interactive Web site; synchronous communication; bulletin board; progress feedback by a facilitator; challenges	Weight loss (0–6 months): 1. 6.8 ± 7.8 kg; 5.1 ± 4.8 kg	n.s.
				2. Internet + in-person: behavior therapy components;* programmatic structure; interactive Web site; synchronous communication; bulletin board; progress feedback by a facilitator; self-monitoring; 12 in-person sessions; challenges	Weight loss (0–12 months): 1. 5.1 ± 7.1 kg; 2. 3.5 ± 5.1 kg	n.s.
Pratt, Jandzio, Tomlinson, Kang, & Smith, 2006	n = 2498; 100% race/ethnicity not specified; 23% Women	1 year of participation over a 4-year period; Aims: 5 fruit/vegetables per day, 10,000 steps per day, healthy BMI	Pre/post; 1 condition	Web chats on nutrition and exercise; self-monitoring; e-mail reminders; challenges	Weight loss amount not reported	Within groups; $p<0.05$

Continued

Table 9.1 Examples of Internet-Based Weight Control Programs for Adults (*Continued*)

Study	Sample	Program Description	Design	Intervention Components	Weight Change	
Rothert et al., 2006	n = 2826; 56% Caucasian, 36% African American, 3% Hispanic, 5% race/ethnicity not specified; 84% Women	Weight loss: 6 weeks	RCT; 2 conditions	1. Tailored expert: automatic progress feedback; a tailored plan in response to questionnaires; could enroll a supportive "buddy" 2. Information only	Weight loss (0–6 month follow-up): 1. 2.8 ± 0.3 kg; 2. 1.1 ± 0.4 kg	Between groups: 1>2; $p<0.001$
Tate, Jackvony, & Wing, 2003	n = 92; 89% Caucasian; 11% race/ethnicity not specified; 90% Women	Weight loss: 12 months	RCT; 2 conditions	1. Basic Internet: bulletin board; weight self-monitoring; e-mail reminders 2. Internet behavioral e-counseling: behavior therapy components;* progress feedback by a facilitator; self-monitoring; e-mail reminders	Weight loss (0–12 months): 1. 2.0 ± 5.7 kg; 2. 4.4 ± 6.2 kg	Between groups: 2>1; $p<0.05$
Tate et al., 2006	n = 192; 90% Caucasian, 10% race/ethnicity not specified; 84% Women	Weight loss: 6 months	RCT; 3 conditions	1. Web site only: behavior therapy components;* interactive Web site; weight self-monitoring; e-mail reminders; meal replacements	Weight loss (0–6 months): 1. 2.6 ± 5.7 kg; 2. 4.9 ± 5.9 kg; 3. 7.3 ± 6.2 kg	Between groups: 3>1 and 2; $p<0.001$

Chapter nine: Obesity

Tate et al., 2001	n = 91; 84% Caucasian, 16% race/ethnicity not specified; 89% Women	Weight loss: 6 months	RCT; 2 conditions	2. Computer-automated feedback: behavior therapy components;* programmatic structure; interactive Web site; bulletin board; computer-based feedback; self-monitoring; e-mail reminders; meal replacements 3. Human e-mail counseling: behavior therapy components;* programmatic structure; interactive Web site; bulletin board; progress feedback by a facilitator; self-monitoring; e-mail reminders; meal replacements 1. Education: self-monitoring; 1 in-person contact	Weight loss (0–6 months): 1. 1.6 ± 3.3 kg; 2. 4.1 ± 4.5 kg	Between groups: 2 >1, $p<0.05$

Continued

Table 9.1 Examples of Internet-Based Weight Control Programs for Adults (*Continued*)

Study	Sample	Program Description	Design	Intervention Components	Weight Change
				2. Behavior therapy: behavior therapy components;* programmatic structure; bulletin board; progress feedback by a facilitator; self-monitoring; 1 in-person contact	
Womble et al., 2004	n = 47; 100% race/ ethnicity not specified; 100% Women	Weight loss: 4 months; Weight maintenance: 8 months	RCT: 2 conditions	1. e-Diets.com: interactive Web site; synchronous communication; food self-monitoring; e-mail reminders; meal plans; 5 in-person contacts	Weight loss (0–4 months): 1. 0.7 ± 2.7 kg; 2. 3.0 ± 3.1 kg	Between groups: 2 >1, $p<0.05$
				2. Manualized treatment: behavior therapy components;* food self-monitoring; 5 in-person contacts	Weight loss (0–12 months): 1. 0.8 ± 3.6 kg; 2. 3.3 ± 4.1 kg	Between groups: 2 >1, $p<0.05$

Notes: RCT: Randomized controlled trial; n.s.: Non-significant difference.

* Behavior therapy components include stimulus control, goal setting, problem solving, social assertion, motivation, cognitive strategies, and relapse prevention.

also help explain their greater weight losses than those achieved in most online programs, including the length of in-person programs, which are typically 3 months to 1 year. Even within the in-person weight control research, longer contact schedules have generally been demonstrated to be more effective for inducing and maintaining weight loss (Perri, Nezu, Patti, & McCann, 1989); thus, it should not be surprising that online programs that are shorter than the typical in-person programs might produce smaller weight losses.

In addition, it appears that incorporating the behavioral therapy components of successful in-person weight management programs (e.g., self-monitoring, goal setting, problem solving, and reinforcement and feedback from a trained facilitator) produces greater weight loss in Internet-based interventions, compared with those programs that merely make health educational resources available (Tate et al., 2001; Tate et al., 2003). However, even with the inclusion of these potentially critical components of programmatic structure and behavior therapy components, only a few studies of Internet-based interventions (Gold et al., 2007; Micco et al., 2007; Tate et al., 2006) have achieved clinically significant weight losses (5% or greater) that are comparable to weight losses that are typical in in-person interventions.

Among the recent studies of Internet-based weight loss interventions that have demonstrated clinically meaningful weight losses, there are several possible factors that may have contributed to the substantial weight losses achieved beyond the inclusion of standard behavioral treatment components. Tate and colleagues (2006) demonstrated some of the highest weight losses in a recent Internet-based program in which they supplemented the online program with partial meal replacement (i.e., Slim-Fast® shakes). Meal replacements have been found to produce greater weight losses among in-person programs as well (Heymsfield, van Mierlo, van der Knaap, Heo, & Frier, 2003), so it is not surprising that an online program that offered them would have enhanced weight loss outcomes. Another beneficial online strategy was identified by Micco and colleagues (2007) who demonstrated that highly successful weight losses could be achieved in a Internet-based weight control program that included synchronous group meetings ("real-time" chat forums), which roughly approximated the interactive nature of in-person classes. This study may point to the advantage of an interactive approach to communication and social support in online programs.

The importance of login frequency

The ability to examine frequency of logins and hits (the number of times a link is chosen) is a clear advantage to Internet technology inasmuch as it allows evaluation of usage patterns of both the Web site overall, as well as individual features. It also provides an estimation of the "dose" of intervention received. In recent research examining login frequency, Micco and colleagues (2007) examined the login frequency to a highly interactive and dynamic Web site (e.g., progress charts/tools, contests, news flashes) that merged the available evidence on crucial components to an Internet-based weight loss program by including a programmatic structure, behavior therapy components, and synchronous group meetings. They found that participants accessed this Web site approximately twice as often over a six-month period (on average, 200 logins) compared to individuals using a less dynamic Web site in a previous study (Tate et al., 2003). Other research (Krukowski et al., 2008) using an interactive Web site found that greater engagement with dynamic Web features that provided participants with progress feedback was a significant predictor of weight loss, consistent with previous research indicating the benefit of individualized feedback (Tate et al., 2001; Tate et al., 2003).

Similarly, Ferney and Marshall (2006) found that focus group participants reported that they preferred interactive Web sites for promoting physical activity—specifically citing their desire for personalized progress charts and e-mail access to experts. In addition, Hurling, Fairley, and Dias (2006) examined the impact of an interactive Web site by comparing a more interactive Internet-based Web site to a less interactive Web site, both of which were designed to increase physical activity. The participants rated the more interactive Web site as more accessible, more useful, more enjoyable, and less impersonal. They found that participants in the interactive condition were more likely to continue logging in to the Web site over the 10-week period and log in more frequently overall in comparison to the participants using the less interactive Web site. Login frequency is a key metric in assessing Internet-based weight control programs because, as with other measures of adherence, it has been consistently related to weight loss success (Krukowski et al., 2008; Tate et al., 2001, 2003, 2006). However, login rates typically tend to decrease over the course of the intervention; thus, developing Web sites that sustain involvement over the entire treatment program is clearly a priority for long-term success of online obesity treatment programs.

In order to encourage login to interactive Web sites, it is important to consider available technology for these interactive and multimedia features. Ferney and Marshall (2006) reported that focus group participants indicated a preference for accessing a quicker Web site over a Web site with complex multimedia features that took longer to download. Importantly, the technological capacity for interactive Web sites in the form of high-speed Internet is growing. Recent research indicates that 70% of home Internet users had high-speed Internet in 2007 (Pew Internet and American Life Project, 2007), which facilitates the use of interactive and multimedia features (e.g., audio and animation) on these Web sites. Nonetheless, developers of online obesity programs should be sensitive to Internet speed availability for the targeted population in the development of interactive Web sites with multimedia features.

Potential advantages of Internet-based weight loss programs

Internet-based weight loss programs may have certain advantages over in-person programs, including self-monitoring adherence, reduced participant burden, convenience, and tailored treatment. First, initial research suggests that participants in Internet-based programs may produce greater adherence to self-monitoring, another fundamental feature of most successful programs, than participants in in-person programs (Harvey-Berino, Pintauro, Buzzell, & Gold, 2004). This may reflect greater ease of self-monitoring through online tools. Self-monitoring is a key dimension of adherence to behavioral treatments and has been shown to be significantly related to weight loss in in-person programs (Boutelle & Kirschenbaum, 1998; Kazdin, 1974) and online programs (Cussler et al., 2008). Therefore, facilitating self-monitoring of dietary and physical activity behavior through online tools may be beneficial for enhancing weight loss success.

Emerging research indicates that daily self-monitoring of weight may facilitate successful long-term weight maintenance (Wing, Tate, Gorin, Raynor, & Fava, 2006), and online tools that prompt for a daily weight and provide interactive feedback on weight changes over time can reinforce and promote this behavior. Some clinicians may understandably be concerned about the validity of self-reported online weights; however, preliminary findings indicate close congruence between self-reported online weight data and observed measures (Harvey-Berino, West, Buzzell, & Ogden, 2008).

Internet-based weight control programs also appear to produce comparable ratings of treatment satisfaction (Haugen, Tran, Wyatt, Barrg, & Hill, 2007) and social support (Harvey-Berino et al., 2004; van den Berg et al., 2007) to in-person programs, and may reduce participant burden (Haugen et al.). Recent research found no significant differences in satisfaction or weight losses between in-person or online weight maintenance programs; however, the individuals in the online program rated their program as more convenient than those in the in-person program (Haugen et al.). Contrary to concerns regarding communication problems in online groups, the majority of participants in one Internet-based physical activity promotion program reported that they were able to develop trust with their online facilitator (85%) and that they did not experience any miscommunication between the facilitator and themselves due to the use of e-mail (89%; van den Berg et al.). Furthermore, comparable levels of perceived social support or therapeutic alliance for in-person and online weight maintenance programs have been demonstrated (Harvey-Berino et al., 2004), suggesting that effective group dynamics can be achieved online. Overall, Internet-based health programs have the potential to address limitations of traditional in-person programs through customized, dynamic, and multimedia Web sites (Neuhauser & Kreps, 2003), while retaining the apparently essential and fundamental features (i.e., programmatic structure, behavior therapy components, and synchronous communication) of in-person weight control programs.

Online weight control programs in managed care and workplace settings

Although most of the studies of online weight control programs have been conducted in community settings, there have been a few key studies that have been conducted within other settings, including the managed care system and the workplace. In a study of a six-week Internet-based weight control program of individuals recruited within a managed care system, Rothert and colleagues (2006) observed that participants assigned to the Internet-based tailored expert system reported significantly greater weight loss at six-month follow-up (−3% of baseline weight) than those assigned to the Internet-based information-only condition (−1.2% of baseline weight). In a worksite wellness program that utilized the Internet to encourage employees to consume 5 fruits and vegetables daily, complete 10,000 steps per day, and achieve and maintain a BMI lower than 25 kg/m^2 (Pratt et al., 2006), researchers found statistically significant improvement in diet, exercise habits, and weight of enrolled participants after one year of participation. The Web site incorporated many elements of other successful online weight control programs, including self-monitoring, challenges, Web chats, and access to professionals. In addition, participants were prompted monthly by e-mail to submit self-monitoring information. These studies demonstrate a few ways that Internet-based weight control interventions may be implemented in specific settings.

Current status of Internet-based weight control research

Thus, Internet-based behavioral weight control programs have been shown to produce significant weight loss, with more recent programs demonstrating larger losses, as the knowledge of essential treatment components grows. To date, the literature on Internet-based weight control programs for adults indicates several program components that may offer greatest efficacy and adherence, including programmatic structure; behavior therapy components; interactive, dynamic Web site features; and synchronous communication. However, this research is in the early stages, and a clear picture of the essential components for the most effective online obesity program remains to be determined. The benefits of tailoring Internet-based interventions to particular settings also merits exploration. Furthermore, because increasing retention and greater Web site utilization would likely improve weight loss and maintenance, the development of Internet-based weight control program features that participants find helpful, attractive, and captivating should clearly be a priority.

Internet-based weight maintenance programs for adults

Research examining Internet-based programs for weight maintenance (i.e., sustaining weight after a successful weight loss) is beginning to emerge as well (Harvey-Berino et al., 2004; Haugen et al., 2007; Stevens et al., 2008; Svetkey et al., 2008; Table 9.2). In one of the largest studies to date examining a tailored weight maintenance Web site targeting successful weight losers who had completed an in-person intervention prior to online maintenance (Stevens et al.; Svetkey et al.), Web site utilization remained high during the first year, with over 80% of the participants still using the Web site at the end of the first year and an average of 35 out of 52 weeks with at least one login. Nonetheless, participants assigned to the Internet-based weight maintenance intervention were not as successful at maintaining their weight loss as those assigned to the in-person counseling sessions (Svetkey et al., 2008). This failure to realize equivalent weight losses online in the Weight Loss Maintenance Trial could be attributed to factors such as insufficient perceived social support in the online intervention, as there was a minimal social support component in the online intervention. However, it may be beneficial to note for future interventions that Stevens and colleagues found that e-mail and telephone prompts (if the participants did not log in once per week as required) were helpful in stimulating participants to sustain their Web site use. This echoes the experiences reported in previous research examining a fruit and vegetable online intervention (Woodall et al., 2007).

In contrast to the unsuccessful weight maintenance outcomes achieved by the Weight Loss Maintenance Trial (Stevens et al., 2008; Svetkey et al., 2008), other researchers have found that some Internet-based weight maintenance programs are as effective in forestalling weight regain as in-person programs (Harvey-Berino et al., 2004; Haugen et al., 2007). One study demonstrating successful weight maintenance featured biweekly lessons, personalized feedback on self-monitoring journals from the facilitator, and synchronous communication with the facilitator and other

Table 9.2 Examples of Internet-Based Weight Maintenance Programs for Adults

Study	Sample	Duration of weight maintenance program	Design	Intervention components	Weight change
Haugen et al., 2007	n = 87; 76% Caucasian, 4% Hispanic, 1% African American, 1% Asian, 18% race/ethnicity not specified; 84% Women; previously participated in an affiliated 6-month weight loss program and lost > 7% of baseline weight	6 months	Program selected by preference; 3 conditions	1. Telehealth: behavior therapy components;* programmatic structure; interactive Web site; progress feedback by a facilitator; self-monitoring; e-mail reminders 2. In-person: behavior therapy components;* programmatic structure; progress feedback by a facilitator; self-monitoring; 3. No treatment	Weight maintenance: 1. −0.6 ± 2.5 kg; 2. −0.5 ± 4.3 kg; 3. + 1.7 ± 3.0 kg Groups 1 and 2 had better maintenance than 3; $p<0.05$

| Harvey-Berino et al., 2004 | n = 255; race/100% ethnicity not specified; 82% Women; previously enrolled in a 6-month interactive television (ITV) behavioral weight control program (M = 7.8 kg) | 12 months | RCT; 3 conditions | 1. Internet (biweekly): behavior therapy components;* programmatic structure; synchronous communication; progress feedback by a facilitator; self-monitoring; incentives
2. Frequent ITV (biweekly): behavior therapy components;* programmatic structure; synchronous communication; progress feedback by a facilitator; self-monitoring; incentives
3. Minimal ITV (monthly for 6 months): behavior therapy components;* programmatic structure; synchronous communication; self-monitoring | Weight loss (0–18 months): 1. 7.6 ± 7.3 kg; 2. 5.1 ± 6.5 kg; 3. 5.5 ± 8.9 kg | n.s. |

Continued

Table 9.2 Examples of Internet-Based Weight Maintenance Programs for Adults (*Continued*)

Study	Sample	Duration of weight maintenance program	Design	Intervention components	Weight change
Stevens et al., 2008; Svetkey et al., 2008	n = 1032; 38% African American, 62% race/ethnicity not specified; 61% Women; successful (>4 kg) weight losers in a 6 month weight loss program (M = 8.5 kg)	30 months	RCT; 3 conditions	1. Interactive technology: behavior therapy components;* unstructured modules; interactive Web site; bulletin board; self-monitoring; weekly e-mail reminders 2. Self-directed: 1 in-person contact 3. Personal contact (monthly): behavior therapy components;* progress feedback by a facilitator; self-monitoring; in-person/phone contacts	Weight loss (0–36 months) (adjusted): 1. −3.3 ± 0.4 kg; 2. −2.9 ± 0.4 kg; 3. −4.2 ± 0.4 kg Weight maintenance (6–36 months) (adjusted): 1. 5.2 ± 0.3 kg; 2. 5.5 ± 0.3 kg; 3. 4.0 ± 0.3 kg Group 3 had better maintenance than Groups 1 and 2; $p<0.01$

Notes: RCT: Randomized controlled trial; n.s.: Non-significant difference.

* Behavior therapy components include stimulus control, goal setting, problem solving, social assertion, motivation, cognitive strategies, and relapse prevention.

participants in a group chat (Harvey-Berino et al.), and therefore was unlike the less interactive and less dynamic Web site utilized by Svetkey and colleagues on several key dimensions. One interesting observation noted by Harvey-Berino et al. was that participants in the Internet-based condition were more likely to report initiating peer support contacts than those in the in-person conditions. In another study demonstrating efficacious online weight maintenance intervention (Haugen et al.), the online intervention also included weekly interactions with a "Healthy Coach" and participants were given weekly personalized feedback and encouragement on their self-monitoring and progress toward goals by trained facilitators. Since both of these successful weight maintenance programs included a social support component and other research has indicated that social support appears to be particularly important for weight maintenance in Internet-based interventions (Krukowski et al., 2008), it is likely that a distinction between the more successful online weight maintenance programs and those that were less successful is the inclusion of meaningful social support. Therefore, it is a strong recommendation that interactive and social support elements be included when developing online weight maintenance interventions.

Using Internet technology for weight loss in children and adolescents

Internet-based weight control programs may be particularly attractive to children and adolescents, as they may be even more comfortable and familiar with Internet technology than adults, because children are often exposed to computers and the Internet at an early age. It is also possible that Internet-based weight management programs may be less stigmatizing than in-person programs for children and adolescents. Adolescents in particular have indicated that minimizing stigma is one of the most important factors to them in obesity treatment programs (Neumark-Sztainer, Martin, & Story, 2000). In addition, transportation to treatment programs may be an even greater concern for children and adolescents due to busy family schedules (Neumark-Sztainer et al.). Furthermore, Internet-based weight control programs for children have the potential to provide social support from peers across wide geographic regions, which may be beneficial for coping with common experiences of bullying among overweight children (Krukowski et al., 2008) or feelings of isolation (Strauss & Pollack, 2003).

Despite the promise, research on Internet-based weight control programs for children has been relatively limited, even in comparison to the rather limited literature on these programs for adults (Table 9.3). A recent Chinese study examined the impact of an interactive Internet-based intervention with adolescents. The researchers utilized an informational and interactive Web site (e.g., games for calculating daily calories, a forum for discussions between participants and facilitators) in combination with an in-person weight loss student group (Hung et al., 2008). The weight loss student group consisted of weekly lessons, exercise classes twice per week, and individual counseling sessions (2 to 3 times over the semester). Over the 14-week study, the researchers observed significant decreases in BMI, waist circumference, and triceps skinfold thickness among the adolescents. The participants also reported high satisfaction with the Web site. Although the effect of the Web site cannot be disentangled from the effect of the in-person weight loss student group, and data on weight maintenance in this sample are not available, this study points to the potential of an interactive Web site in conjunction with a multicomponent comprehensive weight control program for adolescents.

In one of the first and longest studies in duration, Williamson and colleagues (2006) examined weight loss and maintenance among adolescent African American girls and their parents enrolled in a two-year randomized clinical trial, HIPTeens, which compared an online behavior therapy–based weight control program with an online health education program. Consistent with findings from an Internet-based weight control intervention with adults (Tate et al., 2001, 2003), Williamson et al. found that an online behavior therapy–based weight control program for adolescents was significantly more effective than an online health education program in producing reductions in body fat among adolescents and reductions in body weight among the parents at the six-month assessment. Furthermore, the HIPTeens study demonstrated significant associations between adolescent and parent program adherence (i.e., Web site log-in frequency and submission of weekly assignments) and adolescent and parent change in dietary habits (White et al., 2004). Consistent with findings from in-person childhood obesity interventions (Young, Northern, Lister, Drummond, & O'Brien, 2007), these findings suggest that inclusion of the parent may be beneficial for adolescents, perhaps due to modeling of healthy behaviors.

Although these results are encouraging, it is important to note, that Williamson and colleagues (2006) found the changes in percent body fat and body weight in the HIPTeens study were not sustained over the 18-month follow-up period. The investigators reported that the utilization of the Web site decreased significantly after the first year of the study, which may help to explain the failure to demonstrate sustained weight losses. Nonetheless, the findings from the HIPTeens study provide evidence that Internet-based weight control interventions targeting children and adolescents may benefit from including behavior therapy components,

Chapter nine: Obesity

Table 9.3 Examples of Internet-Based Weight Management Programs for Children and Adolescents

Study	Sample	Program Description	Design	Intervention components	Weight/adiposity outcomes	
Hung et al., 2008	n = 37; 100% race/ethnicity not specified; 43% Girls	Weight loss: 14 weeks	Pre/post; 1 condition	Programmatic structure; interactive Web site; synchronous communication; self-monitoring; incentives; weekly in-person classes; 2–3 in-person counseling sessions; twice weekly in-person exercise classes	BMI change (0–14 weeks): -0.23 kg/m^2	Within groups: $p<0.05$
Jones et al., 2008	n = 105; 64% Caucasian, 8% African American, 21% Hispanic, 7% race/ethnicity not specified; 70% Girls	Weight gain prevention/reduction of binge eating: 16 weeks	RCT; 2 conditions	1. Internet: behavior therapy components; programmatic structure; interactive Web site; asynchronous discussions; self-monitoring; reminder letters; in-person contact in cohort 2 2. Wait-list control	BMI change (0–9 month follow-up): 1. -0.7 kg/m^2; 2. 0.44 kg/m^2	Between groups: 1>2; $p<0.05$

Continued

Table 9.3 Examples of Internet-Based Weight Management Programs for Children and Adolescents

Study	Sample	Program Description	Design	Intervention components	Weight/adiposity outcomes	
White et al., 2004; Williamson et al., 2006	n = 57; 100% African American; 100% Girls, and their parents	Weight loss: 6 months; Weight maintenance: 18 months	RCT; 2 conditions	1. Behavioral: behavior therapy components;* programmatic structure; interactive Web site; progress feedback by a facilitator; self-monitoring; 4 in-person sessions	BMI change (0–6 months): 1. −0.24 ± 1.38 kg/m^2; 2. +0.71 ± 1.19 kg/m^2	Between groups: 1>2; $p<0.01$
				2. Educational: information about nutrition and physical activity; 4 in-person counseling sessions	BMI change (0–24 months): 1. +0.73 ± 0.66 kg/m^2; 2. +1.2 ± 0.65 kg/m^2	n.s.

Notes: RCT: Randomized controlled trial; n.s.: Non-significant difference.

* Behavior therapy components include stimulus control, goal setting, problem solving, social assertion, motivation, cognitive strategies, and relapse prevention.

adopting a family-based approach, and developing an intervention Web site that sustains high login rates.

Using Internet technology for weight gain prevention and weight maintenance in children and adolescents

In a study aimed at preventing weight gain and reducing binge eating among adolescents, Jones and colleagues (2008) demonstrated successful weight maintenance in a 16-week Internet-based intervention (StudentBodies) compared with those in the wait-list control condition. Despite successful weight maintenance, Jones et al. found that sustaining engagement of adolescent participants was a significant challenge; the majority of the adolescents actively participated in the online program for less than eight weeks. Although the program was based in cognitive-behavioral theory and included self-monitoring recommendations, the program was self-directed and included only asynchronous communication with a facilitator for social support. Although there has not been a systematic investigation of the impact of programmatic structure of online weight loss programs designed for adolescents and children, research from Internet-based programs among adults (Gold et al., 2007; Tate et al., 2001; Womble et al., 2004) and observations based on in-person weight control programs for children (Snethen, Broome, & Cashin, 2006) suggest that Internet-based programs for adolescents and children might be more effective if they were to follow a programmatic behavioral structure. In addition, drawing upon the adult literature on Internet-based weight control programs indicating the importance of social support (Krukowski et al., 2008), it could be hypothesized that if the StudentBodies program had incorporated a greater degree of social support from the facilitator or other participants, better adherence to the program might have been observed.

Current status of online obesity prevention and treatment research for children and adolescents

Research on online weight control programs is clearly more limited for children and adolescents than for adults, particularly in the area of weight maintenance. The extent to which existing knowledge of effective Internet-based weight control programs from the adult literature will be pertinent for effective programs for younger populations is unknown, but it appears that there are some commonalities, including the importance of behavior therapy components, social support, and developing a Web site that encourages frequent traffic. Nonetheless, there may be some important differences in components of successful online treatment programs for youths, particularly as children have less control over the home food environment and the younger generation may require a strong entertainment component to the intervention. Further research to identify the elements necessary for an effective online approach to weight control for children and adolescents, as well as which aspects will produce the best outcomes, is necessary to fully exploit the potential of interactive technologies for addressing the growing problem of childhood obesity.

Future directions for research that uses technology to treat obesity

There are many frontiers for expanding existing research examining technology to treat obesity. Research examining the cost-effectiveness of Internet-based weight control programs is one of the most important future directions to explore to guide dissemination of these interventions. Limited research (McConnon et al., 2007) has compared the cost-effectiveness of conducting Internet-based weight control programs in comparison to in-person interventions. When conducting economic evaluations of Internet programs compared to in-person programs, it will be important to examine a broad range of factors, including development costs of the programs, structural costs (e.g., a Web server for online compared with printed intervention materials and meeting space for in-person programs), personnel costs (e.g., program facilitators for either delivery modality, with the addition of a technology specialist or Webmaster for online programs), and participant costs (e.g., time and transportation). Considerations of greater potential reach with online programs may outweigh minor differences in costs, should they emerge.

The qualifications (and associated cost) of the program facilitator may be another important direction for exploration in the dissemination of low-cost, Internet-based obesity treatment. The importance of personalized feedback and social support has been demonstrated, and as a result, the preferred program would be staffed by trained personnel. Program personnel in Internet-based weight control studies are typically trained professionals in psychology, nutrition, exercise physiology, or health education (Gold et al., 2007; Micco et al., 2007; Tate et al., 2001, 2003, 2006). However, the capacity for broad dissemination and cost-effectiveness of these programs may be increased by training lay health educators to facilitate the programs. To date, research has not yet examined factors such as essential content for training, optimal training duration, and feasibility for training lay health educators to facilitate Internet-based weight control programs, but these questions represent a promising direction to pursue.

Tate and colleagues (2006) recently investigated computer-automated tailored counseling, which has the potential to reduce personnel time for conducting Internet-based weight control programs. Although they found that participants in the computer-automated condition were not as successful in losing weight at six months compared to those who had human e-mail counseling, 34% of participants in the computer-automated condition achieved a 5% weight loss or greater. Refinement of computer-automated counseling or a combination of computer-automated and human counseling could allow programs utilizing computer-automated methodology to produce clinically significant weight losses in a more wide-reaching and cost-effective manner. In addition, the impact of supplementing an online program with periodic in-person sessions is worthy of exploration. Although a small pilot study examining the addition of monthly in-person support to Internet-based treatment did not indicate a benefit over Internet-based treatment alone (Micco et al., 2007), the potential for this innovative hybrid design is intriguing and worthy of more definitive study. Moreover, further research may wish to examine whether Internet-based weight control programs in combination with individual or group interventions would provide advantages (e.g., social support, between-session resources, a modality for assignments).

An additional consideration for future research is whether matching individuals to a preferred technological intervention format will enhance outcomes. For example, technologically naïve individuals may perform best in in-person programs, while other individuals who prefer a technology-based approach may have better engagement and weight losses in an Internet-based program. A single obesity treatment modality for all individuals may not be the best way to conceptualize treatment delivery (Brownell & Wadden, 1991). Treatment preference may have a significant impact on outcome within a treatment modality; yet to date, no studies comparing online programs to other treatment modalities have included information on the participants' treatment modality preferences. It is conceivable that published outcomes of online programs underestimate the efficacy that would be achieved among individuals who indicate a strong preference for the format or convenience of an online program. Preference for the treatment modality would be an important variable to consider in future research.

Other types of technology might also be considered in efforts to enhance obesity treatment outcomes. Although initial evidence indicates that personal digital assistants (PDAs) do not appear to improve the validity or frequency of dietary self-monitoring (Yon, Johnson, Harvey-Berino, & Gold, 2006, 2007), novel technologies such as PDAs may have implications for the concept of treatment preference matching. Some participants may have a preference for self-monitoring using a PDA or other types of technology for reasons of convenience or social acceptability (i.e., many individuals use handheld devices in public settings), which may facilitate program adherence. Devices such as PDAs or cellular phones could also facilitate self-monitoring if features were utilized such as reminders to record meals. Technology will certainly continue to change, and it will be important for researchers to seize technological opportunities that offer promise to refine behavioral weight control programs so that they are more effective, accessible, and convenient.

Conclusions

Research evaluating online obesity intervention outcomes has demonstrated that delivery by Internet is a promising avenue to promote weight loss and maintenance. Internet-based weight control programs could be critically important in curtailing the obesity epidemic, given the possibilities for extended reach and cost-effectiveness. As Cassell, Jackson, and Cheuvront (1998) state, Internet-based health programs can merge the advantages of an individually tailored approach with the public health advantages of a wide-reaching intervention. A developing picture of the program parameters for both adults and children indicates that interventions that incorporate behavioral strategies, progress feedback, strong social support, and interactive Web site features are likely to achieve the most significant weight losses among adults and children.

However, on balance, Internet-based programs have not achieved weight losses of the magnitude typically produced by in-person behavioral weight control programs. Furthermore, the most readily accessible Internet interventions are commercial programs that have not produced weight losses commensurate with those seen in research-based Internet programs (Gold et al., 2007; Womble et al., 2004). However, by continuing to incorporate features as they are demonstrated efficacious in the research literature, it may be possible to develop and disseminate widely accessible evidence-based online weight management programs that are more successful than currently available commercial versions. Further research to enhance the weight losses produced with Internet-based programs and to expand research on weight maintenance, particularly among youths, is undoubtedly necessary. Finally, research that evaluates the cost-effectiveness and the training needs for facilitators of Internet-based obesity treatment, as well as research on the implications for matching based on treatment preference, would be beneficial in informing the development and dissemination of Internet-based weight control interventions.

Acknowledgment

Funded in part by the National Institute of Diabetes and Digestive and Kidney Diseases (RO1 DK056746).

References

Boutelle, K. N., & Kirschenbaum, D. S. (1998). Further support for consistent self-monitoring as a vital component of successful weight control. *Obesity Research, 6*(3), 219–224.

Bray, G. A. (2004). Medical consequences of obesity. *The Journal of Clinical Epidemiology & Metabolism, 89*(6), 2583–2589.

Brownell, K. D., & Wadden, T. A. (1991). The heterogeneity of obesity: Fitting treatments to individuals. *Behavior Therapy, 22*, 153–177.

Cassell, M. M., Jackson, C., & Cheuvront, B. (1998). Health communication on the Internet: An effective channel for health behavior change? *Journal of Health Communication, 3*, 71–79.

Cooper, S., Valleley, R. J., Polaha, J., & Evans, J. H. (2006). Running out of time: Physician management of behavioral health concerns in rural pediatric primary care. *Pediatrics, 118*(1), 132–138.

Cussler, E. C., Teixeira, P. J., Going, S. B., Houtkooper, L. B., Metcalfe, L. L., Blew, R. M., et al. (2008). Maintenance of weight loss in overweight middle-aged women through the Internet. *Obesity, 16*(5), 1052–1060.

DeBell, M., & Chapman, C. (2003). *Computer and Internet use by children and adolescents in 2001*. Washington, DC: National Center for Education Statistics.

Diabetes Prevention Program Research Group. (2002). Reduction in the incidence of type 2 diabetes with lifestyle intervention or metformin. *The New England Journal of Medicine, 346*(6), 393–403.

Epstein, L. H., Valoski, A., Wing, R. R., & McCurley, J. (1994). Ten-year outcomes of behavioral family-based treatment for childhood obesity. *Health Psychology, 13*(5), 373–383.

Ferney, S. L., & Marshall, A. L. (2006). Web site physical activity interventions: Preferences of potential users. *Health Education Research, 21*(4), 560–566.

Gold, B. C., Burke, S., Pintauro, S., Buzzell, P., & Harvey-Berino, J. (2007). Weight loss on the Web: A pilot study comparing a structured behavioral intervention to a commercial program. *Obesity, 15*(1), 155–164.

Gottschalk, A., & Flocke, S. A. (2005). Time spent in face-to-face patient care and work outside the examination room. *Annals of Family Medicine, 3*(6), 488–493.

Harvey-Berino, J., Pintauro, S., Buzzell, P., & Gold, E. C. (2004). Effect of Internet support on the long-term maintenance of weight loss. *Obesity Research, 12*, 320–329.

Harvey-Berino, J., West, D. S., Buzzell, P., & Ogden, D. (2008). Weight reported on the Web: Is it accurate?[Abstract]. *Obesity, 16(Suppl.), 168*.

Haugen, H. A., Tran, Z. V., Wyatt, H. R., Barry, M. J., & Hill, J. O. (2007). Using telehealth to increase participation in weight maintenance programs. *Obesity, 15*(2), 3067–3077.

Heymsfield, S., van Mierlo, C., van der Knaap, H., Heo, M., & Frier, H. (2003). Weight management using a meal replacement strategy: Meta and pooling analysis from six studies. *International Journal of Obesity 27*, 537–549.

Hung, S.-H., Hwang, S.-L., Su, M.-J., Lue, S.-H., Hsu, C.-Y., Chen, H.-L., et al. (2008). An evaluation of a weight-loss program incorporating e-learning for obese junior high students. *Telemedicine and e-Health, 14*(8), 783–792.

Hurling, R., Fairley, B. W., & Dias, M. B. (2006). Internet-based exercise intervention systems: Are more interactive designs better? *Psychology & Health, 21*(6), 757–772.

Jones, M., Luce, K. H., Osborne, M. I., Taylor, K., Cunning, D., Celio, D. A., et al. (2008). Randomized, controlled trial of an Internet-facilitated intervention for reducing binge eating and overweight in adolescents. *Pediatrics, 121*(3), 453–462.

Kazdin, A. E. (1974). Self-monitoring and behavior change. In M. J. Mahoney & C. F. Thorsen (Eds.), *Self control: Power to the person* (pp. 218–246). Monterey, CA: Brooks/Cole.

Krukowski, R., Harvey-Berino, J., Ashikaga, T., Thomas, C. S., & Micco, N. (2008). Internet-based weight control: The relationship between Web features and weight loss. *Telemedicine and e-Health, 14*(8), 775–782.

Krukowski, R. A., West, D. S., Siddiqui, N. J., Bursac, Z., Phillips, M. M., & Raczynski, J. M. (2008). No change in weight-based teasing when school-based obesity policies are implemented. *Archives of Pediatric and Adolescent Medicine, 162*, 936–942.

Look AHEAD Research Group. (2007). Reduction in weight and cardiovascular disease risk factors in individuals with type 2 diabetes. *Diabetes Care, 30*(6), 1374–1383.

McConnon, A., Kirk, S. F. L., Cockroft, J. E., Harvey, E. L., Greenwood, D. C., Thomas, J. D., et al. (2007). The Internet for weight control in an obese sample: Results of a randomised controlled trial. *BMC Health Services Research, 7*, 206.

Micco, N., Gold, B., Buzzell, P., Leonard, H., Pintauro, S., & Harvey-Berino, J. (2007). Minimal in-person support as an adjunct to Internet obesity treatment. *Annals of Behavioral Medicine, 33*(1), 49–56.

National Heart, Lung, and Blood Institute. (1998). *Clinical guidelines on the identification, evaluation, and treatment of overweight and obesity in adults: The evidence report.* Washington, DC: U.S. Government Press.

Nemet, D., Barkan, S., Epstein, Y., Friedland, O., Kowen, G., & Eliakim, A. (2005). Short- and long-term beneficial effects of a combined dietary-behavioral-physical activity intervention for the treatment of childhood obesity. *Pediatrics, 115*(4), e443–e449.

Neuhauser, L., & Kreps, G. L. (2003). Rethinking communication in the e-health era. *Journal of Health Psychology, 8*(1), 7–23.

Neumark-Sztainer, D., Martin, S. L., & Story, M. (2000). School-based programs for obesity prevention: What do adolescents recommend? *American Journal of Health Promotion, 14*(4), 232–235.

Ogden, C. L., Carroll, M. D., Curtin, L. R., McDowell, M. A., Tabak, C. J., & Flegal, K. M. (2006). Prevalence of overweight and obesity in the United States. *Journal of the American Medical Association, 295*(13), 1549–1555.

Perri, M. G., Nezu, A. M., Patti, E. T., & McCann, K. L. (1989). Effect of length of treatment on weight loss. *Journal of Consulting and Clinical Psychology, 57*(3), 450–452.

Pew Internet & American Life Project. (2000). The online health care revolution: How the Web helps Americans take better care of themselves. Retrieved November 20, 2008, from http://www.pewinternet.org/pdfs/PIP_Health_Report.pdf

Pew Internet & American Life Project. (2005a). Generations online. Retrieved August 20, 2008, from http://www.pewinternet.org/pdfs/PIP_Generations_Memo.pdf

Pew Internet & American Life Project. (2005b). Health information online. Retrieved August 20, 2008, from http://www.pewinternet.org/pdfs/PIP_Healthtopics_May05.pdf

Pew Internet & American Life Project. (2006). Teens and the Internet. Retrieved November 20, 2008, from http://www.pewinternet.org/pdfs/PIP_Teens_Tech_July2005web.pdf

Pew Internet & American Life Project. (2007). Broadband Adoption in 2007. Retrieved August 20, 2008, from http://www.pewinternet.org/press_release.asp?r=141

Pew Internet & American Life Project. (2008). Demographics of Internet users. Retrieved August 19, 2008, from http://www.pewinternet.org/trends/User_Demo_2.15.08.htm

Pratt, D. S., Jandzio, M., Tomlinson, D., Kang, X., & Smith, E. (2006). The 5-10-25 Challenge: An observational study of a Web-based wellness intervention for a global workforce. *Disease Management, 9*(5), 284–290.

Reinehr, T., & Andler, W. (2004). Changes in the atherogenic risk factor profile according to the degree of weight loss. *Archives of Disease in Childhood, 89*, 419–422.

Reinehr, T., Kiess, W., Kapellen, T., & Andler, W. (2004). Insulin sensitivity among obese children and adolescents, according to the degree of weight loss. *Pediatrics, 114*(6), 1569–1573.

Rothert, K., Strecher, V. J., Doyle, L. A., Caplan, W. M., Joyce, J. S., Jimison, H. B., et al. (2006). Web-based weight management programs in an integrated health care setting: A randomized, controlled trial. *Obesity, 14*(2), 266–272.

Savoye, M., Shaw, M., Dziura, J., Tamborlane, W. V., Rose, P., Guandalini, C., et al. (2007). Effects of a weight management program on body composition and metabolic parameters in overweight children. *Journal of the American Medical Association, 297*, 2697–2704.

Snethen, J. A., Broome, M. E., & Cashin, S. E. (2006). Effective weight loss for overweight children: A meta-analysis of intervention studies. *Journal of Pediatric Nursing, 21*(1), 45–56.

Spear, B. A., Barlow, S. E., Ervin, C., Ludwig, D. S., Saelens, B. E., Schetzina, K. E., et al. (2007). Recommendations for treatment of child and adolescent overweight and obesity. *Pediatrics, 120*, S254–288.

Stevens, V. J., Funk, K. L., Brantley, P. J., Erlinger, T. P., Myers, V. H., Champagne, C. M., et al. (2008). Design and implementation of an interactive Web site to support long-term maintenance of weight loss. *Journal of Medical Internet Research, 10*(1), e1.

Strauss, R. S., & Pollack, H. A. (2003). Social marginalization of overweight children. *Archives of Pediatric and Adolescent Medicine, 157,* 746–752.

Svetkey, L. P., Stevens, V. J., Brantley, P. J., Appel, L. J., Hollis, J. F., Loria, C. M., et al. (2008). Comparison of strategies for sustaining weight loss: The Weight Loss Maintenance randomized controlled trial. *Journal of the American Medical Association, 299,* 139–1148.

Tate, D., Jackvony, E., & Wing, R. (2003). Effects of Internet behavioral counseling on weight loss in adults at risk of type 2 diabetes: A randomized trial. *Journal of the American Medical Association, 289,* 1833–1836.

Tate, D., Jackvony, E., & Wing, R. (2006). A randomized trial comparing human e-mail counseling, computer-automated tailored counseling, and no counseling in an Internet weight loss program. *Archives of Internal Medicine, 166,* 1620–1625.

Tate, D., Wing, R., & Winett, R. (2001). Using Internet technology to deliver a behavioral weight loss program. *Journal of the American Medical Association, 285,* 1172–1177.

Tsai, A. G., Asch, D. A., & Wadden, T. A. (2006). Insurance coverage for obesity treatment. *Journal of the American Dietetic Association, 106*(10), 1651–1655.

van den Berg, M., Ronday, H., Peeters, A., Voost-van der Harst, E., Munneke, M., Breedveld, F., et al. (2007). Engagement and satisfaction with an Internet-based physical activity intervention in patients with rheumatoid arthritis. *Rheumatology, 46,* 545–552.

Wang, Y., Beydoun, M., Liang, L., Caballero, B., & Kumanyika, S. (2008). Will all Americans become overweight or obese? Estimating the progression and cost of the U.S. obesity epidemic. *Obesity, 16,* 23–30.

Weiss, R., Dziura, J., Burgert, T., Tamborlane, W. V., Taksali, S. E., Yeckel, C. W., et al. (2004). Obesity and the metabolic syndrome in children and adolescents. *New England Journal of Medicine, 350,* 2362–2374.

White, M. A., Martin, P. D., Newton, R. L., Walden, H. M., York-Crowe, E. E., Gordon, S. T., et al. (2004). Mediators of weight loss in a family-based intervention presented over the Internet. *Obesity Research, 12*(7), 1050–1059.

Wilfley, D. E., Tibbs, T. L., Van Buren, D. J., Reach, K. P., Walker, M. S., & Epstein, L. H. (2007). Lifestyle interventions in the treatment of childhood overweight: A meta-analytic review of randomized controlled trials. *Health Psychology, 26*(5), 521–532.

Williamson, D. A., Walden, H. M., White, M. A., York-Crowe, E. E., Newton, R. L., Alfonso, A., et al. (2006). Two-year Internet-based randomized controlled trial for weight loss in African-American girls. *Obesity, 14*(7), 1231–1243.

Wing, R., & Jeffery, R. (1999). Benefits of recruiting participants with friends and increasing social support for weight loss and maintenance. *Journal of Consulting and Clinical Psychology, 67,* 132–138.

Wing, R. R., Tate, D. F., Gorin, A. A., Raynor, H. A., & Fava, J. L. (2006). A self-regulation program for maintenance of weight loss. *New England Journal of Medicine, 355,* 1563–1571.

Womble, L., Wadden, T., McGuckin, B., Sargent, S., Rothman, R., & Krauthamer-Ewing, E. (2004). A randomized controlled trial of a commercial Internet weight loss program. *Obesity, 12*(6), 1011–1018.

Woodall, W. G., Buller, D. B., Saba, L., Zimmerman, D., Waters, E., Hines, J. M., et al. (2007). Effect of e-mailed messages on return use of a nutrition education Web site and subsequent changes in dietary behavior. *Journal of Medical Internet Research, 9*(3), e27.

Yon, B. A., Johnson, R. K., Harvey-Berino, J., & Gold, B. C. (2006). The use of a personal digital assistant for dietary self-monitoring does not improve the validity of self-reports of energy intake. *Journal of the American Dietetic Association, 106,* 1256–1259.

Yon, B., Johnson, R. K., Harvey-Berino, J., & Gold, B. C. (2007). Use of a personal digital assistant for dietary self-monitoring during a weight loss program. *Journal of Behavioral Medicine, 30*(2), 165–175.

Young, K. M., Northern, J. J., Lister, K. M., Drummond, J. A., & O'Brien, W. H. (2007). A meta-analysis of family-behavioral weight-loss programs. *Clinical Psychology Review, 27,* 240–249.

chapter ten

Diabetes management

Jun Ma, Sarah B. Knowles, and Sandra R. Wilson

Diabetes mellitus is a growing, complex, and costly chronic disease in the United States. The disease is increasingly prevalent, is frequently associated with comorbidities, and requires multiple medications and intensive behavior change in its management. Evidence-based standards of diabetes care exist, which emphasize coordinated medical care and patient self-management. Coordinated care management includes identifying patient health care needs, customizing medical care and patient self-management support, and coordinating ongoing care across providers and settings and with patients to achieve long-term, optimal control.

However, clinical implementation of diabetes care recommendations remains inadequate and varied for a sizable proportion of patients. Continued challenges for most health care systems (e.g., constrained resources, the acute-care orientation) and persistent barriers to patient self-management (e.g., lack of information, skill training, social support) make the growing need for diabetes services an untenable situation without reformative interventions.

Advances in health information technology (IT) have provided a partial solution, and evidence of their effectiveness in diabetes care has accumulated over the past two decades. Technologies such as diabetes registries and clinical decision support systems can accelerate the clinical implementation of recommended diabetes care by providing diabetes-specific treatment guidelines and best-practice reminders or prompts at the point of care and by generating population and individual patient health status reports for ongoing quality measurement and reporting.

Patient self-management technologies empower patients in self-care of diabetes by offering patient educational resources, providing interactive behavior change tools, facilitating self-monitoring of changes in disease control and self-care needs, and enhancing patient–provider communication and social support. More recently, integrated IT systems have been developed and evaluated that provide a suite of technologies to support a full range of diabetes management activities by the medical care team and the patient.

In this chapter, we first illustrate that diabetes presents an opportune context for the incorporation of IT in care management due to the complexity of diabetes management, the persistence of treatment gaps, and the existence of quality measures. We then describe various technologies in view of their use to enhance diabetes medical management and patient self-management. We also summarize the evidence to date regarding their impacts on processes and outcomes of diabetes care. A thorough literature review is beyond the scope of this chapter. Instead, we highlight individual well-designed studies while referring the reader to reviews published by others that more comprehensively synthesize the available scientific evidence. We conclude by discussing directions for future research.

Burden of diabetes in the United States

Diabetes is a high-risk, high-cost chronic disease. Since the middle of the last century, chronic disease has been a leading cause of death and disability in the United States, accounting for disproportionate health care utilization and cost (Hoffman, Rice, & Sung, 1996; Murray & Lopez, 1996). Diabetes alone affects nearly 23.6 million, or 7.8% of Americans, a number expected to increase to 39 million by 2050 (Boyle et al., 2001; National Institute of Diabetes and Digestive and Kidney Diseases, 2005). Most of the expected increase will be in type 2 diabetes, which accounts for 90 to 95% of all diagnosed cases of diabetes each year (Boyle et al.; National Institute of Diabetes and Digestive and Kidney Diseases). The microvascular complications of diabetes make it the leading cause of preventable blindness, renal failure, and lower extremity amputation (National Institute of Diabetes and Digestive and Kidney Diseases). Also, people with diabetes have 2 to 4 times the risk of heart disease and stroke, compared to nondiabetics, and are 2 to 3 times more likely to die from these macrovascular complications (Czyzyk, Muszynski, & Krolewski, 1977; Kannel & McGee, 1979; National Institute of Diabetes and Digestive and Kidney Diseases, 2005). The total estimated cost of diabetes was $174 billion in 2007 and is expected to grow as diabetes prevalence increases (American Diabetes Association, 2008a).

Complexity and challenges of quality diabetes care

Diabetes care is complex and clinical standards of care have been established based on a large body of evidence (American Diabetes Association, 2008b).
Optimal control of diabetes requires routine screening and monitoring, intensive glycemic control (measured by HbA1c and patient self-monitoring of blood glucose), assessment and treatment of concomitant risk factors (e.g., blood pressure, lipids, and body weight), and

prevention and control of microvascular and macrovascular complications (Table 10.1). The quality of diabetes care is measured based on appropriate receipt of recommended processes (e.g., HbA1c tests, eye exams, and foot exams) and control of surrogate outcomes (e.g., HbA1c, cholesterol, and blood pressure levels) that are expected to predict the risks of microvascular and macrovascular disease outcomes (Agency for Healthcare Research and Quality, 2008a).

In addition, optimal diabetes care is dependent on persistent patient self-management (Table 10.1). Patients with chronic conditions, such as diabetes, are necessarily responsible for the day-to-day management of the factors that affect their risk. The 2001 Institute of Medicine (IOM) report *Crossing the Quality Chasm* acknowledges patient-centeredness as a key attribute of quality health care and emphasizes ongoing self-management

Table 10.1 Key Components of Recommended Diabetes Care

Screening and monitoring	Medical treatment	Patient self-management (5-A Model)

Glycemic control
  HbA1c
  Fasting blood glucose
Blood pressure
Lipids
Retinopathy
Peripheral neuropathy
Microalbuminuria
Weight
	Glycemic control
	Hypertension
	Dyslipidemia
	
	Cardiovascular disease
	Nephropathy
	Neuropathy
	Retinopathy
	Immunization
	Use of angiotensin-converting enzyme inhibitors
	Use of antiplatelet agents
		Assess periodically the patient's disease-related knowledge, attitudes, and behaviors and provide personalized results and feedback.
		Advise on personally relevant, specific recommendations for behavior change and risk reduction.
		Agree on specific, measurable, feasible self-management goals through shared decision making.
		Assist with development of personalized self-management action plans that include effective behavior change and problem-solving strategies.
		Arrange for follow-up on action plans and referrals and for provider–patient communication between visits.

support by health professionals in recognition of the centrality of the informed, activated patient to a productive patient–provider relationship. Patient well-being is positively influenced by systematic efforts to foster patient self-management through increased patient knowledge, skills, and confidence to manage their condition (Bodenheimer, Lorig, Holman, & Grumbach, 2002; Holman & Lorig, 2004; Lorig & Holman, 2003).

Diabetes self-management education (DSME) has been considered an important aspect of diabetes care since the 1930s (Bartlett, 1986), and national standards exist to guide its implementation (Mensing et al., 2006). DSME has evolved over time from primarily didactic presentations to collaborative, comprehensive interventions that focus not only on knowledge, but on lifestyle, skills, and self-efficacy, which are essential to achieving behavior change and metabolic control in diabetics (Norris, Engelgau, & Narayan, 2001; Peyrot, 1999). Decades of research in patient self-management has shown that the fundamental elements of effective self-management interventions include ongoing health risk assessment, sustained behavior change counseling and reinforcement, individualized goal setting and action planning, shared problem definition and decision making, and continued self-management training and support (Glasgow et al., 2002; Von Korff, Gruman, Schaefer, Curry, & Wagner, 1997). The 5-A model—Assess, Advise, Agree, Assist, and Arrange (Table 10.1)—which stems from the behavioral change counseling literature, provides a useful framework for conceptualizing and delivering diabetes self-management interventions (Whitlock, Orleans, Pender, & Allan, 2002).

Despite the wealth and strength of evidence about effective management of diabetes and established guidelines for providers and patients, the quality of diabetes care remains suboptimal and varied across subpopulations of patients with diabetes (Saaddine et al., 2002, 2006; Saydah, Fradkin, & Cowie, 2004). Improvements in standard measures of diabetes processes of care and surrogate outcomes were modest in the 1990s and the rate of improvement since 2000 has slowed (Agency for Healthcare Research and Quality, 2008a; Saaddine et al., 2006). It is estimated that 33% of U.S. adults with diagnosed diabetes have high blood pressure, 37% have inadequate glycemic control, and 40% have elevated levels of low-density lipoprotein cholesterol (LDL-C; Saaddine et al., 2006). Because of these uncontrolled risk factors, millions of Americans remain at high risk for diabetes-related complications. Moreover, the continuing gaps in the quality of care for diabetes disproportionately affect racial and ethnic minorities and people of low socioeconomic status where health disparities have remained unchanged or, in some cases, worsened in recent years (Agency for Healthcare Research and Quality, 2008b). Innovative interventions are needed to continue to improve health care quality and reduce disparities for populations of patients with diabetes.

The emergence of information technology in U.S. health care

Health IT emerged in the center of the national health care debate in 1991 when an IOM committee called to the nation's attention the slow adoption of automated patient records (Dick & Steen, 1991). It is the IOM's vision that "information technology, including the Internet, holds enormous potential for transforming the health care delivery system" (Institute of Medicine, 2001, p. 5).

Effective care of chronic illness, including diabetes, arguably presents the greatest challenge to the U.S. health care system. The U.S. health care system has evolved primarily to react to acute episodes that patients present with, and it remains largely so even though chronic disease has increasingly dominated patient demands and health care expenditures since the middle of the last century (Hoffman et al., 1996). As a result, the current system leaves the majority of chronically ill patients inadequately treated, and many opportunities are missed for quality chronic and preventive care (Agency for Healthcare Research and Quality, 2008a; Grumbach & Bodenheimer, 2002; Institute of Medicine, 2001).

With the advent of chronic disease dominance, a new patient–provider relationship is required in which informed, internally motivated patients interact with their providers in partnership. Face-to-face interactions remain fundamental to many functions of a productive patient–provider partnership. Nonetheless, the quality and continuity of care may profoundly benefit from integrating in-person visits with less labor-intensive delivery venues, for example, through the use of health IT (Piette, 2002).

The importance of IT to chronic disease management is explicitly embraced in the new chronic care paradigm. The Chronic Care Model put forth by Wagner and others (Bodenheimer, Lorig et al., 2002; Bodenheimer, Wagner, & Grumbach, 2002a, 2002b; Wagner, 1998; Wagner et al., 2001) is a population-based, proactive, and patient-centered approach to treating chronic diseases, such as diabetes, and preventing disease complications. The model emphasizes the need for health IT systems that can (1) provide disease-specific registries for population and individual health status tracking and reporting, (2) enable point-of-care decision support for providers, (3) empower active patient self-management, and (4) enhance patient–provider communication to facilitate collaborative decision making and patient-centered, information-rich longitudinal care (Wagner, 1998; Wagner et al., 2001). These functional capabilities go far beyond the basic documentation, retrieval, and billing functions of traditional electronic health record (EHR) systems.

A 2006 systematic review conducted for the Agency for Healthcare Research and Quality (AHRQ) found that health IT systems, particularly those that are multifunctional and include clinical decision support, are efficacious in increasing adherence to evidence-based guidelines, enhancing disease surveillance and monitoring, and reducing medical errors (Chaudhry et al., 2006). Other potential benefits include greater continuity of care, greater efficiency in tracking changes to patients' individual treatment, and the potential of reduced health care costs.

Diabetes management technologies

A variety of health ITs have been studied in the context of diabetes care. There are two broad categories of ITs depending on the intended user: (1) technologies used by providers (e.g., computerized disease registries and clinical decision support systems), and (2) technologies used by patients (e.g., electronic self-monitoring devices, clinic-based CD-ROM systems, automated calling systems, and the Internet). Such technologies facilitate and support the delivery of the key components of recommended diabetes care described above, including screening and monitoring, medical treatment, and patient self-management.

Provider-centered technologies

Clinical inertia, or the failure to intensify therapy when clinically indicated, occurs at well over 60% of office visits with diabetic patients who are not at goal for HbA1c, blood pressure, or LDL-C (Phillips et al., 2001). This phenomenon, and the resulting consequences of suboptimal medical treatment, is partly attributable to the limitations of "unaided human minds" in medical decision-making processes. For example, providers often lack adequate knowledge of current evidence-based care, information about their own performance, and timely, accurate, and detailed patient health information (Greco & Eisenberg, 1993).

Two examples of technologies designed to address provider-level barriers to optimal care are computerized disease registries and clinical decision support systems, which provide clinicians with patient-centric assessments and recommendations to improve adherence to screening, monitoring, and medical treatment guidelines. Computerized disease registries are electronic lists of patients with chronic conditions such as diabetes that assist in the identification of patients and tracking and reporting on their disease-specific information. Registries may interface with administrative systems to facilitate scheduling recommended lab test ordering (e.g., HbA1c tests) and health maintenance examinations (e.g., eye and foot exams). Registries may also generate individual and population health status summaries, in relation to guideline-based treatment targets and quality improvement metrics, for ongoing monitoring. A number of computerized disease registry products are in use today; some are stand-alone applications and others are an integral component of a multifunctional EHR system. Two California HealthCare Foundation reports provide a thorough overview of the function and use of various computerized disease registry products and outline issues for consideration in product selection (Metzger, 2004; Simon & Powers, 2004).

The second provider-level IT, clinical decision support, is defined as "providing clinicians or patients with clinical knowledge and patient-related information, intelligently filtered or presented at appropriate times, to enhance patient care" (Osheroff, Pifer, Teich, Sittig, & Jenders, 2005, p. x). Clinical decision support systems compare patient information to a set of rules developed according to evidence-based practice guidelines, generate best practice reminders/prompts, and present medical treatment options to providers at the point of care (Institute of Medicine, 2003). Research suggests that a high level of clinical decision support is essential to the ability of an EHR system to improve clinical practice (Chaudhry et al., 2006). State-of-the-art EHRs include increasingly sophisticated clinical decision support features and functions (Kawamoto & Lobach, 2005). Several recent articles review features of clinical decision support systems important for improving clinical practice, as well as pressing challenges in their implementation in routine clinical use (Kawamoto, Houlihan, Balas, & Lobach, 2005; Sittig et al., 2008).

The impact of these technologies on diabetes management has been specifically examined. In a group randomized controlled trial (RCT) of 12 intervention and 14 control staff providers and 307 intervention and 291 control patients with type 2 diabetes in a hospital-based internal medicine clinic, Meigs and colleagues (2003) evaluated a Web-based information management/decision support system. The system displayed patient-specific clinical data that were interactively linked to evidence-based treatment recommendations and provided treatment advice at the time of patient contact. Changes in evidence-based processes and outcomes of diabetes care from the year preceding the study through the year of the study were compared between the intervention and control groups. The authors reported modest but significant improvements in rates of testing for levels of HbA1c and LDL-C and screening for foot disease. There was no evidence, however, that these improved processes of care led to better glycemic or lipid control in the intervention group. Similarly, other recent studies have shown that diabetes guideline–based reminders/prompts and decision support aids targeting providers are associated with improved compliance with recommended diabetes care procedures, for example, higher rates of routine diabetes care visits, recommended lab testing, and health maintenance examinations (Balas et al., 2004; Montori et al., 2002; O'Connor et al., 2005; Weber, Bloom, Pierdon, & Wood, 2007). However, evidence is thin regarding the impact of these technologies on physiological outcomes of diabetes control, for example, control of HbA1c, blood pressure, and cholesterol.

Patient-centered technologies

Even the best informed, highly motivated, self-confident patients with diabetes may fail in their efforts at day-to-day self-management of the disease due to the complexity of information that must be obtained, processed, and tracked in order to make good self-management decisions. Information technology has led to major advances in diabetes self-management research and practice (Piette, 2007). A growing number of technologies allow for active involvement in health care by patients who can interact in some fashion with a portable device, a phone, or a computer and receive individually tailored information, advice, or access to resources. Examples include, but are not limited to, personal digital assistants (PDAs) or other handheld, portable devices, clinic-based CD-ROM systems, automated calling systems, the Internet, and a variety of rapidly emerging interventions that combine these different technologies. These technologies can effectively support one or more dimensions of the 5-A model and hence enhance the delivery of diabetes self-management interventions (Table 10.2).

Handheld, portable, and mobile devices

The typical capabilities of a PDA (a small handheld computer) include tracking appointments and tasks with built-in alarms or reminders, storing phone numbers and addresses, saving memos and messages, and functioning as a calculator. Newer PDAs also have audio capabilities, enabling them to be used as mobile phones (smartphones) or portable media players. Many PDAs can access the Internet and be used as Web browsers. The device can be customized to support diabetes self-management through installation of diabetes self-care activity monitoring software programs such as HealthEngage Diabetes (http://www.healthengage.com) and Diabetes Pilot (http://www.diabetespilot.com). These software programs have the capabilities of recording and tracking self-monitored blood glucose measurements, test results, medications taken and prescribed; accepting automatic uploads from glucose meters; planning and recording daily meals, activities, and exercises; tracking intake of calories, carbohydrates, and other nutrients in the foods consumed using an integrated food database; graphing and reporting trends

Table 10.2 Technologies with Strong Functional Capabilities Supporting Each 5-A Dimension

5-A Model	Technologies	Functional capabilities
	Assess	
Knowledge, attitudes, behaviors	PDA CD-ROM Automated calling systems Internet	Computerized data collection, acquisition, and transmission Automated data analysis, contextualization, and abstraction
	Advise	
Personally relevant, guideline-based recommendations	CD-ROM Internet	Computerized patient education materials in multimedia Computer-generated, instant feedback Professional (a)synchronous e-counseling
	Agree	
Personalized self-management goals	CD-ROM Internet	Graphic user interfaces for eliciting goals and tracking and displaying progress over time
	Assist	
Action planning and problem solving	CD-ROM Internet	Interactive action plan platforms Data acquisition and abstraction about barriers to behavior change
	Arrange	
Follow-up care and communication Automated calling systems	PDA Internet	Automated follow-up care scheduling and reminders Tracking of progress on action plans Synchronous or asynchronous communication modalities

in blood glucose levels; and exporting data in various formats for further analysis. While PDA technology has been found to improve patient compliance with diabetes self-care activity monitoring and patient satisfaction (Forjuoh, Reis, Couchman, & Ory, 2008; Sevick et al., 2008; Tsang et al., 2001), rigorous studies evaluating its impacts on guideline-based processes and outcomes of diabetes care are lacking.

Clinic-Based CD-ROM systems

CD-ROM-based IT systems, such as touch-screen kiosks, have been used in clinic waiting rooms by patients before or after their scheduled appointments. These systems are easy to use, can reach a large number of patients, and require minimum staffing. CD-ROM applications allow for display of self-paced and disease-specific educational materials in large multimedia (audio and video) files, and thus are amenable to use by persons with low literacy. They also allow for use of complex programming algorithms to assist patients in health risk assessment, goal setting and action planning, identification of barriers to goal attainment, and receiving immediate, automated, tailored feedback. This information is then provided to the provider in a summary form and can be used to facilitate shared decision making regarding treatment options during the patient's visit.

Studies evaluating such applications for self-management in diabetes reported modest improvements in patients' attitudes (e.g., perceived susceptibility to diabetes complications) and health behaviors (e.g., dietary intake) but failed to demonstrate significant intervention effects on self-efficacy for self-care, physiological outcomes, diabetes-related health care utilization, or health-related quality of life (Gerber et al., 2005; Glasgow & Bull, 2001; Glasgow et al., 1997; Glasgow & Toobert, 2000).

Automated calling systems

Automated calling systems use specialized computer technology to conduct health status assessments (e.g., self-monitored blood glucose readings, self-care activities, and symptoms) and deliver tailored self-care education. Calls are placed with a predetermined frequency (e.g., weekly) and at times indicated by patients to be most convenient for them. During automated monitoring calls, patients use their touch-tone keypad or voice recognition technology to respond to hierarchically structured messages composed of statements and queries recorded in a human voice. The patient's responses are recorded and determine the subsequent content of the message. The automated systems generate reports to identify patients with health and self-care problems and these reports can then be used by providers to prioritize their patient contacts.

These systems represent a pragmatic and inexpensive way to conduct frequent follow-up between visits or for patients who have difficulty accessing clinic-based services or who lack the resources to access other IT-based disease management interventions. Indeed, Piette and colleagues have shown that these systems are acceptable and can be efficiently used by low-income, non-English-speaking patients with diabetes (Piette, 1999; Piette, McPhee, Weinberger, Mah, & Kraemer, 1999; Piette, Weinberger, McPhee et al., 2000). These same researchers have conducted several RCTs to evaluate automated calls with telephone nurse follow-up as a strategy for improving diabetes treatment processes and outcomes in English- and Spanish-speaking adults with diabetes (Piette, Weinberger, Kraemer, & McPhee, 2001; Piette, Weinberger, & McPhee, 2000; Piette, Weinberger, McPhee et al., 2000). They demonstrated that the intervention had a significant impact on patient self-efficacy, self-care behaviors (e.g., self-monitoring of blood glucose), medication adherence, glycemic control, and symptoms of depression for up to 12 months. However, long-term maintenance data are lacking. Few studies have evaluated automated telephonic diabetes assessment and self-care education alone, though some preliminary evidence suggests feasibility and potential benefits (Boren et al., 2006; Mollon et al., 2008).

The Internet

The Internet is increasingly used as a way to access health information, with more than 60 million American residents (Berland et al., 2001) or 40% of Internet users (Baker, Wagner, Singer, & Bundorf, 2003) seeking such information. Those with ongoing health needs (e.g., the chronically ill) are among the individuals who tend to look to the Internet for health information (Bundorf, Wagner, Singer, & Baker, 2006). In addition, the Internet provides an innovative and versatile tool for delivering continuous, patient-centered care because of its multimedia capability, instantaneous interactivity, complex tailoring capacity, continuous availability, wide accessibility, automated data uploading and sharing, improved openness of communication, and low maintenance costs (Robinson, Patrick, Eng, & Gustafson, 1998). Hence, even though the Internet is only the latest IT available to help address challenges inherent in effective diabetes self-management, it is perhaps the one with the greatest potential.

Internet-based diabetes management applications have garnered a lot of research and development attention in recent years. Bull, Gaglio, McKay, and Glasgow (2005) have identified and evaluated 87 publicly available diabetes Web sites (the article provides the names and URLs of the sites) hosted by government, health plan, commercial, pharmaceutical, and not-for-profit organizations. The authors conclude that most sites clearly fail to utilize the full potential of the Internet and instead merely provid didactic information, often using an electronic newspaper or pamphlet format and being accessible only to well-educated English speakers. In addition, site content was rarely based on theories of behavior change. A few sites did allow users to track personal data and receive automated, tailored feedback; however, the tailoring was limited to biomedical variables and seldom included behavioral variables (e.g., dietary intake, exercise, and medication adherence). A small number of sites offer social support features, such as a chat room or bulletin board.

Below, we review three Web-based diabetes management applications that have been most rigorously evaluated to date.

MyCareTeam Web site. MyCareTeam™ (https://mycareteam.georgetown.edu) is an interactive Web-based diabetes management intervention designed and hosted at the Imagine Science and Information Systems Center at Georgetown University Medical Center. It was developed to bridge the length of time between clinic visits (usually every 2 to 3 months per American Diabetes Association recommendations), providing access to more frequent 2-way communication via the Web that allows more consistent monitoring of glucose levels and appropriate medication or behavioral changes when necessary. The Web site accepts uploads from glucose and blood pressure monitoring devices and displays these data in graphic and tabular form for the patient and provider to view. Patients can send comments with the data that providers can then respond to directly. There is also a separate secure messaging component for patient–provider communication. Finally, the Web site contains diabetes education modules and links to other Web-based diabetes resources.

An initial 6-month feasibility study was conducted in a small sample of adults with poorly controlled type 1 or type 2 diabetes (Smith et al., 2004). One of the purposes of the feasibility study was to evaluate whether patients with poorly controlled diabetes would respond to more frequent interactions with their care team. Sixteen patients aged 19 to 65 years with poorly controlled diabetes (HbA1c > 8.5% in past 6 months) were enrolled and consented to add the MyCareTeam program to their standard of care. Patients attended 1 clinic visit at baseline and 2 follow-up visits at 3 and 6 months. Each week during the 6-month study, patients were instructed to upload their self-monitoring blood glucose and exercise. A significant reduction of 2.22 percentage points in mean HbA1c was observed at 6 months follow-up, and the magnitude of reduction was positively associated with the frequency of Web site use.

McMahon et al. (2005) evaluated the MyCareTeam program in a 12-month RCT among 104 veterans with diabetes (type not specified) whose baseline HbA1c was greater than or equal to 9.0%. The study used broad inclusion criteria to maximize eligibility, including recruiting patients with little computer experience. The primary outcomes were HbA1c and systolic and diastolic blood pressure. All patients attended a diabetes education session and were then randomized to either Web-based care or the control group (n = 52 each). Both groups continued with usual care. In addition, the Web-based group received a laptop computer, a blood glucose monitor, a blood pressure monitor, and access to the intervention Web site. The computer was programmed to connect to the MyCareTeam Web site, using complimentary dial-up Internet access. Patients also received computer training and instructions about how to use the glucose and blood pressure devices. Patients were asked to monitor their blood pressure on their own and upload the results via laptop to the MyCareTeam Web site a minimum of 3 times per week. The frequency of glucose monitoring and uploads varied depending on each patient's own treatment plan. At 12-months follow-up, the intervention group had a significant net decline in HbA1c, compared with the control group, and the frequency of Web site logins and the frequency of data uploads were both significant predicators for improvement in HbA1c. The intervention also resulted in a significant net decrease in systolic blood pressure and triglyceride levels, and a significant net increase in high-density lipoprotein cholesterol (HDL-C), compared with the control group.

Informatics for Diabetes Education and Telemedicine (IDEATel) project. The IDEATel project was a large one-year randomized trial of 1,665 Medicare patients with diabetes (type unspecified) living in federally identified medically underserved areas throughout New York and receiving care from private primary practices (Shea et al., 2002, 2006; Starren et al., 2002). The IDEATel intervention consisted of installing a specially designed home medicine unit that included a computer with Internet access to the study Web site (via modem) as well as self-monitoring devices for blood glucose and blood pressure. The home unit was designed for four specific functions: (1) videoconferencing, (2) remote glucose and blood pressure monitoring using mobile devices, (3) online access to the patient's EHR and secure messaging for communication between the patient and care manager, and (4) online access to an educational Web site designed specifically for the IDEATel study (Shea et al., 2006).

Participants were Medicare beneficiaries aged at least 55 years who currently were being treated for diabetes via diet, insulin, or oral hypoglycemic medication. The primary outcomes were HbA1c, blood pressure, and LDL-C. Intervention patients received training on how to use the home telemedicine unit, including its videoconferencing capabilities to interact with care managers. The care managers were also trained to use the study's diabetes management software and on how to tailor videos and computer alerts to their individual patients. The intervention protocols regarding the number of uploads and frequency of contacts were not reported.

Compared with usual care, the intervention group had significantly improved HbA1c, systolic and diastolic blood pressure, and LDL-C. The authors also noted that there was a greater intervention effect on HbA1c among patients whose baseline levels were ≥ 7.0%.

The IDEATel study had several strengths. The RCT was one of the largest in the published literature, incorporated an existing EHR system and used multimedia, did not require computer experience, included diabetes patients with both controlled and poorly controlled glycemic levels, and was delivered to medically underserved patients, a large portion of whom were minorities (approximately 50%). Despite these positive characteristics, the expense of the intervention, reported at $3,425 per home medicine unit, and the necessity of home installation highlight important obstacles to widespread implementation of this intervention.

Internet-Based Blood Glucose Monitoring System (IBGMS). The IBGMS was designed and evaluated in a series of randomized trials by Cho, Kwon, and colleagues (Cho et al., 2006; Kwon et al., 2004). The primary feature of the IBGMS includes a study Web site (http://www.biodang.com) where patients send and receive messages from their physician. The Web site allows patients to upload self-monitored blood glucose data, as well as diabetes medication information (type of medication, dose, and frequency); that information is then added to the patient's EHR. Patients are also able to input other relevant information such as weight changes, diet, or other information relevant to their diabetes care. Based on those results, physicians send their patients medication changes or other recommendations, in compliance with the national Korean diabetes management guidelines.

In an initial 3-month study of the IBGMS, 110 patients with type 2 diabetes (mean age 54 years; 61% men) were recruited from the outpatient clinic at a hospital-based diabetes center in Seoul, South Korea (Kwon et al., 2004). Intervention patients were trained how to use the IBGMS and asked to upload blood glucose levels before and after meals for 1 to 3 times per day for at least 3 days per week, along with medication information and other information relevant to diabetes care (diet, exercise, and weight). Patients who did not send information for more than a week were sent a reminder message. Patients who did not send information for more than 3 weeks were considered lost to follow-up. In response to self-monitoring data the patient submitted and after reviewing the patient's EHR, a coordinated care team of endocrinologists, dietitians, and nurses sent, via the study Web site, their recommendations regarding any medication changes and lifestyle modification. Intervention patients did not have any clinic visits during the follow-up period. Control patients received the

Chapter ten: Diabetes management

standard care, which included monthly outpatient clinic visits. Similar to the intervention visits, these clinic visits could also include recommendations about medication and/or lifestyle changes.

At 3 months follow-up, the intervention group had significantly lower HbA1c than it had at baseline and the control group had significantly higher HDL-C levels. There were no other significant pre/post changes for any other biomarkers for either group. In comparison to the control group, the intervention group had significantly lower HbA1c at 3 months; especially among patients whose HbA1c baseline levels were ≥ 7.0%.

The IBGMS was subsequently evaluated during a 30-month RCT to evaluate the potential long-term effectiveness of the intervention (Cho et al., 2006). Eighty patients with type 2 diabetes (mean age 53 years; 60% men) were randomized to participate in the IBGMS or control group. As in the 3-month study, intervention participants were trained to use the IBGMS and were provided a glucometer for blood glucose self-monitoring. They were asked to upload their glucose levels, along with other relevant information (weight, medication use, etc.) at their convenience. They were also able to electronically send specific questions to the care team. The care team responded to the patient's information with any necessary medication changes or other recommendations. Intervention patients had clinic visits every 3 months for the 30-month follow-up period. Control patients received the standard of care for diabetics at the outpatient clinic, including treatment changes when necessary.

For the entire 30-month follow-up, the mean HbA1c of the intervention group was significantly lower than that of the control group. Additionally, the fluctuation of HbA1c levels (measured as the standard deviation of HbA1c values for each patient over the course of the study) was notably less for the intervention group than the control group.

The MyCareTeam, IDEATel, and IBGMS projects are examples of well-conducted studies that have demonstrated the ability of Web-based care management interventions to significantly improve diabetes control. Together, these studies indicate that Web-based care management interventions are applicable to a wide range of diabetic populations (e.g., patients with type 1 or type 2 diabetes, patients with good or poor glycemic control, older adults, ethnic minorities, and persons with limited computer experience) and in various health care settings (e.g., Veterans Affairs health care system, private practices in medically underserved communities, and hospital-based specialty clinics). As one of few long-term studies, the IBGMS project suggests that these interventions can have sustainable effects to improve diabetes control (Cho et al., 2006).

Several common functional capabilities have emerged from the interventions tested in the MyCareTeam, IDEATel, and IBGMS projects that directly support the implementation of the 5-A model as applied to diabetes self-management. The functional capabilities include (1) data collection and acquisition for remote monitoring of self-management behaviors and success; (2) data storage and tracking for longitudinal care planning; (3) information management through data analysis, contextualization, and abstraction; and (4) patient–provider communication via synchronous (e.g., videoconferencing) and/or asynchronous (e.g., secure text messaging) mechanisms that link objective information about the patient with evidence-based treatment and personalized behavioral counseling by the medical care team, as well as collaborative goal setting and decision making. It is expected that as ITs advance and research in Web-based interventions proceeds, additional capabilities and their application to the 5-A model for enhanced diabetes self-management support will continue to evolve.

Future research recommendations

The studies highlighted in this chapter represent the growing number of creative, health IT–based approaches to improving the effectiveness and efficiency of diabetes care, but important gaps in knowledge remain. There is a critical need for additional rigorously designed studies to address these gaps and provide evidence-based solutions.

Some already operating EHR systems have an integrated personal health record (PHR) component (e.g., Epic Systems' MyChart® module) that gives patients greater access to their own medical records and a means of secure electronic communication with their physicians. Such integrated EHR/PHR systems have the ability to provide a suite of technologies supporting both physicians and patients, such as the disease registry, clinical decision support, remote monitoring, and Web-mediated self-management support techniques discussed above. Research is needed to determine the most effective, mutually supportive combinations of these technologies to deliver comprehensive IT-facilitated diabetes management interventions in real-world practice settings.

Diabetes care processes and intermediate outcomes have been improving over recent years, as a result of growing quality improvement initiatives in many health care systems (Saaddine et al., 2006; Shojania et al., 2006). Though the traditional approach of limiting study eligibility to patients with measurements above a certain threshold indicating poor control (e.g., HbA1c > 9.0%) can address the problem of overall improvement in the population, such a continued approach will cause interventions to be tested in an increasingly small proportion of the general diabetic population. This can have undesired implications for both internal and external validity because of reduced sample size (and thus statistical power) and generalizability. In addition, the emerging quality improvement initiatives targeting standard diabetes quality indicators make "usual care" a moving target and can also make it difficult to demonstrate incremental intervention effects on the same measures without accruing large samples. These methodological challenges will need to be carefully addressed in future research on novel diabetes interventions, including those that incorporate use of technological innovations.

There is a continuing lack of understanding of the mechanisms by which IT-based interventions may impact health behavior and clinical outcomes for patients with diabetes. Increased knowledge about the mechanisms or components of IT-based applications will maximize their benefit and facilitate adoption. Established models from the information systems literature (e.g., the Unified Theory of Acceptance and Use of Technology, Venkatesh, Morris, Davis, & Davis, 2003) can provide important information about such mechanisms.

Depending on the type of technology, the cost of acquiring, implementing, and maintaining IT-based diabetes care can be substantial (Adler-Milstein et al., 2007). Yet there are few data to address the question whether the incremental efficacy of IT-based diabetes care justifies the financial investment by the payer, the health care system, and the patient. Future studies of IT-based diabetes interventions should carefully report costs associated with the interventions including both direct and indirect health care costs and incremental costs per additional patient, which few studies to date have done. Rigorous research on the cost-effectiveness of such interventions is also needed.

Although some studies have produced encouraging results regarding the applicability of IT-based diabetes interventions, particularly automated telephonic and Web-based diabetes management, the generalizability of such interventions and their potential to widen (or narrow) the "digital divide" (Warschauer, 2003) and/or health disparities remain an ongoing controversy and the subject of continuing scientific investigation. Computer access is an important barrier, although it has been suggested that lack of experience or access does not impede patients' willingness to use the technology (Feil, Glasgow, Boles, & McKay, 2000). Research grounded in a sound conceptual framework such as RE-AIM (Glasgow, Vogt, & Boles, 1999) is needed to inform translation of IT-based diabetes care into real-world settings.

Given the chronic nature of diabetes and resulting need for long-term self-management and behavior change, sustainability becomes a critical issue. IT interventions are also susceptible to participant desensitization to computer-generated messages and lack of motivation to continue using the technology. Some authors noted that patient participation (number of logons, number of data uploads) tended to decline as soon as 3-months post-IT intervention (Piette, 2007). Notwithstanding some early evidence of potential sustainability of IT-based diabetes management from the few long-term studies (Cho et al., 2006), further investigation is needed.

Conclusion

The disease complexity of diabetes and the failure to achieve optimal management of the disease in the United States have resulted in an increased burden in terms of prevalence and economic cost. As part of ongoing research efforts to identify effective and feasible management strategies, IT-based diabetes care has shown enormous potential.

A number of IT tools have been tested, with varying degrees of success, and research has helped clarify the types of IT with the greatest potential to improve care. Provider-level technologies (e.g., disease registries, clinical decision support systems) have been shown to improve provider adherence to diabetes screening, monitoring, and medical treatment guidelines, but their ability to change patient behaviors and improve outcomes has not been demonstrated. Patient-centered technologies, particularly Web-based diabetes management interventions, have shown great promise in improving glycemic control and stability, as well as other physiological outcomes. Automated monitoring calls with telephone nurse follow-up are an effective alternative to reach diverse populations, including patients without computers, non-English speakers, and those with health literacy deficits. These technologies can be used to implement and sustain the behavioral changes necessary for diabetes self-management, as illustrated by the 5-A framework.

In conclusion, the key functional capacities of an effective IT-facilitated diabetes management intervention include (1) data collection and acquisition for remote monitoring, (2) data storage and tracking for longitudinal care planning, (3) data processing and analysis for information sharing and tailored feedback, and (4) (a)synchronous patient–provider communication for follow-up contacts and care between visits. Comprehensive EHR systems that contain an integrated PHR component have the ability to support a suite of technologies targeting both providers and patients and hence provide a functional platform to deliver comprehensive IT-based diabetes management interventions. With additional research that focuses on sustainability, generalizability, and cost-effectiveness, IT-based interventions offer great opportunity to improve the quality of diabetes care.

References

Adler-Milstein, J., Bu, D., Pan, E., Walker, J., Kendrick, D., Hook, J. M., et al. (2007). The cost of information technology-enabled diabetes management. *Disease Management, 10*(3), 115–128.
Agency for Healthcare Research and Quality. (2008a). *2007 National Healthcare Quality Report* (No. AHRQ Pub. No. 08-0040). Rockville, MD: U.S. Department of Health and Human Services, Agency for Healthcare Research and Quality.
Agency for Healthcare Research and Quality. (2008b). *2007 National Healthcare Disparities Report* (No. AHRQ Pub. No. 08-0041). Rockville, MD: U.S. Department of Health and Human Services, Agency for Healthcare Research and Quality.
American Diabetes Association. (2008a). Economic costs of diabetes in the U.S. in 2007. *Diabetes Care, 31*(3), 596–615.
American Diabetes Association. (2008b). Standards of medical care in diabetes—2008. *Diabetes Care, 31* (Suppl 1), S12–54.
Baker, L., Wagner, T. H., Singer, S., & Bundorf, M. K. (2003). Use of the Internet and e-mail for health care information: Results from a national survey. *Journal of American Medical Association, 289*(18), 2400–2406.
Balas, E. A., Krishna, S., Kretschmer, R. A., Cheek, T. R., Lobach, D. F., & Boren, S. A. (2004). Computerized knowledge management in diabetes care. *Medical Care, 42*(6), 610–621.
Bartlett, E. E. (1986). Historical glimpses of patient education in the United States. *Patient Education and Counseling, 8*(2), 135–149.
Berland, G. K., Elliott, M. N., Morales, L. S., Algazy, J. I., Kravitz, R. L., Broder, M. S., et al. (2001). Health information on the Internet: Accessibility, quality, and readability in English and Spanish. *Journal of American Medical Association, 285*(20), 2612–2621.
Bodenheimer, T., Lorig, K., Holman, H., & Grumbach, K. (2002). Patient self-management of chronic disease in primary care. *Journal of American Medical Association, 288*(19), 2469–2475.
Bodenheimer, T., Wagner, E. H., & Grumbach, K. (2002a). Improving primary care for patients with chronic illness. *Journal of American Medical Association, 288*(14), 1775–1779.
Bodenheimer, T., Wagner, E. H., & Grumbach, K. (2002b). Improving primary care for patients with chronic illness: The chronic care model, Part 2. *Journal of American Medical Association, 288*(15), 1909–1914.
Boren, S. A., De Leo, G., Chanetsa, F. F., Donaldson, J., Krishna, S., & Balas, E. A. (2006). Evaluation of a Diabetes Education Call Center intervention. *Telemedical Journal of Electronic Health, 12*(4), 457–465.
Boyle, J. P., Honeycutt, A. A., Narayan, K. M., Hoerger, T. J., Geiss, L. S., Chen, H., et al. (2001). Projection of diabetes burden through 2050: Impact of changing demography and disease prevalence in the U.S. *Diabetes Care, 24*(11), 1936–1940.
Bull, S. S., Gaglio, B., McKay, H. G., & Glasgow, R. E. (2005). Harnessing the potential of the Internet to promote chronic illness self-management: Diabetes as an example of how well we are doing. *Chronic Illness, 1*(2), 143–155.
Bundorf, M. K., Wagner, T. H., Singer, S. J., & Baker, L. C. (2006). Who searches the Internet for health information? *Health Services Research, 41*(3, Pt 1), 819–836.

Chaudhry, B., Wang, J., Wu, S., Maglione, M., Mojica, W., Roth, E., et al. (2006). Systematic review: Impact of health information technology on quality, efficiency, and costs of medical care. *Annals of Intern Medicine, 144*(10), 742–752.

Cho, J. H., Chang, S. A., Kwon, H. S., Choi, Y. H., Ko, S. H., Moon, S. D., et al. (2006). Long-term effect of the Internet-based glucose monitoring system on HbA1c reduction and glucose stability: A 30-month follow-up study for diabetes management with a ubiquitous medical care system. *Diabetes Care, 29*(12), 2625–2631.

Czyzyk, A., Muszynski, J., & Krolewski, A. S. (1977). Lactate coma in diabetes. *Pol Tyg Lek, 32*(21), 793–795.

Dick, R. S., & Steen, E. B. (Eds.). (1991). *The computer-based patient record: An essential technology for health care.* Washington, DC: Institute of Medicine.

Feil, E. G., Glasgow, R. E., Boles, S., & McKay, H. G. (2000). Who participates in Internet-based self-management programs? A study among novice computer users in a primary care setting. *Diabetes Education, 26*(5), 806–811.

Forjuoh, S. N., Reis, M. D., Couchman, G. R., & Ory, M. G. (2008). Improving diabetes self-care with a PDA in ambulatory care. *Telemedical Journal of Electronic Health, 14*(3), 273–279.

Gerber, B. S., Brodsky, I. G., Lawless, K. A., Smolin, L. I., Arozullah, A. M., Smith, E. V., et al. (2005). Implementation and evaluation of a low-literacy diabetes education computer multimedia application. *Diabetes Care, 28*(7), 1574–1580.

Glasgow, R. E., & Bull, S. S. (2001). Making a difference with interactive technology: Considerations in using and evaluating computerized aids for diabetes self-management education. *Diabetes Spectrum, 14*(2), 99–106.

Glasgow, R. E., Funnell, M. M., Bonomi, A. E., Davis, C., Beckham, V., & Wagner, E. H. (2002). Self-management aspects of the improving chronic illness care breakthrough series: Implementation with diabetes and heart failure teams. *Annals of Behavior Medicine, 24*(2), 80–87.

Glasgow, R. E., La Chance, P. A., Toobert, D. J., Brown, J., Hampson, S. E., & Riddle, M. C. (1997). Long-term effects and costs of brief behavioural dietary intervention for patients with diabetes delivered from the medical office. *Patient Education and Counseling, 32*(3), 175–184.

Glasgow, R. E., & Toobert, D. J. (2000). Brief, computer-assisted diabetes dietary self-management counseling: Effects on behavior, physiologic outcomes, and quality of life. *Medical Care, 38*(11), 1062–1073.

Glasgow, R. E., Vogt, T. M., & Boles, S. M. (1999). Evaluating the public health impact of health promotion interventions: The RE-AIM framework. *American Journal of Public Health, 89*(9), 1322–1327.

Greco, P. J., & Eisenberg, J. M. (1993). Changing physicians' practices. *New England Journal of Medicine, 329*(17), 1271–1273.

Grumbach, K., & Bodenheimer, T. (2002). A primary care home for Americans: Putting the house in order. *Journal of American Medical Association, 288*(7), 889–893.

Hoffman, C., Rice, D., & Sung, H. Y. (1996). Persons with chronic conditions. Their prevalence and costs. *Journal of American Medical Association, 276*(18), 1473–1479.

Holman, H., & Lorig, K. (2004). Patient self-management: A key to effectiveness and efficiency in care of chronic disease. *Public Health Report, 119*(3), 239–243.

Institute of Medicine. (2001). *Crossing the quality chasm: A new health system for the twenty-first Century*. Washington, DC: National Academy Press.
Institute of Medicine. (2003). *Key capabilities of an electronic health record system*. Washington, DC: Committee on Data Standards for Patient Safety Board on Health Care Services.
Institute of Medicine. Crossing the quality chasm: A new health system for the twenty-first century. Retrieved July 10, 2009 from http://www.iom.edu/object.file/master/27/184/chasm-8pager.pdf
Kannel, W. B., & McGee, D. L. (1979). Diabetes and cardiovascular disease. The Framingham study. *Journal of American Medical Association, 241*(19), 2035–2038.
Kawamoto, K., Houlihan, C. A., Balas, E. A., & Lobach, D. F. (2005). Improving clinical practice using clinical decision support systems: A systematic review of trials to identify features critical to success. *British Medical Journal, 330*(7494), 765.
Kawamoto, K., & Lobach, D. F. (2005). Design, implementation, use, and preliminary evaluation of SEBASTIAN, a standards-based Web service for clinical decision support. *AMIA Annu Symp Proc*, 380–384.
Kwon, H. S., Cho, J. H., Kim, H. S., Song, B. R., Ko, S. H., Lee, J. M., et al. (2004). Establishment of blood glucose monitoring system using the Internet. *Diabetes Care, 27*(2), 478–483.
Lorig, K. R., & Holman, H. (2003). Self-management education: History, definition, outcomes, and mechanisms. *Annals of Behavior Medicine, 26*(1), 1–7.
McMahon, G. T., Gomes, H. E., Hicksonhohne, S., Hu, T. M., Levine, B. A., & Conlin, P. R. (2005). Web-based care management in patients with poorly controlled diabetes. *Diabetes Care, 28*(7), 1624–1629.
Meigs, J. B., Cagliero, E., Dubey, A., Murphy-Sheehy, P., Gildesgame, C., Chueh, H., et al. (2003). A controlled trial of Web-based diabetes disease management: The MGH diabetes primary care improvement project. *Diabetes Care, 26*(3), 750–757.
Mensing, C., Boucher, J., Cypress, M., Weinger, K., Mulcahy, K., Barta, P., et al. (2006). National standards for diabetes self-management education. *Diabetes Care, 29 Suppl 1*, S78–85.
Metzger, J. (2004). Using computerized disease registries in chronic disease care [Electronic Version]. Retrieved December 9, 2008, from http://www.chcf.org/documents/chronicdisease/ComputerizedRegistriesInChronicDisease.pdf
Mollon, B., Holbrook, A. M., Keshavjee, K., Troyan, S., Gaebel, K., Thabane, L., et al. (2008). Automated telephone reminder messages can assist electronic diabetes care. *Journal of Telemedical Telecare, 14*(1), 32–36.
Montori, V. M., Dinneen, S. F., Gorman, C. A., Zimmerman, B. R., Rizza, R. A., Bjornsen, S. S., et al. (2002). The impact of planned care and a diabetes electronic management system on community-based diabetes care: The Mayo Health System Diabetes Translation Project. *Diabetes Care, 25*(11), 1952–1957.
Murray, C. J., & Lopez, A. D. (1996). Evidence-based health policy—Lessons from the Global Burden of Disease Study. *Science, 274*(5288), 740–743.

National Institute of Diabetes and Digestive and Kidney Diseases. (2005). *National Diabetes Statistics fact sheet: General information and national estimates on diabetes in the United States, 2005*. Bethesda, MD: U.S. Department of Health and Human Services, National Institute of Health.

Norris, S. L., Engelgau, M. M., & Narayan, K. M. (2001). Effectiveness of self-management training in type 2 diabetes: A systematic review of randomized controlled trials. *Diabetes Care, 24*(3), 561–587.

O'Connor, P. J., Crain, A. L., Rush, W. A., Sperl-Hillen, J. M., Gutenkauf, J. J., & Duncan, J. E. (2005). Impact of an electronic medical record on diabetes quality of care. *Annals of Family Medicine, 3*(4), 300–306.

Osheroff, J. A., Pifer, E. A., Teich, J. M., Sittig, D. F., & Jenders, R. A. (2005). *Improving outcomes with clinical decision support: An implementer's guide*. Chicago, IL: Health Information Management and Systems Society.

Peyrot, M. (1999). Behavior change in diabetes education. *Diabetes Education, 25*(6 Suppl), 62–73.

Phillips, L. S., Branch, W. T., Cook, C. B., Doyle, J. P., El-Kebbi, I. M., Gallina, D. L., et al. (2001). Clinical inertia. *Annals of Internal Medicine, 135*(9), 825–834.

Piette, J. D. (1999). Patient education via automated calls: a study of English and Spanish speakers with diabetes. *American Journal of Preventative Medicine, 17*(2), 138–141.

Piette, J. D. (2002). Enhancing support via interactive technologies. *Current Diabetes Report, 2*(2), 160–165.

Piette, J. D. (2007). Interactive behavior change technology to support diabetes self-management: Where do we stand? *Diabetes Care, 30*(10), 2425–2432.

Piette, J. D., McPhee, S. J., Weinberger, M., Mah, C. A., & Kraemer, F. B. (1999). Use of automated telephone disease management calls in an ethnically diverse sample of low-income patients with diabetes. *Diabetes Care, 22*(8), 1302–1309.

Piette, J. D., Weinberger, M., Kraemer, F. B., & McPhee, S. J. (2001). Impact of automated calls with nurse follow-up on diabetes treatment outcomes in a Department of Veterans Affairs Health Care System: A randomized controlled trial. *Diabetes Care, 24*(2), 202–208.

Piette, J. D., Weinberger, M., & McPhee, S. J. (2000). The effect of automated calls with telephone nurse follow-up on patient-centered outcomes of diabetes care: A randomized, controlled trial. *Medical Care, 38*(2), 218–230.

Piette, J. D., Weinberger, M., McPhee, S. J., Mah, C. A., Kraemer, F. B., & Crapo, L. M. (2000). Do automated calls with nurse follow-up improve self-care and glycemic control among vulnerable patients with diabetes? *American Journal of Medicine, 108*(1), 20–27.

Robinson, T. N., Patrick, K., Eng, T. R., & Gustafson, D. (1998). An evidence-based approach to interactive health communication: A challenge to medicine in the information age. Science Panel on Interactive Communication and Health. *Journal of American Medical Association, 280*(14), 1264–1269.

Saaddine, J. B., Cadwell, B., Gregg, E. W., Engelgau, M. M., Vinicor, F., Imperatore, G., et al. (2006). Improvements in diabetes processes of care and intermediate outcomes: United States, 1988–2002. *Annals of Internal Medicine, 144*(7), 465–474.

Saaddine, J. B., Engelgau, M. M., Beckles, G. L., Gregg, E. W., Thompson, T. J., & Narayan, K. M. (2002). A diabetes report card for the United States: Quality of care in the 1990s. *Annals of Internal Medicine, 136*(8), 565–574.

Saydah, S. H., Fradkin, J., & Cowie, C. C. (2004). Poor control of risk factors for vascular disease among adults with previously diagnosed diabetes. *Journal of American Medical Association, 291*(3), 335–342.

Sevick, M. A., Zickmund, S., Korytkowski, M., Piraino, B., Sereika, S., Mihalko, S., et al. (2008). Design, feasibility, and acceptability of an intervention using personal digital assistant–based self-monitoring in managing type 2 diabetes. *Contemporary Clinical Trials, 29*(3), 396–409.

Shea, S., Starren, J., Weinstock, R. S., Knudson, P. E., Teresi, J., Holmes, D., et al. (2002). Columbia University's Informatics for Diabetes Education and Telemedicine (IDEATel) project: Rationale and design. *Journal of American Medical Information Association, 9*(1), 49–62.

Shea, S., Weinstock, R. S., Starren, J., Teresi, J., Palmas, W., Field, L., et al. (2006). A randomized trial comparing telemedicine case management with usual care in older, ethnically diverse, medically underserved patients with diabetes mellitus. *Journal of American Medical Information Association, 13*(1), 40–51.

Shojania, K. G., Ranji, S. R., McDonald, K. M., Grimshaw, J. M., Sundaram, V., Rushakoff, R. J., et al. (2006). Effects of quality improvement strategies for type 2 diabetes on glycemic control: A meta-regression analysis. *Journal of American Medical Association, 296*(4), 427–440.

Simon, J., & Powers, M. (2004). Chronic disease registries: A product review [Electronic Version]. Retrieved December 9, 2008, from http://www.chcf.org/documents/chronicdisease/ChronicDiseaseRegistryReview.pdf

Sittig, D. F., Wright, A., Osheroff, J. A., Middleton, B., Teich, J. M., Ash, J. S., et al. (2008). Grand challenges in clinical decision support. *Journal of Biomedical Information, 41*(2), 387–392.

Smith, K. E., Levine, B. A., Clement, S. C., Hu, M. J., Alaoui, A., & Mun, S. K. (2004). Impact of MyCareTeam for poorly controlled diabetes mellitus. *Diabetes Technological Therapy, 6*(6), 828–835.

Starren, J., Hripcsak, G., Sengupta, S., Abbruscato, C. R., Knudson, P. E., Weinstock, R. S., et al. (2002). Columbia University's Informatics for Diabetes Education and Telemedicine (IDEATel) project: Technical implementation. *Journal of American Medical Information Association, 9*(1), 25–36.

Tsang, M. W., Mok, M., Kam, G., Jung, M., Tang, A., Chan, U., et al. (2001). Improvement in diabetes control with a monitoring system based on a handheld, touch-screen electronic diary. *Journal of Telemedical Telecare, 7*(1), 47–50.

Venkatesh, V. V., Morris, M. G., Davis, G. B., & Davis, F. D. (2003). User acceptance of information technology: Toward a unified view. *Mis Quarterly, 27*(3), 425–478.

Von Korff, M., Gruman, J., Schaefer, J., Curry, S. J., & Wagner, E. H. (1997). Collaborative management of chronic illness. *Annals of Internal Medicine, 127*(12), 1097–1102.

Wagner, E. H. (1998). Chronic disease management: What will it take to improve care for chronic illness? *Effects of Clinical Practice, 1*(1), 2–4.

Wagner, E. H., Austin, B. T., Davis, C., Hindmarsh, M., Schaefer, J., & Bonomi, A. (2001). Improving chronic illness care: Translating evidence into action. *Health Affairs (Millwood), 20*(6), 64–78.

Warschauer, M. (2003). *Technology and social inclusion. Rethinking the digital divide.* Cambridge, MA: The MIT Press.

Weber, V., Bloom, F., Pierdon, S., & Wood, C. (2007). Employing the electronic health record to improve diabetes care: A multifaceted intervention in an integrated delivery system. *J General Intern Med, 23*(4), 379–382.

Whitlock, E. P., Orleans, C. T., Pender, N., & Allan, J. (2002). Evaluating primary care behavioral counseling interventions: An evidence-based approach. *American Journal of Preventative Medicine, 22*(4), 267–284.

section two

Issues concerning implementation and evaluation

chapter eleven

Implementation

Michael A. Cucciare

Implementing technology-based behavioral health applications into health care systems has become increasingly popular over the last two decades (Curran, Mukherjee, Allee, & Owen, 2008; Doebbeling, Chou, & Tierney, 2006; Weingardt, 2004). This is at least partly due to a desire to implement evidence-based behavioral health practices in a cost-effective manner (Gruber, Moran, Roth, & Taylor, 2001). Technologies such as personal and handheld computers (Gruber et al., 2001), virtual reality (Maltby, Kirsch, Mayers, & Allen, 2002), and audio and visual media such as CD-ROM/DVD (Whitfield, Hinshelwood, Pashely, Campsie, & Williams, 2006) are increasingly being used to facilitate and support clinicians in the delivery of evidence-based practices (EBPs). However, despite the growing evidence to support the use of such applications, successful implementation (e.g., sustained adoption) of these tools remains a significant challenge.

Before we continue, it is important to distinguish the term *implementation* from terms with possibly similar meaning such as *dissemination*. We use the term *implementation* in this chapter to refer to the actual use of evidence-based procedures and techniques in practice, which differs from the sole act or goal of providing information to providers about EBPs (i.e., dissemination).

The challenges to implementing new clinical practices into health care settings are well documented (Cabana et al., 1999; Grol, 1992). Potential barriers to the sustained adoption of new practices are many and can range from a host of patient factors (e.g., patient preferences) to organizational factors (e.g., limited resources, clinician factors). Researchers have responded to these challenges by developing theoretical frameworks (e.g., PARIHS framework, Rycroft-Malone et al., 2002) and testing the effectiveness of specific strategies (see Grol & Grimshaw, 2003, for a review) in maximizing the success of implementing clinical interventions. A major objective of this chapter is to provide an overview of these general frameworks and specific strategies by discussing them in the broad context of implementing new clinical practices. Although we emphasize how these strategies can be used to support the successful implementation of technology-based behavioral health interventions, we borrow liberally from the evidence-based medical literature (e.g., Grol, 1992; Grol & Grimshaw, 2003; Grol & Wensing, 2004). Furthermore, to illustrate how these frameworks and strategies can be used to guide implementation initiatives, we conclude this chapter with a case example. However, we begin this chapter with a discussion of the potential advantages and disadvantages of implementing technology-based behavioral health interventions in health care systems to provide some context for why it might be beneficial (and perhaps in some cases not so beneficial) to do so.

Advantages and disadvantages

A sizeable literature has emerged regarding the potential disadvantages and advantages of integrating technology into behavioral health service delivery (see Cucciare, Weingardt, & Villafranca, 2008; Emmelkamp, 2005; Marks, Shaw, & Parkin, 1998; Przeworski & Newman, 2006; Tate & Zabinski, 2004; Taylor & Luce, 2003). Some of the potential advantages include increased cost-effectiveness of behavioral health services, extended reach of behavioral health interventions, and increased facilitation of client disclosure. Some of the potential disadvantages include higher treatment dropout rates, the elimination of common factors of therapeutic success (e.g., therapeutic alliance), and reduced need for face-to-face psychotherapy. We briefly discuss these below.

Advantages

Increased cost-effectiveness. Recent studies suggest that integrating technology-based behavioral health interventions into the treatment process can reduce the costs of treatment delivery and the amount of time spent conducting treatment, without compromising treatment efficacy (Gruber et al., 2001; Marks, Kenwright, McDonough, Whittaker, & Mataix-Cols, 2004; Wright et al., 2005). For example, Gruber et al. showed that a 12-week cognitive-behavioral therapy (CBT) group for social phobia can be reduced to 8 weeks when using a handheld cognitive restructuring computer program. Furthermore, this reduction in clinician time produced $133 in cost savings per client, which is modest, but considered over the course of several groups, may lead to substantial savings in treatment delivery without compromising treatment efficacy. This advantage will increase the demand for technology-based behavioral health services for clinics, practitioners, and consumers for which financial issues are the primary barrier to adoption, implementation, and access to treatment (Przeworski & Newman, 2006). Furthermore, developing technological adjuncts to clinician-delivered behavioral health treatments may be especially important as the supply of therapists trained to deliver EBPs is not currently adequate to meet the demand for such services (Emmelkamp, 2005).

Extended reach. Lack of time, money, and transportation are often cited as barriers preventing clients from accessing behavioral health care. Persons with physical and/or sensory disabilities and/or those living in geographically remote areas may find these barriers to be particularly burdensome. One potential benefit of technology is that it can serve as a mechanism for delivering behavioral health services to those who might not otherwise be able to access them by decreasing the amount of time needed for face-to-face contact with a provider, while maintaining the efficacy of many interventions (Marks et al., 1998).

Facilitates disclosure. Computers can make it easier for some clients to disclose important details about their emotional problems. Research shows that in some instances, primary care clients feel more comfortable disclosing details about emotional problems to a computer than when speaking face-to-face with a clinician (Marks et al., 1998). For example, the use of a computer has been found to facilitate an increased openness to discussing important issues concerning substance abuse (Erdman, Klein, & Greist, 1985; Supple, Aquilino, & Wright, 1999) and sexual experiences (Chinman, Young, Schell, Hassell, & Mintz, 2004; Gilbert et al. 2008; Hassell, & Mintz, 2004; Lapham, Henley, & Skipper, 1997). Integrating technology into behavioral health service delivery may allow some clients, who might otherwise feel uncomfortable disclosing details about substance use or other emotional problems, to talk more comfortably about these issues.

Disadvantages

Higher dropout rates. Several authors have voiced concern that the integration of technology into behavioral health treatment will lead to disproportionate numbers of clients dropping out of treatment (see Przeworski & Newman, 2006, for a review). This concern appears to be valid. A recent study comparing the efficacy of a computer-based and clinician-delivered exposure treatment did find higher dropout rates in the former group (43% compared to 23%; Marks et al., 2004). The literature on this topic appears to be mixed, with some research supporting this finding (Emmelkamp, 2005), while other research shows similar dropout rates and equal levels of treatment satisfaction among the two mechanisms (computer and face-to-face) of treatment delivery (Carlbring, Ekselius, & Andersson, 2003). We hypothesize that the type and degree of clinician involvement may affect dropout rates. For example, one might predict that the more tightly integrated the technology is with a face-to-face therapeutic relationship, the higher the likelihood that clients will complete assigned computer-based tasks.

Elimination of common factors of therapeutic success. Some critics argue that technology-based behavioral health services will exclude factors that constitute the interpersonal *common factors* of effective behavioral health treatment, namely, the therapeutic alliance, empathy, and important nonverbal behavior such as facial cues and body language. Interestingly, it appears that interpersonal factors important to treatment outcome can be simulated, to some extent, in computer-based treatments. For example, participants exposed to computer-based therapies have reported that computer-generated responses can have an empathic quality (Ghosh, Marks, & Carr, 1988), and have reported higher relationship ratings with their online therapist when compared to individuals meeting with a therapist on a face-to-face basis (Cook & Doyle, 2002). These findings may help explain why some individuals tend to disclose more to a computer than during face-to-face interactions with a provider. The point here is not to suggest that the use of technology in the delivery of EBPs can completely replicate the experience of live, face-to-face contact between a client and provider, but to illustrate that interpersonal aspects that are valuable to the therapeutic process are not necessarily lost when technology is used to deliver behavioral health services.

Reduced need for face-to-face psychotherapy. Some critics have argued that integrating technology (e.g., personal computer) into the delivery of behavioral health care will reduce the need for face-to-face clinician delivered psychotherapy and may therefore reduce the demand for clinicians (Marks et al., 1998). This does not appear to be the case, and in fact, there is some evidence to suggest that integrating technology into behavioral health treatment delivery may actually increase demand for face-to-face treatment. A consumer satisfaction survey on this issue showed that of a convenience sample of 619 visitors to an Internet site, 452 (73%) stated that they had consulted with a psychotherapist on the Internet and 307 (68%) of those individuals had never sought face-to-face psychotherapy. Most importantly, of the 307 individuals not in psychotherapy at the time of consultation, 196 (64%) sought out face-to-face psychotherapy after their consultation, with an additional 43 (14%) reporting plans to seek out face-to-face psychotherapy in the future (Ainsworth, 2001). This survey was posted on a Web site devoted to Internet therapy, which may account for the large number of people reporting to have consulted with an online psychotherapist.

Implementation frameworks

Should it be determined that implementing technology-based behavioral health applications would be beneficial to a health care setting, the question remains, how does one go about systematically implementing the

Chapter eleven: Implementation

technology? The next sections of this chapter describe two comprehensive implementation frameworks that can help guide the user in the implementation process, followed by a discussion of more specific evidence-based implementation strategies that can be used independently or as components of a more comprehensive systematic approach.

PARIHS

The PARIHS framework (Promoting Action on Research Implementation in Health Services) was developed out of several implementation projects conducted by researchers at the Royal College of Nursing (Rycroft-Malone et al., 2002). The major premise behind PARIHS is that successful implementation occurs when there is clarity about the *evidence* available that supports the implementation of a new practice, the *context* (e.g., environment or setting) in which health care practice occurs, and the level and availability of *facilitation* or factors that support the change process (Rycroft-Malone et al., 2002; Rycroft-Malone, 2004). These three components sit on a continuum of low to high, with those judged to be high increasing the likelihood of successful implementation. For example, new clinical practices supported by well-designed research studies, matching clinical experiences and expertise, patient health needs, and that are supported by local data (perhaps in the form of a pilot study) are deemed to be supported by high evidence. The following section outlines some of the key elements of these three components of PARIHS (see Rycroft-Malone et al., 2002; Rycroft-Malone, 2004 for more thorough descriptions of the PARIHS framework).

Evidence. The first component of the PARIHS approach is to determine the level of evidence (e.g., efficacy) available for a clinical practice for a potential implementation initiative. Three domains of evidence are considered important in the implementation process—empirical research evidence, and clinical and patient experiences. A high level of research evidence, in this context, was originally defined as practices supported by randomized controlled trials (RCTs) or systematic reviews, while low evidence was characterized by anecdotal and descriptive data, for example, clinical experience. However, over the last decade, researchers have broadened these "evidence anchors" to include not only evidence from randomized trials but also clinical and patient experiences and preferences. Regardless of evidence type, the defining factor that determines whether research evidence is high or low is the extent to which the source of knowledge is subjected to scrutiny and determined to be credible.

Context. Context refers to the environment or setting in which the new practice is to be implemented. There are three sub-elements of context—culture, leadership, and evaluation. Cultural contextual elements that are consistent with successful implementation efforts include environments that value individual staff and clients, value and promote continued learning, have available needed resources (e.g., human, financial, and equipment), promote consistency of provider roles in terms of relationships with others, teamwork, power and authority, and processes for rewards and recognition. Effective leadership characteristics associated with successful implementation initiatives include leaders who have clearly defined roles, effective teamwork among staff, democratic decision-making processes, and leaders who promote and provide opportunities for continued teaching and learning. The third component of context is evaluation. Researchers of the PARIHS model (see Rycroft-Malone et al., 2002) argue that consistent use of feedback (e.g., clinical, economic) for both providers and teams engaged in new clinical practices can help support the implementation of new clinical practices. (See Chapter 12 for a more thorough discussion of the evaluation process.)

Facilitation. The process of facilitating implementation often involves designating a person to function as a liaison between administrative and clinical staff who can support and influence the change process with his or her reputation and influence. A facilitator may be appointed from within or chosen from outside the target organization. The idea behind the facilitator concept is to appoint a person (or persons) who can help support and encourage behavioral health providers to adopt new clinical practices. Developing sustained partnerships with key stakeholders, providing counseling and feedback, and supporting critical evaluation are characteristics of facilitation associated with successful implementation. Implementation researchers point out the importance of involving staff and potential facilitators of initiatives as early as possible in the implementation process (Hagedorn et al., 2006). In the case of a clinic setting, this might involve interviewing clinic staff to identify both administrative leaders and clinical staff members who have considerable influence over day-to-day clinic operations. Attempts should be made to recruit these individuals to function as part of the implementation team and to be potential facilitators of the implementation process should they have the time available and interest in doing so (Hagedorn et al.).

In summary, the PARIHS model provides criteria for evaluating the evidence available for choosing a specific health care intervention to implement, and provides guidance regarding the successfully implementation of that intervention in a health care system. In contrast, the PRECEDE-PROCEED framework, which we will discuss next, is arguably more comprehensive in that this approach focuses on the identification of health care needs and the development of health interventions, as well as the implementation and evaluation of health care interventions into a system of care. In addition, administrators or providers might follow the PARIHS framework when the intervention being considered for implementation is being adopted from the research literature, as in the case of an evidence-based medical or psychological practice, while the PRECEDE-PROCEED framework might be more appropriate for circumstances that require the development of new practices to meet an emerging health care need. The following section presents the main elements of the PRECEDE-PROCEED approach to intervention development and implementation.

PRECEDE

The PRECEDE framework (Predisposing, Reinforcing, and Enabling Constructs in Educational Diagnosis and Evaluation) outlines an approach to identifying health needs with the purpose of designing, developing, and implementing public health interventions (e.g., educational health interventions). This entire implementation framework consists of nine phases (see Table 11.1). PRECEDE consists of the first five phases, which are focused primarily on identifying factors contributing to a target population's quality of life and any potential health problems that might impact their well-being. The premise behind these five phases is that a person's well-being is influenced by a variety of factors and that it is therefore important to identify these factors in order to effectively develop and implement health interventions.

The *first three steps* of PRECEDE center on identifying social, epidemiological, behavioral, and environmental factors that contribute to the well-being and general health of a target group of individuals. *Steps four and five* focus on assessing possible contributing/maintaining factors such as educational, organization, administrative, and policy resources

Table 11.1 Description of the Steps Involved in the PRECEDE-PROCEED Implementation Framework

PRECEDE
Step 1: Social Diagnosis
Step 2: Epidemiological Diagnosis
Step 3: Behavioral & Environmental Diagnosis
Step 4: Education & Organizational Diagnosis
Step 5: Administrative & Policy Diagnosis

PROCEED
Step 6: Implementation
Step 7: Process Evaluation
Step 8: Impact Evaluation
Step 9: Outcome Evaluation

(and barriers) that are important to consider for successful implementation. Accordingly, step 1 might involve identifying social factors (e.g., low socioeconomic status (SES), lack of information on healthy diet) that impact the quality of life of a group of individuals in a community. A link between these factors and specific health problems (e.g., obesity, diabetes) is the next phase in step 1. Methods for identifying social factors impacting quality of life within a population of individuals may be gathered from community forums, focus groups, and surveys. Steps 2 and 3 focus on the epidemiological/behavioral issues associated with important health problems identified in step 1. Mortality, disability, incidence and prevalence, and morbidity data would be included in this analysis, along with behavioral or environmental factors such as age, gender, adequacy of health care facilities, and workplace environments. Steps 4 and 5 apply information gathered in the first three steps to identify and develop strategies for addressing the health problem(s) and identifying methods for altering (if necessary) current health care resources (step 5) to maximize the implementation of interventions identified in step 4 (see Green & Kreuter, 2005, for a more thorough discussion of the PRECEDE steps).

PROCEED

The second part of this model consists of PROCEED (or Policy, Regulatory, and Organizational Constructs in Educational and Environbehavioral Development). It is designed to guide the implementation process and provide a framework for evaluation of new programs developed by the PRECEDE portion of this approach. PROCEED consists of four steps (six through nine of the entire process): (6) implementation of a developed program, (7) evaluation of the implementation process, (8) program effectiveness with respect to short-term objectives (e.g., change in predisposing factors contributing to an identified problems such as diabetes), and (9) long-term examination of the overall program efficacy.

Step 6, the first of the PROCEED steps, is the actual implementation of a program in a target community or system of care. The remaining three steps (7 through 9) are focused on evaluating different aspects of the implementation process from a funnel-down approach. Accordingly, step 7 is used to evaluate the implementation process at a broad level to determine whether the process was useful and to what extent it was effective. Step 8 begins to narrow the evaluation process to examine changes in immediate objectives (e.g., at the system, provider, and/or patient level), while step 9 is focused on patient health indices (e.g., improved diabetes management and decreased morbidity/mortality).

Chapter eleven: Implementation 233

Evidence-based implementation strategies

We next present specific implementation strategies for which there is empirical support (Grimshaw et al., 2006; Grol, 1992; Grol & Grimshaw, 2003; Weingardt, 2004). These strategies can be used as elements of a comprehensive implementation model or individually. Before we begin, it is important to point out that there is no one ideal tool to ensure successful implementation; therefore, it is recommended that using multiple tools along with an understanding of barriers that might exist in a particular organization is likely the best approach (Grol, 1992). The following discussion was derived from the evidence-based medical practice literature, but we believe it applies equally well to the implementation of technology-supported behavioral health practices.

Education. Educational materials help promote the sustained adoption of Web-based behavioral health applications (see Grimshaw et al., 2006). Educational tools might include *written educational materials* (e.g., printed practice guidelines, journal articles, and brochures) and in-person *educational outreach visits* to groups of providers and/or administrators to provide information about new practices that are being promoted. Reviews have concluded that one-on-one, in-person visits with health care providers and small group discussions that promote interactive education are particularly effective in promoting new practices among providers (Grol, 1992; Grol & Grimshaw, 2003). In addition, *educational and training activities* such as Continuing Medical Education courses, didactic workshops, and Web-based training can be effective in promoting adoption of new practices, and in particular, educating providers and administrators in new clinical applications and techniques. In their review of the specific implementation strategies, Grol and Grimshaw (2003) report that studies utilizing interactive educational content were most effective in promoting the adoption of new practices. We partially affirm this finding in a recent randomized trial in which we compared training outcomes obtained in 147 substance abuse counselors who completed 8 self-paced online modules on cognitive-behavioral therapy (CBT), and attended a series of four weekly group supervision sessions using Web conferencing software (Weingardt, Cucciare, Bellotti, & Lai, in press). Participants were randomly assigned to two conditions that systematically varied the degree of interactivity of the protocol. We found that counselors in both conditions demonstrated similar improvements in CBT knowledge and self-efficacy with respect to delivering CBT components. However, counselors in the more interactive condition demonstrated reductions in job-related burnout. This study offers some support (albeit limited) for interactive educational protocols which can not only help promote the adoption of evidence-based behavioral health practices but have additional beneficial effects on other factors such as job-related burnout.

Feedback. Feedback in the context of implementation involves collecting data on one or more indicators of job performance and providing that information to relevant stakeholders (e.g., providers, patients, administrators). In his review, Grol (1992) draws several conclusions about the nature of feedback in the implementation process. For example, when delivered directly (or close as possible) after the target behavior, and when used in conjunction with other evidence-based implementation strategies (e.g., education and/or reminders), feedback can be effective for changing clinical practice, especially when given by respected colleagues/peers. Dr. Grol also suggests that the effectiveness of feedback is enhanced when it is provided to individuals in a one-on-one fashion rather than in groups.

An example of integrating feedback into the implementation process can be illustrated by a recent initiative in the Veterans Health Administration (VHA). The VHA recently adopted the practice of evidence-based alcohol screening for all primary care patients. Patients who screen positive are offered brief interventions or counseling (Bradley et al., 2006). The VHA incorporates computerized clinical reminders to prompt clinicians to screen patients for alcohol misuse and provide brief alcohol counseling when appropriate. When screening and brief counseling are conducted, this information is written back to the electronic medical record (and provider note) and stored in an electronic database. This allows administrators, health managers, and researchers to access this information and to provide performance-oriented feedback to clinicians.

Clinical reminders. Clinical reminders are often automated (e.g., delivered via computer) and have the purpose of prompting a provider or patient to engage in a treatment-related behavior (e.g., conduct an assessment, provide a brief intervention, or check the status of test results). The example provided in the previous section demonstrates how clinical reminders may be useful in the implementation process and particularly the provider side of health care delivery. The VHA has been at the forefront of using computerized clinical reminders to implement evidence-based practices in all types of medical settings. Findings from one relatively large-scale study (12 VA medical centers, 275 resident physicians caring for over 18,000 patients) showed that computerized clinical reminders could increase physician compliance to well-accepted standards of care for patients presenting to outpatient medical clinics (Demakis et al., 2000).

Computers are often used for implementing clinical reminders because they provide a very quick, reliable method for storing data and prompting providers (or patients) to engage in a behavior or set of behaviors. Health care organizations are increasingly using clinical reminders as part of their electronic medical records to prompt both providers and patients about important aspects of health care. For example, Kaiser allows patients to access their electronic medical record (see http://www.kp.org), which gives them the ability to securely e-mail their providers, access test results, find a physician, and access other health-related educational information via the Web. When new information (e.g., test results, provider messages) becomes available, Kaiser prompts registered patients to access their electronic health record through a personal e-mail address provided by the patient.

Marketing and mass media. Marketing consists of identifying customer needs and wants, targeting markets that an organization can serve most effectively, and developing products to meet the needs and wants of customers in those markets. Marketing also has to do with developing customer satisfaction by building relationships between those developing or implementing products and customers (American Marketing Association, http://www.marketingpower.com, 2009). For our purposes, we use the term *customers* to refer to those interested in implementation initiatives such as a business entity, health care administrators of a system of care, or even providers in a clinic interested in adopting new practices. The term *customer* may also be used to refer to those for which new applications or practices were designed. Customers include providers for whom new practices support efficient and/or consistent delivery of evidence-based practices and patients for whom the new practices might lead to improvements in care and overall well-being. Marketing strategies may include persuading health care providers and/or patients to engage in (or utilize) a particular practice by communicating its value (e.g., enhancement of the efficiency at a system or practice level and/or potential health benefits).

The strategy of marketing is reflected in models of implementation such as PRECEDE-PROCEED (see Green & Kreuter, 2005), which utilize needs assessments of a target population and target problems to design educational interventions. For example, PRECEDE consists of five phases, the first three of which target the identification of a particular health problem and needs of a target population impacted by that problem, as well as identifying factors that contribute to these issues (see "Precede-Proceed Model," 2002). This process is designed to inform marketing strategies that can be tailored to the specific problems and needs of health care stakeholders such as providers and patients.

Marketing through mass media campaigns can be effective for promoting the adoption of new clinical practices. A recent review conducted by the Cochrane Collaboration concluded that mass media campaigns are effective for promoting the utilization of effective health care practices among providers, patients, and the general public (Grilli, Ramsay, & Minozzi, 2001).

A common example of this strategy in health care is the promotion of flu shots during the fall months. Posters, brochures, word of mouth, the Web, and television advertisements are some of the ways that this information is advertised to the public.

Incentives. Perhaps one of the most commonsense strategies for promoting the adoption of new clinical practices among both providers and patients is through the use of incentives. A common example of an incentive is the use of a financial reward contingent on a provider engaging in a new clinical practice (Grol, 1992; Grol & Grimshaw, 2003). Studies show that forms of payment can alter the utilization of health services in health care systems. For example, the use of salary or capitation versus fee-for-service results can impact the amount of referrals to certain types of inpatient and outpatient services (Hillman, Pauly, & Kerstein, 1989). Similarly, financial incentives can be structured to promote the identification of important health issues and increase the utilization of relevant health services. For example, research shows that providing financial payouts for identifying smokers in a primary care clinic and referring them to appropriate counseling services can increase both of these behaviors among providers (Roski et al., 2003).

Barriers to implementation

Implementation efforts can be hindered by various provider-, patient-, and system-level factors. A systematic review published in the *Journal of the American Medical Association* (Cabana et al., 1999) conducted a thorough examination of barriers that limit the implementation of clinical practice guidelines in medical settings. Although the focus of this systematic review was on identifying barriers that limit physicians' adherence to practice guidelines, our hope is that we might generalize these findings to the implementation of technologies to support evidence-based behavioral health practices. We briefly discuss some of the major findings of this review below. We have organized the findings into provider- and patient-level barriers.

Provider barriers

Barriers to adopting new practices include simply being unaware of new clinical guidelines or tools, as well as a host of attitudinal factors regarding the quality and applicability of a particular practice (Cabana et al., 1999). Accordingly, *providers who are unaware of new clinical practices* may understandably have difficulty implementing them into clinical practice.

Cabana et al. report that as many 84% of physicians identify lack of awareness as a barrier to implementation of clinical practice guidelines. This is undoubtedly an important issue with respect to implementing computer-based behavioral health applications. Although literature reviews on computer-aided psychotherapy yield over 150 studies demonstrating varying degrees of evidence for these applications (Marks, Cavanagh, & Gega, 2007), providers are unlikely to be familiar with such a large literature as well as how to effectively integrate such applications into their clinical practice. Therefore, educating providers on the availability and potential merits (and perhaps drawbacks) of computer-based behavioral health applications may be an important component of the initial phase of the implementation process.

Second, research reviews have concluded that *provider attitudes* may interfere with clinicians' willingness to adopt new practices (Grol & Wensing, 2004). Cabana et al. (1999) report that at least 1 out of 10 physicians demonstrates some resistance to clinical practice guidelines because of disagreement with the specific approach to patient care outlined in the guidelines. Some of these concerns include differences in interpretation of evidence supporting a guideline, beliefs that the benefits of engaging in a new practice are not worth the risks to patients, and feeling that guidelines are oversimplified.

External barriers

Barriers to implementation may also exist from outside the control of providers. External barriers to successful implementation include concerns that patients may react negatively to a particular practice, the influence of colleagues, as well as organization or clinic factors (e.g., lack of resources). We will briefly describe some of these potential external barriers in this section.

First, *patient treatment preferences* are important to consider when implementing new treatment practices such those that are technology based (Cabana et al., 1999; Grol, 1992; Rycroft-Malone, 2004). Accordingly, it is important to determine, prior to large-scale implementation efforts, the feasibility (e.g., acceptability of treatment application among patients) of particular treatment strategies. For example, we recently conducted a pilot study in a large primary care clinic at the Veterans Affairs Palo Alto Health Care System to determine the feasibility of implementing a brief motivational intervention for treating alcohol misuse. One important purpose of this study was to determine the acceptability of using the computer to (a) briefly screen for alcohol use and (b) provide brief motivational counseling for patients reporting risky drinking behavior. Despite some initial concerns about using the computer with older veterans, we found that our application was very well received as veterans reported a high degree of satisfaction with the program. Using pilot data to determine local response to an implementation initiative can be useful in identifying potential resistance and/or lack of need for an intervention among a target patient population. Thus, collecting pilot data to determine the feasibility of implementing a clinical practice is a critical component of some implementation models (see PARIHS model discussed above).

Colleagues and organizational staff may also influence the types of interventions that we accept into our clinical repertoire. Grol (1992) suggests that colleagues, practice staff, and other important opinion leaders may exert a powerful influence on the adoptability of new clinical practices by resisting or accepting new initiatives. He suggests that solo practitioners (e.g., those working in private practice or in isolation) may be particularly resistant to change as they are likely to have less information about new practices (e.g., current evidence, advantages/disadvantages) than providers working closely with colleagues on a day-to-day basis. Fortunately, most models of implementation, such as the PARIHS model, have approaches for addressing this potential barrier. For example, the facilitation component of the PARIHS model provides some guidance on how to approach resistance to implementation initiatives through the use of facilitators or liaisons who have been identified as key opinion leaders in the target organization.

Finally, *organizations that have limited resources* may have more difficulty successfully implementing new clinical practices. This is a particularly important issue when considering implementing computer-based behavioral health applications. Computer equipment, relevant software, printers, computer carts, and appropriate training may be required to support such initiatives. These prerequisite resources may exist in an organization but may be currently used by other providers and/or for other programs limiting their use for these purposes. In such cases, implementation frameworks may provide guidance on both identifying resource deficiencies and developing plans for remedying the situations (see steps outlined in the PRECEDE model).

Example implementation initiative using the PRECEDE-PROCEED framework

The focus of this chapter is centered on providing a broad overview of tools that can guide the implementation process and perhaps increase the success of such efforts. We started our discussion of implementation with a broad overview of implementation frameworks, followed by a discussion of specific strategies and potential barriers to the implementation process. This section presents an example of how one might apply these strategies to implement technology-based EBPs into a health care setting. In the following section, we apply the PRECEDE-PROCEED implementation framework (see Table 11.1) to an example initiative—the implementation of CBT in an outpatient substance abuse setting. As mentioned earlier, PRECEDE is mainly concerned with the identification of health needs in a patient population and factors that give rise to and/or maintain those health issues. Since substance abuse in many patient populations (e.g., veterans, primary care), along with relevant predictors, correlates, and epidemiology, is well documented in the research literature, we will skip steps 1 through 4 and advance directly to step 5, which begins the implementation process.

The PRECEDE-PROCEED implementation framework occurs on a continuum with one end consisting of the identification of health problems in a community or patient population, while the other end consists of evaluating the objectives of an implementation initiative. For illustrative purposes, we will start in the middle of this spectrum to highlight how we might begin the actual implementation process once we have identified initial relevant factors such as a target health problem (e.g., substance abuse), chosen an intervention (e.g., CBT), and chosen a health care clinic (e.g., outpatient).

Determining feasibility prior to implementation

Step 5 of the PRECEDE-PROCEED framework helps guide the initial part of the implementation process by outlining organizational issues to consider prior to launching an implementation project. The availability of resources (e.g., computers, providers, and physical space), budget allocation, clinic coordination with other departments, and an implementation timetable are all important issues to consider prior to large-scale implementation efforts. It is also helpful, in terms of evaluating the "success" of implementation efforts at a later time, to further clarify the objectives of the implementation process (e.g., what are the specific clinical outcomes we are aiming for?). Small pilot studies can be helpful in providing program planners with a preliminary sense of potential barriers that might hinder large-scale implementation efforts, and whether the intervention will have the desired effect on patient outcomes (e.g., increasing patient satisfaction with treatment, increasing appropriate health care utilization, reducing substance use and frequency).

We might start this process by first developing interdisciplinary collaborations with providers and/or administrative stakeholders who are familiar with important feasibility issues (e.g., availability of providers, physical space, computers, and necessary software, as well as the work flow of the clinic). In the case of implementing a computer-based CBT intervention, there will be several important questions to consider, such as:

- How will this intervention fit into current processes of care?
- Are the necessary resources (e.g., computers and printers) available?
- Will providers require training to operate relevant software?
- How will patients interact with this intervention (e.g., will there be a room(s) designated specifically for patients to complete computer-based CBT tasks or will providers incorporate this intervention into their face-to-face sessions with patients)?

Should the necessary resources be deemed available, we might proceed next with a small pilot study to test the feasibility of implementing this intervention. To proceed, we will need to address important issues, such as securing relevant permissions and identifying space to house computers, and conduct any necessary provider training. This latter issue will almost certainly depend on the degree to which the computerized CBT program supports face-to-face clinical practice. For example, clinicians might use this program to provide patients with CBT-based psychoeducation (e.g., helping patients complete CBT-related tasks such as cognitive restructuring, completing homework assignments, learning about relapse prevention and relaxation techniques, and assessing and tracking symptoms; see Cucciare & Weingardt, 2007, for a review; Marks et al., 2004; Robinson, 2003). Once we have "set up shop," we might recruit a small sample (e.g., 20) of patients to interact with the program in the manner we envisioned in the full implementation project. This will help us identify feasibility issues such as the extent to which the software is user friendly, patient (and provider) satisfaction with how the software is integrated into the face-to-face treatment process, and how the intervention fits into the general clinic flow (e.g., is it facilitating or disrupting the treatment process?).

Determining feasibility can be a tricky issue. Unfortunately, we may not always have a clear sense of whether an implementation idea is feasible, even after the completion of a pilot project. The good news is that any pilot testing that we conduct will provide us with information about potential concerns that need to be addressed prior to full-scale implementation efforts. Should we determine based on pilot study results that it is generally feasible to implement this intervention, we can, proceed to the next step—implementation.

Implementation

The first step of PROCEED is implementation. The implementation process consists of many elements. Given restrictions on space, we will only briefly discuss some of these. For example, launching the full-scale implementation of computer-based CBT in an outpatient substance abuse treatment clinic will require practical steps such as installing software on all computers, ensuring that patients have privacy while interacting with the software, determining how (and if) patients will interact with the program at home, and ensuring providers have received any relevant training as well as have opportunities to receive updated training when questions or concerns arise. More conceptual issues will also need to be addressed during the implementation phase. These include determining the extent to which the software will support the delivery of face-to-face CBT and training providers in how to integrate this method of delivery into their practice.

Beyond these practical and conceptual issues, it is also useful to consider utilizing the strategies outlined earlier in this chapter to enhance the sustained adoption of the target intervention. For example, we might consider setting goals for using and incorporating tools, such as automated clinical reminders and provider feedback, and setting up an incentive system to encourage the use of a technology-based CBT intervention among providers and patients. We might also make a special effort to educate and market the intervention to patients highlighting the advantages of engaging in this treatment component (e.g., is convenient, can be accessed at home, increases privacy, enhances efficacy of substance abuse treatment).

Evaluation

The remaining steps of PROCEED guide program planners through the evaluation process. We note that Chapter 12 in this volume is devoted to the role(s) of evaluation in the implementation process; therefore, we will discuss it only briefly in this section. In the PROCEED framework, evaluation can be conducted to answer questions about the implementation process itself (e.g., was the implementation process efficient?) and as a tool to examine overall objectives of the implementation process such as measuring various patient outcomes (e.g., reduction in drinking, improvements in overall well-being and quality of life).

Conclusion

The growing popularity of technology-based behavioral health applications is at least partially fueled by the persuasive advantages of integrating these strategies into health care systems. Some of these potential advantages were presented in this chapter and include enhancing the cost-effectiveness of evidence-based behavioral health practices, extending the reach of these services to a wide variety of patient populations, and in some cases facilitating greater comfort on the part of clients to disclose issues concerning potentially uncomfortable health and emotional topics. To be balanced, we also presented potential disadvantages (and common criticisms) of integrating technology into behavioral health practices including the possibility of higher dropout rates (when compared to face-to-face treatment), the elimination or reduction of interpersonal factors that can contribute to therapeutic success, and the criticism that these technologies may reduce the need for face-to-face care.

We then presented strategies for implementing technology-based behavioral health services into health care systems. Two comprehensive frameworks were presented—PARIHS and PRECEDE-PROCEED, along with a discussion of evidence-based strategies for addressing common barriers to the implementation process. Our hope is that this top-down presentation approach to the implementation process will enhance the success of the implementation process by first providing readers with a comprehensive overview of critical issues to consider (e.g., criteria for selecting an intervention to implement, strategies for determining feasibility, and methods for promoting the sustained adoption of such practices post implementation), followed by a discussion of more specific evidence-based strategies (e.g., educating providers, provider feedback, clinical reminders, mass marketing, and provider incentives) for addressing common barriers (e.g., provider and patient preferences, lack of knowledge, and organizational resources) to the implementation process.

Finally, in an attempt to put some of the strategies into practice, we provided a case example implementation project. Specifically, we applied the PRECEDE-PROCEED framework to the objective of implementing a computer-based CBT intervention into an outpatient substance abuse clinic. We also discussed how some of the evidence-based implementation strategies outlined in this chapter might be used to enhance the sustained adoption of such an intervention. Due to space restrictions and the complexities involved in the implementation process, we of course were unable to discuss all of the possible issues that require consideration, but instead hope that we have provided readers with a foundation from which to begin implementing technology-based interventions into health care settings.

References

Ainsworth, M. (2001). E-therapy history and survey. Retrieved June, 2007, from http://www.metanoia.org/imhs/history.htm

American Marketing Association. (2009). About AMA: Definition of marketing. Retrieved July 1, 2009, http://www.marketingpower.com/AboutAMA/pages/DefinitionsofMarketing.aspx

Bradley, K. A., Williams, E. C., Achtmeyer, C. E., Volpp, B., Collins, B. J., & Kivlahan, D. R. (2006). Implementation of evidence-based alcohol screening in the Veterans Health Administration. *The American Journal of Managed Care, 12*(10), 597–606.

Cabana, M. D., Rand, C. S., Powe, N. R., Wu, A. W., Wilson, M. H., et al. (1999). Why don't physicians follow clinical practice guidelines? A framework for improvement. *Journal of the American Medical Association, 282*(15), 1458–1465.

Carlbring, P., Ekselius, L., & Andersson, G. (2003). Treatment of panic disorder via the Internet: a randomized trial of CBT vs. applied relaxation. *Journal of Behavior Therapy and Experibehavioral Psychiatry, 34,* 129–140.

Chinman, M., Young, A. S., Schell, T., Hassell, J., & Mintz, J. (2004). Computer-assisted self-assessment in persons with severe behavioral illness. *Journal of Clinical Psychiatry, 65,* 1343–1351.

Cook, J. E., & Doyle, C. (2002). Working alliance in online therapy as compared to face-to-face therapy: Preliminary results. *Cyberpsychology and Behavior, 5,* 95–105.

Cucciare, M. A., & Weingardt, K. (2007). Integrating technology into the delivery of evidence-based psychotherapies. *Clinical Psychologist, 11*(2), 1–10.

Cucciare, M. A., Weingardt, K., & Villafranca, S., (2008). Using blended learning to implement evidence-based psychotherapies. *Clinical Psychology: Science and Practice, 15,* 299–307.

Curran, G. M., Mukherjee, S., Allee, E., & Owen, R. R. (2008). A process for developing an implementation intervention: QUERI Series. *Implementation Science, 3*(17), 1–11.

Demakis, J. G., Beauchamp, C., Cull, W. L., Denwood, R., Elsen, S. A., et al. (2000). Improving residents' compliance with standards of ambulatory care: Results from the VA cooperative study on computerized reminders. *Journal of the American Medical Association, 284*(11), 1411–1416.

Doebbeling, B. N., Chou, A. F., & Tierney, W. M. (2006). Priorities and strategies for the implementation of integrated informatics and communications technology to improve evidence-based practice. *Journal of General Internal Medicine, 21,* S50–57.

Emmelkamp, P. M. G. (2005). Technological innovations in clinical assessment and psychotherapy. *Psychotherapy and Psychosomatics, 74,* 336–343.

Erdman, H. P., Klein, M. H., & Greist, J. H. (1985). Direct patient computer interviewing. *Journal of Consulting and Clinical Psychology, 53,* 760–773.

Ghosh, A., Marks, I. M., & Carr, A. C. (1988). Therapist contact and outcomes of self-exposure treatment for phobias: A controlled study. *British Journal of Psychiatry, 152,* 234–238.

Gilbert, P., Ciccarone, D., Gansky, S. A., Bangsberg, D. R., Clanon, K., McPhee, S. J., et al. (2008). Interactive "video doctor" counseling reduces drug and sexual risk behavior among HIV-positive patients in diverse settings. *PLoS ONE, 3*(4), e1988.

Green, L. W. & Kreuter, M. W. (2005) *Health promotion planning: An educational and environbehavioral approach* (4th ed.). Mountain View, CA: Mayfield.

Grilli, R., Ramsay, C., & Minozzi, S. (2001). Mass media interventions: Effects on health services utilisation. *Cochrane Database of Systematic Reviews, 4,* Art. No.: CD000389. DOI: 10.1002/14651858.CD000389.

Grimshaw J., Eccles, M., Thomas, R., MacLennan, Ramsay, C., et al. (2006). Toward evidence-based quality improvement. *Journal of General Internal Medicine, 21,* S14–20.

Grol, R. (1992). Implementing guidelines in general practice care. *Quality in Health Care, 1,* 184–191.

Grol, R., & Grimshaw, J. (2003). From best evidence to best practice: Effective implementation of change in patients' care. *Lancet, 362,* 1225–1230.

Grol, R., & Wensing, M. (2004). What drives change? Barriers to and incentives for achieving evidence-based practice. *Medical Journal of Australia, 180,* S57–S60.

Gruber, K., Moran, P. J., Roth, W. T., & Taylor, C. B. (2001). Computer-assisted cognitive behavioral group therapy for social phobia. *Behavior Therapy, 32,* 155–165.

Hagedorn, H., Hogan, M., Smith, J. L., Bowman, C., Curran, G. M., et al. (2006). Lessons learned about implementing research evidence into clinical practice. Experiences from VA QUERI. *Journal of General Internal Medicine, 21,* S21–24.

Hillman, A. L., Pauly, M. V., & Kerstein, J. J. (1989). How do financial incentives affect physicians' clinical decisions and the financial performance of health maintenance organizations? *Journal of the American Medical Association, 321,* 86–92.

Lapham, S. C., Henley, H., & Skipper, B. J. (1997). Use of computerized prenatal interviews for assessing high-risk behaviors among American Indians. *American Indian and Alaska Native Behavioral Health Research, 8,* 11–23.

Maltby, N., Kirsch, I., Mayers, M., & Allen, G. J. (2002). Virtual reality exposure therapy for the treatment of fear of flying: A controlled investigation. *Journal of Consulting and Clinical Psychology, 70*(5), 1112–1118.

Marks, I. M., Cavanagh, K., & Gega, L. (2007). *Hands-on help: Computer aided psychotherapy.* New York: Psychology Press.

Marks, I. M., Kenwright, M., McDonough, M. Whittaker, M., & Mataix-Cols, D. (2004). Saving clinicians' time by delegating routine aspects of therapy to a computer: A randomized controlled trial in phobia/panic disorder. *Psychological Medicine, 34,* 9–18.

Marks, I., Shaw, S., & Parkin, R. (1998). Computer-aided treatments of behavioral health problems. *Clinical Psychology: Science and Practice, 5,* 151–170.

Precede-Proceed Model. (2002). In Lester Breslow (Ed.), *Encyclopedia of Public Health.* New York: Macmillan Reference/Gale Cengage; eNotes.com, 2006. Retrieved October 17, 2008, from http://www.enotes.com/public-health-encyclopedia/precede-proceed-model

Przeworski, A., & Newman, M. G. (2006). Efficacy and utility of computer-assisted cognitive behavioral therapy for anxiety disorders. *Clinical Psychologist, 10*(2), 43–53.

Robinson, P. (2003). Homework in cognitive behavior therapy. In W. O'Donohue, J. E. Fisher, & S. C. Hayes (Eds.). *Cognitive behavior therapy: Applying empirically supported techniques in your practice* (pp. 202–211). Hoboken, NJ: John Wiley & Sons, Inc.

Roski, J. R., Jeddeloh, R., An, L., Lando, H., Hannan, P., Hall, C., et al. (2003). The impact of financial incentives and a patient registry on preventive care quality: Increasing provider adherence to evidence-based smoking cessation practice guidelines. *Preventative Medicine, 36*(3), 291–299.

Rycroft-Malone, J. (2004). The PARIHS framework—A framework for guiding the implementation of evidence-based practice. *Journal of Nursing Care Quality, 19*(4), 297–304.

Rycroft-Malone, J., Kitson, A., Harvey, G., McCormick, B., Seers, K., Titchen, A., et al. (2002). Ingredients for change: Revisiting a conceptual framework. *Quality and Safety in Health Care, 11,* 174–180.

Supple, A. J., Aquilino, W. S., & Wright, D. L. (1999). Collective sensitive self-report data with laptop computers: Impact on the response tendencies of adolescents in a home interview. *Journal of Research on Adolescents, 9,* 467–488.

Tate, D. F., & Zabinski, M. F. (2004). Computer and Internet applications for psychological treatment: Update for clinicians. *Journal of Clinical Psychology, 60*(2), 209–220.

Taylor, C. B., & Luce, K. H. (2003). Computer- and Internet-based psychotherapy interventions. *Current Directions in Psychological Science, 12,* 18–24.

Weingardt, K. R. (2004). The role of instructional design and technology in the dissemination of empirically supported, manual-based therapies. *Clinical Psychology: Science & Practice, 11,* 313–331.

Weingardt, K., Cucciare, M., Bellotti, C., & Lai, W. P. (in press). A randomized trial comparing two models of Web-based training in cognitive behavioral therapy for substance abuse counselors. *Journal of Substance Abuse Treatment.*

Whitfield, G., Hinshelwood, R., Pashely, A., Campsie, L., & Williams, C. (2006). The impact of a novel computerized CBT CD-ROM (overcoming depression) offered to patients referred to clinical psychology. *Behavioural and Cognitive Psychotherapy, 34,* 1–11.

Wright, J. H., Wright, A. S., Albano, A. M., Basco, M. R., Goldsmith, L. J., Raffield, T., et al. (2005). Computer-assisted cognitive therapy for depression: Maintaining efficacy while reducing therapist time. *American Journal of Psychiatry, 162* (6), 1158–1164.

chapter twelve

Evaluation

Kenneth R. Weingardt

As the number and variety of technology applications designed to support evidence-based behavioral health practices have grown dramatically in recent years, so too has the emerging research literature in this area. This newfound empirical attention is most welcome. However, a natural consequence has been the proliferation of published studies that have used different research designs, sampling strategies, and outcome measures to address a host of different research questions. This variability, along with the exponential increase in the sheer volume of publications in this area, makes it particularly challenging to draw generalizable conclusions about the effectiveness of technology to support evidence-based behavioral health practices.

The present chapter begins with a brief primer on the fundamentals of research design and methodology, which may help the reader learn to critically evaluate the primary research in this area. As the chapters in this volume attest, the vast majority of the studies published in peer review journals consist of summative evaluations, meaning that the evaluation activities focus on summarizing the outcomes or impacts of an intervention after it has been delivered. Methodologically rigorous summative evaluations, such as randomized controlled trials (RCTs), are critically important in order to accurately and reliably measure the clinical impact of an intervention.

The second part of this chapter explores the equally important role of formative evaluation. In contrast to summative evaluation, which typically occurs after the intervention has been completed, formative evaluation takes place throughout the design and development process, providing critical information about the intervention or product as it forms and evolves. It involves frequent, iterative, and incremental review of the intervention by a variety of different stakeholder groups, which typically include the end users as well as project sponsors. Each cycle of feedback is then used to inform the next version, draft, or prototype.

Although formative evaluations appear to be underrepresented in the scientific literature, formative evaluation plays a central role in many of the processes that are used by industry to guide the development of new technologies. This includes (a) the Instructional Systems Design (ISD) model (Dick & Carey, 1996; Morrison, Ross, & Kemp, 2001), which is commonly used to guide the development of technology-based training; (b) contemporary "agile" software development processes such as the Rational Unified Process (RUP; Kroll & Kruchten, 2003), which are frequently used to guide enterprise (large-scale) software development projects; and (c) the New Products Management (NPM) framework (Crawford & DiBenedetto, 2003), which is used to guide the development of new products ranging from electronics and automobiles to insurance and financial services. Whether one's business is developing Web-based customer service training for new employees, designing an enterprise software system for a large federal agency, or dreaming up the next iPod, it seems that one key to success is getting feedback from key stakeholders early and often, listening to what they have to say, and revising one's prototype accordingly.

The third section of this chapter explores how technology-based behavioral health applications can (and, the authors argue, should) be evaluated in context. On the micro level, we consider an individual practitioner striving to implement evidence-based practices (EBPs) within the context of his/her own clinical expertise and the client's preferences and expectations. On the macro level, we discuss how various emerging models of implementation science such as PRECEDE-PROCEED (Green & Kreuter, 2005) and PARIHS (Rycroft-Malone, 2004) can inform the evaluation of technology-based applications in the context of the clinic, hospital, or health care system.

We conclude with a discussion of some of the challenges faced by researchers who are interested in evaluating technologies to support evidence-based behavioral health practices, and suggest some innovative approaches to overcoming these barriers. These challenges include (a) balancing the competing demands of internal and external validity, (b) conducting ongoing formative evaluation to guide the design, development, implementation, and evaluation of technologies throughout the project life cycle, and (c) the need to understand and evaluate technology applications within the context of individual client–therapist relationships, as well as the broader context of the clinic or health care system in which behavioral health services are delivered.

Research 101

This section is intended to provide those readers who have not had formal research training with some of the fundamental knowledge required to become a more informed consumer of the empirical literature. It covers such topics as how randomized control trials can support causal inferences, why larger samples are preferable to smaller ones, and why it is important to use measures that are both valid and reliable. If you are already familiar with such research basics, please feel free to skip ahead to the next section on formative evaluation.

One of the guiding principles underlying sound research design is the desire to minimize the biases inherent in the process of human reasoning under uncertainty. In solving problems, making predictions, and making decisions, humans (including clinicians and researchers!) typically rely on heuristics, which are templates, models, or mental rules of thumb (Kahneman & Tversky, 1983). For example, research has found that people often judge the probability of an event by the ease with which an instance of that event comes to mind. Although this so-called availability heuristic can sometimes result in accurate estimates of probability, it often leads us astray as it is easily biased by factors such as salience, vividness, or personal relevance—factors that may be entirely unrelated to the frequency of the event. For example, when I think of an online stop-smoking intervention, I may readily think of Marge, my flamboyant and engaging client who had a particularly successful experience with it. However, Marge's positive response may not accurately reflect the experience of the majority of people who have completed the program.

Rigorous research designs strive to minimize the influence of such biases in human decision making, as well as other sources of error, in establishing a causal relationship. Researchers would typically like to conclude that an independent variable that they experimentally manipulate (e.g., completion of an online treatment intervention) causes improvement in some dependent or measured variable (e.g., reduction in psychiatric symptoms). When the researcher may confidently attribute the observed changes or differences in patient outcomes to the intervention, and when he can rule out other explanations, then his or her causal inference is said to have internal validity (Liebert & Liebert, 1995).

In order to support a causal inference (i.e., to be able to conclude that an online intervention caused a reduction in psychiatric symptoms or improvement in functioning), the researcher must carefully design his or her study so as to rule out any other possible explanations for the results. Other explanations for the results are commonly referred to as threats to internal validity or sources of systematic error. Confounding is one such source of error and refers to changes in the dependent variables that are caused by a third variable, which is related to the independent variable. For example, if we are evaluating a technology intervention that involves daily visits to a Web site where clients are prompted to enter information regarding their private thoughts, any improvements in depressive symptoms that are subsequently observed may actually be the result of engaging in a process of regular self-monitoring rather than accessing the specific online depression content. In this example, self-monitoring is a confounding variable that would need to be controlled in the research design—perhaps by creating a control condition in which participants are asked to enter daily self-monitoring data that is unrelated to the depression program content, for example, completing a daily food diary.

Other common threats to internal validity include history, when historical events outside the research study (e.g., natural disasters, policy changes) influence participants' responses; maturation, when participants naturally grow and change during the course of the study; selection bias, when participants with certain characteristics related to the phenomenon of interest are unequally distributed among the experimental and control groups; and differential attrition, when relatively more participants from one condition drop out of the study. Good experimental design allows researchers to rule out these and other possible alternative explanations of the results, thereby increasing our confidence in the causal inference that the measured improvement in patient outcomes (dependent variable) was caused by participation in the intervention (independent variable).

Researchers' most powerful weapon against all of these threats to internal validity is randomization. Randomization is a process that assigns research participants by chance, rather than by choice, to either the experimental or control group. Each study participant has a fair and equal chance of receiving either the new intervention being studied or the control intervention. In an RCT, researchers use a computer program or a table of random numbers to assign each study participant to a group. As long as the number of participants in the study is sufficiently large, random assignment to condition ensures that both known and unknown threats to internal validity are evenly distributed between the experimental groups.

The size of the sample, or the number of individuals who agree to be randomly assigned to the study conditions, is itself a critically important feature of good experimental design. Generally speaking, the larger the sample size, the more confidence we have that any observed improvement in client outcomes were caused by the intervention. Although the statistical explanation for this relationship is beyond the scope of this chapter, the general principle is that the more people a researcher recruits to participate in her study, (a) the more accurately the sample will represent the entire population and (b) the more confident we are that any observed differences in client outcomes between the experimental and control groups are the result of the intervention and not simply due to chance or random error.

In the hierarchy of evidence that influences health care policy and practice, large RCTs are considered the highest level—the "gold standard" research design (Lachin, Matts, & Wei, 1988). As the "gold standard" label implies, they are also very expensive to conduct, and typically require grant funding from government or industry. In the emerging field of technology in mental health, much of the published literature reports the results of studies that are not true RCTs (Marks, Cavanagh & Gega, 2007). Although alternate designs do not support the level of causal inference allowed by an RCT, such quasi-experimental designs provide important information regarding the real-world effectiveness of technology-based mental health applications as they are implemented in the context of real-world clinical practice. Although lower in internal validity than RCTs, quasi-experimental designs are generally higher in external validity, which means that the results are more likely to apply to or generalize to the population as a whole. The issue of external validity is critically important for improving public health, and a growing chorus of researchers has exhorted the field to do more to increase the applicability of research findings to clinical practice (e.g., Green & Glasgow, 2006; Glasgow et al., 2006)

Some quasi-experimental designs that appear frequently in the literature include the nonequivalent control group design and time series designs (Campbell & Stanley 1968). In the nonequivalent control group design, comparison groups are naturally occurring rather than randomly selected. Here control groups are matched to the experimental group; they are chosen to be as similar as possible in all experimentally relevant aspects. For example, researchers interested in evaluating the effectiveness of an intervention delivered to clients in crisis (e.g., highly stressed individuals) may choose not to randomly assign participants to an experimental and control condition due to ethical concerns (e.g., it is likely not ethical to withhold treatment to clients in crisis). Thus, all participants seeking services at a clinic delivering such services may be included, while those who refuse or delay treatment may be assigned to a control condition.

Perhaps the most widely used quasi-experimental design is the pre/post design, or time series design. In the pre/post design, the researchers establish a baseline measurement of some phenomenon of interest (e.g., client symptoms, functioning, or quality of life), deliver the intervention, and then administer the same set of measures again. Although many studies use a single measurement before and after the intervention (pre/post design), conducting multiple observations over time, both before and after the intervention, helps to rule out history as a potential threat to internal validity. More frequent observations and a long follow-up window between the baseline and final assessment are generally more desirable.

The observations, interviews, assessment instruments, or other types of data that are collected during a study, constitute another important indicator of its methodological quality. Studies that employ measures that have been demonstrated to be reliable and valid are preferable to those that have not. *Reliable* means that a set of measurements will remain consistent over time and over repeated administration (e.g., will I get the same score when I complete the measure the next day?) *Validity* reflects how well the instrument measures what it is supposed to measure (i.e., are people who meet the cutoff score on the Beck Depression Inventory diagnosed as having major depressive disorder by an independent clinical interview?). Standardized measures, such as the Beck Depression Inventory (BDI; Beck, Steer, & Brown, 1996); the Alcohol Use Disorders Identification Test (AUDIT; Babor, Higgins-Biddle, Saunders, & Monteiro, 2001); and the Posttraumatic Stress Disorder Checklist–Military Version (PCL-M; Weathers, Litz, Herman, Huska, & Keane, 1993) have been thoroughly tested to document their reliability and validity, and have amassed a large body of normative data that allow researchers to accurately interpret the scores that result.

This concludes our brief primer on research design and methodology. Readers who are interested in learning more about research design are referred to the classic text by Donald Campbell and Julian Stanley (1968), or more recent textbooks on conducting research such as Rosnow and Rosenthal (2008) or Bordens and Abbott (2008).

Formative evaluation

Summative evaluation is a systematic process of collecting data on the impacts, outputs, products, or outcomes hypothesized in a study (Stetler et al., 2006). Summative evaluations, which include RCTs and the various quasi-experimental designs described in the previous section, allow us to accurately and reliably measure the clinical impact of an intervention after it has been delivered. At the conclusion of a trial, we can conclude with some level of certainty that any improvement in measured outcomes (e.g., symptom reduction) can be attributed to the intervention. Such studies provide us with invaluable scientific evidence regarding the efficacy of technology-based interventions in improving behavioral health outcomes.

However, summative evaluation data are often silent on the issue of process, for example:

- By what mechanism did these changes occur?
- If the desired changes were not obtained, what might have gone wrong?
- How can we manipulate the internal dynamics and actual operations of the intervention in order to understand the changes that occur in patient outcomes over time?

- How can we evaluate the intervention as it forms, during the design and pretesting process, in order to better understand its strengths and weaknesses and improve it as a result?

Such questions fall under the domain of formative evaluation (Bhola, 1990; Rossi & Freeman, 1993; Stetler et al., 2006).

Formative evaluation can be entirely consistent with the purposes of a rigorous summative evaluation. In fact, the two are quite complementary. Many published RCTs began life as a series of smaller pilot studies that were used to inform the design and development of the intervention that ultimately finds its way into print in an academic journal. It is tempting to think of formative evaluation as being primarily qualitative, and summative evaluation as being primarily quantitative. However, this is not necessarily the case. Investigators can use focus groups and qualitative interviews as part of rigorous summative evaluations, and have also used standardized questionnaires and online surveys as elements of formative evaluation efforts.

Although academic journal articles describing formative evaluations are not common, the process of iterative review and incremental improvement is a centerpiece of conceptual models that are widely used to guide the development of new technologies. Formative evaluation plays a critical role in the Instructional Systems Design model, which is used throughout both the public and private sectors to guide the development of interactive Web-based training (Morrison et al., 2003). Ongoing formative evaluation is also an important feature of most contemporary "agile" approaches to enterprise (i.e., large-scale) software development such as the RUP (Kroll & Kruchten, 2003). Finally, formative evaluation is a key element in the New Products Management framework, which is commonly used to guide the development of innovative products ranging from technical devices and software, to automobiles and consumer goods (Crawford & DiBenedetto, 2003). In the section that follows, we provide a brief description of these seemingly disparate models and explore how they converge on the importance of formative evaluation.

Instructional Systems Design (ISD). ISD refers to a systematic process of creating, developing, and testing instructional materials to effectively facilitate learning (Dick & Carey, 1996; Gagne, Briggs, & Wagner, 1992; Morrison et al., 2003). Although there are some minor variations in ISD models, they share the core elements of analysis, design, development, implementation, and evaluation (or ADDIE, see Figure 12.1; see Weingardt, 2004, for more detailed discussion). For present purposes, the most important feature of this model is the relationship between evaluation and the other activities. Note that evaluation activities inform each stage of the instructional design process, from analysis through implementation.

ADDIE Model

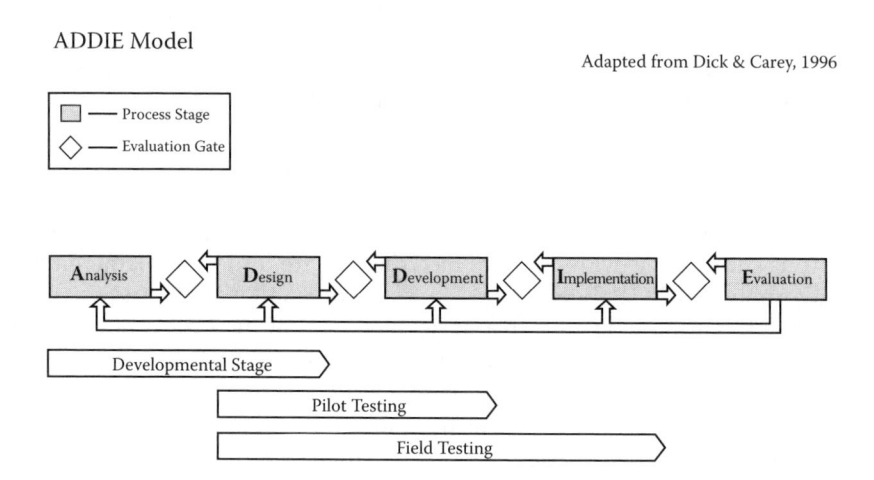

Adapted from Dick & Carey, 1996

Figure 12.1 Formative evaluation in the ADDIE Instructional Systems Design model.

The progenitors of the ADDIE framework developed a three-stage model to describe the various applications of formative evaluation (FE) in their model of ISD. The first stage, which occurs toward the beginning of the process, consists of an individual interview known as *developmental testing* (Thiagarjan, Semmel, & Semmel, 1974). In developmental testing, the instructional designer tries out the instruction with individual learners (Brenneman, 1989; Flagg, 1990). The goal at this stage is to obtain early feedback regarding the impact, clarity, and overall appeal of the instruction. The second stage of this approach to FE typically consists of pilot tests in which a more developed version of the instruction is reviewed. These groups of 8 to 25 people are asked to provide observational, attitudinal, and performance data regarding the instructional materials. The evaluator integrates the group feedback into a final form of the instruction, which is then tested in a field trial (stage 3). The field trial then evaluates the instruction with a full-sized group under real-world conditions (Morrison et al., 2003).

Our team has followed the ISD process in developing technology-based training for nurses (http://www.detoxguideline.org; Weingardt & Villafranca, 2005), and substance abuse counselors (http://www.nidatoolbox.org; see Weingardt, Villafranca, & Levin, 2006; Weingardt, Cucciare, Bellotti, & Lai, in press). At each phase of the process we obtained feedback on our work in progress from a variety of different stakeholders. Prior to the formal evaluation (in this case an RCT), we conducted developmental and pilot testing with small samples of the target population, and conducted several rounds of review on the course content and its presentation by asking subject matter experts to review the prototype materials and provide feedback for improvement.

Iterative software development

The waterfall model of enterprise (large-scale) software development sees development as flowing steadily downward (like a waterfall) through a fixed sequence of discrete phases, from initial conception, initiation, analysis, design (validation), construction, and testing to maintenance. In this model, a software development effort must proceed from one phase to the next in a rigid series. For example, design and development cannot begin until the requirements (a detailed functional description of the planned system) are completely specified and approved by all project stakeholders. The requirements document is thus intended to serve as a blueprint for the programmers to follow. Interaction between those who drew the blueprint and those building the application is intentionally kept to a minimum. Large government agencies (e.g., U.S. Department of Defense, NASA, Department of Veterans Affairs) have historically used the waterfall model.

The waterfall process frustrated many developers, as it provided no means of adapting to changing requirements or responding to new information as it is discovered. Furthermore, clients often had a very difficult time of thinking through all the possible requirements up front, and could not answer important design questions until they were able to review a working prototype. A new generation of so-called agile software development processes in reaction to the limitations of the waterfall process. Agile methodologies generally promote a project management process that encourages frequent inspection and adaptation; in other words, they embrace the importance of formative evaluation.

The Rational Unified Process (RUP) is an agile software engineering process framework developed by the Rationale Software division of IBM (Kroll & Kruchten, 2003; http://www.ibm.com/developerworks/rational/library/05/wessberg/). RUP is essentially a set of evidence-based best practices for software developers (think clinical guidelines). Interestingly, many of the best practices in the RUP were identified by a group of expert software developers conducting a detailed review of the root causes responsible for previous failed software projects. This analysis found that many failures were attributable to the waterfall process where each phase is completed in sequence. In contrast, the RUP and other agile development processes use an *iterative* approach. *Iterate* means literally "to repeat," and iteration is defined as "a procedure in which repetition of a sequence of

The Rational Unified Process (RUP)

Developed by IBM

PHASES	Inception	Elaboration		Construction			Transition	
INTERATIONS	Initial	#1	#2	#1	#2	#N	#1	#2
Business Modeling								
Requirements								
Analysis & Design								
Implementation								
Test								
Deployment								
Configuration & Change Management								
Project Management								
Environment								

(DISCIPLINES on left axis)

Figure 12.2 Formative evaluation in the Rational Unified Process.

operations yields results successively closer to a desired result" (Iteration, 2009). In software, this means that development proceeds in a sequence of incremental steps or iterations.

As depicted in Figure 12.2, each iterative development cycle involves all of the various related professional disciplines (e.g., requirements analysis, design, development) and results in the deployment of a partial working prototype of the final system. During each cycle or iteration, the prototype is subjected to extensive testing that evaluates everything from user acceptance of a new interface design, to performance testing, which measures the demands that the prototype places on the technical infrastructure (e.g., server capacity, processing speed, and bandwidth). The results of each cycle of tests are used to inform the development of the next version or prototype. In this manner, each successive iteration builds upon the work of previous iterations to evolve and refine the system (Kroll & Kruchten, 2003).

The shift away from the waterfall process, and toward more adaptive or agile development processes, reflects an increased appreciation of the important role of formative evaluation in software development. This field has clearly embraced periodic feedback and iterative review as an industry best practice. The next section briefly describes how the New Products Management framework has also come to place a similar emphasis on formative evaluation and iterative feedback in the development of innovative new products.

Chapter twelve: Evaluation 257

New product management

Variously referred to as product planning, product innovation management, or New Product Development (Kahn, 2004) the New Product Management framework (NPM; Crawford & DiBenedetto, 2003) is widely used in business and engineering to guide the development of a dizzying array of products in virtually all sectors of the economy. This includes such consumer products as groceries, electronics, and automobiles, but also many less tangible products in such industries as financial services, technology, insurance, and communications. There are generally five phases in the new products process: (1) opportunity identification and selection; (2) concept generation; (3) concept evaluation; (4) development; and (5) launch (Crawford & DiBenedetto, 2003, Chapter 2).

As illustrated in Figure 12.3, the New Product Management process incorporates formative evaluation throughout, from the very idea of a new product through to its successful launch. Note the multiple evaluation gates indicated by diamond shapes in the figure. This term comes from the Stage-Gate Process and refers to evaluation tasks that must occur between stages (Cooper, 1993). Evaluation gates typically include the Concept Test, which is used to determine if the intended user really needs the proposed product; the Product-in-Use Test, which is used to see if the item developed actually meets that need; and the Market Test, which is used to see if the marketing plan is effective. In industry, each of these evaluation gates is often considered a "go/no go" decision regarding the product's future. If the prototype fails to pass through any of the required gates, it may be

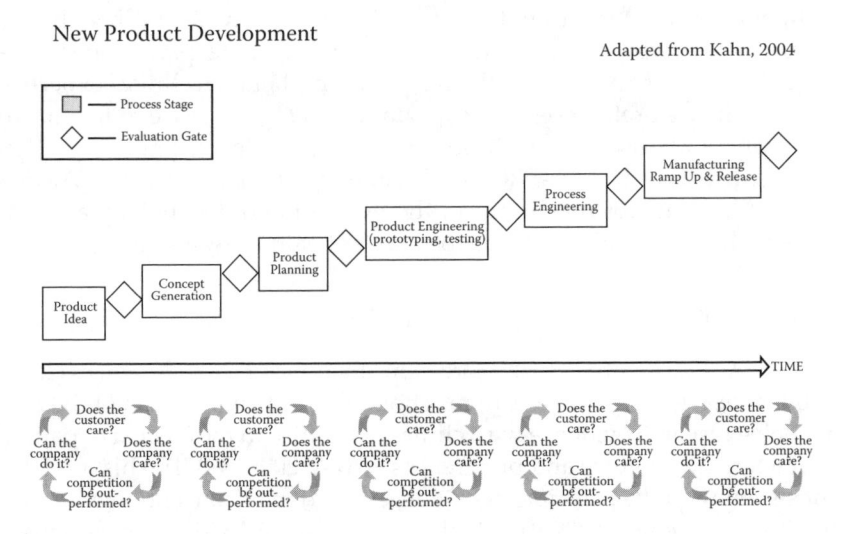

Figure 12.3 Formative evaluation in the New Products Management framework.

dropped and resources diverted to the development of other, potentially more promising new products. As in the RUP process, each of the relevant professional disciplines involved in New Products Management is engaged to varying degrees in all phases of development.

An important common element linking these seemingly disparate models is the critical and central role of formative evaluation. Whatever the product, intervention, or endeavor, it is clearly a good idea to test early and test often. However, putting the intervention in front of your potential students, clients, patients, or customers and repeatedly asking them "Is this it?" "How about now?" and acting on their feedback requires a tremendous amount of financial and institutional support, not to mention resilience and cognitive flexibility. This can be difficult to do under the best of circumstances. As we discuss in the conclusion of this chapter, grant-funded research programs may constitute a particularly challenging environment in which to embrace a genuinely formative evaluation approach.

Evaluation of technology applications in context

Social psychologists have long exhorted our field to embrace the importance of context in understanding human behavior. For example, Kurt Lewin's influential Field Theory (Lewin, 1951) explains behavior as a function of the "field of forces" that arises from the interaction of the person and his or her social and physical environment. In this section, we discuss the importance of attending to the interpersonal, social, organizational, and technical context in evaluating the feasibility and effectiveness of new technologies to support evidence-based practices in mental health. We begin at the micro level, where we consider an individual practitioner striving to implement EBPs within the context of his/her own clinical expertise and the client's preferences and expectations. We then move to the macro level, where we discuss how various emerging models of implementation science such as PRECEDE-PROCEED (Green & Kreuter, 2005) and PARIHS (Rycroft-Malone, 2004) can inform the evaluation of technology applications in the context of the clinic, hospital, or health care system.

Evaluation in the context of the individual practitioner

The APA Presidential Task Force on Evidence-Based Practice (2006) defines *evidence-based practice* in psychology as "the integration of the best available research evidence with clinical expertise in the context of patient characteristics, culture and preferences" (APA, 2006, p. 271). This model is consistent with the principle of evidence-based medicine endorsed by the Institute of Medicine (2001), which conceptualizes evidence-based practice as a stool resting upon three "legs": (a) research evidence, (b) clinical expertise of the health professional, and (c) the unique preferences and expectations of the client (APA, 2006; Spring, 2007).

Evidence-based practice does not stand on research alone, but rather on the application of specific research evidence within the context of a therapeutic relationship with an individual client or group of clients. From this perspective, evaluation of a particular intervention (technology based or otherwise) must begin and end with a sound understanding of the interpersonal context in which it is deployed. Just as manual-based therapies are not equally appropriate for all clients who meet a certain set of diagnostic criteria, technologies to support evidence-based practices are unlikely to be universally applicable to broad categories of clients. Mental health professionals must rely upon their training and experience, as well as their understanding of the desires and capabilities of the individual client, in their decision to incorporate a particular technology into their practice.

Perhaps the most context-sensitive evaluation design that can be used by clinician is a single subject, multiple-time-series design. In this design, which has long been advocated by proponents of cognitive-behavioral therapy (CBT), clients regularly complete objective, standardized outcome measures and submit them to the therapist who tracks them over time and uses them to continuously monitor the client's response to treatment (e.g., Lambert et al., 2003). The systematic collection of outcomes data provides the evidence-based practitioner with perhaps the best measure of whether his/her interventions are helping a client within the context of his or her individual circumstances. The process of obtaining regular outcomes data and using this clinical feedback to inform the course of therapy can be understood as a type of formative evaluation. Much as the developers of instructional technology, software, and new products use iterative feedback to refine and evolve prototype systems, therapists can use formative evaluation of a client's symptoms and level of functioning to continuously adjust and improve the process of treatment. Such ongoing evaluation may be particularly helpful in the context of a clinician who has decided to implement a new technology application into his or her practice. Regular feedback about the client's level of symptoms and functioning might be combined with an ongoing collaborative discussion about ways in which the technology application is helpful and ways in which it can be improved.

Evaluation in the context of implementation science

Much as evaluations of client progress must attend to the individual's interpersonal, social, and environmental context, efforts designed to promote the systematic update and sustained adoption (i.e., implementation) of evidence-based practices in routine clinical care must also attend to the larger organizational and interpersonal context in which such efforts occur. As this section illustrates, ongoing evaluation plays a central role in several of the conceptual frameworks that are widely used in the nascent field of implementation science.

Researchers who have proposed the Promoting Action on Research Implementation in Health Services framework (or PARIHS model; Rycroft-Malone et al., 2002) argue that consistent use of feedback for both providers and teams engaged in new clinical practices is critical in supporting the implementation of new practices. The PARIHS framework consists of three main elements (i.e., evidence, context, and facilitation) that, when maximized, promote the successful implementation of new practices. The second main element or context has many sub-elements including the process of evaluation. Rycroft-Malone (2004) argues that evaluation plays a critical role in the implementation process in that it provides opportunities to frequently measure the impact of new practices on outcomes of interest (e.g., are the change processes effective for patients, providers, and overall clinic operations?). The role of evaluation in the PARIHS models is broad and may include the assessment of a wide variety of factors related to not only the impact of new clinical practices (e.g., on patient outcomes), but information on the effectiveness of provider training, efforts to improve work flow (e.g., to enhance the efficiency of a clinic or practice), and the degree to which providers are engaging in new practices (e.g., sustained adoption).

The PRECEDE-PROCEED implementation framework offers a "packaged" approach to the implementation process by first providing a strategy for identifying health concerns in a population of individuals and (if necessary) developing new clinical practices to address those concerns. This part of the implementation process is packaged in the PRECEDE or (or Predisposing, Reinforcing, and Enabling Constructs in Educational Diagnosis and Evaluation) portion of the framework, while the second portion of this framework (PROCEED or Policy, Regulatory, and Organizational Constructs in Educational and Environmental Development) provides a model for the implementation of new programs that have been deemed necessary based on the diagnostic information provided by PRECEDE (Green & Kreuter, 2005).

Similar to the broad role of evaluation in the PARIHS framework, PROCEED focuses on evaluating various aspects of the implementation process ranging from macro-level concerns (e.g., is the implementation process effective?) to more micro-level questions such as "Is the new clinical practice effectively addressing the health issues identified in PRECEDE?" and "What is the impact of the new interventions on patients' quality of life?" For more discussion of the PARIHS and PRECEDE-PROCEED implementation frameworks, see Chapter 11 in this volume.

Discussion

The evaluation of technology applications to support evidence-based behavioral health interventions presents many challenges. As in any area of inquiry, researchers in this area must balance the need for judicious use of random assignment and experimental control to establish causality (internal validity), with the need to ensure that whatever results are discovered can generalize to the world outside the study (i.e., external validity). If researchers overemphasize internal validity, for example, by conducting a randomized control trial that includes only those participants who have a single psychiatric diagnosis or those who have a stable residence with high-speed Internet access, they may inadvertently be excluding some of the most vulnerable individuals from their study (Humphreys, Weingardt, & Harris, 2007). On the other hand, if researchers place relatively greater emphasis on external validity, for example, by recruiting a sample that includes all of the individuals who are seeking treatment, or by using a less rigorous quasi-experimental design, they may end up with results that generalize to the real world, but not be able to rule out alternative explanations of the results and allow the conclusion that the intervention caused any observed improvement in patient outcomes.

Another significant challenge for those evaluating technologies designed to support evidence-based behavioral health practices is the need for a process of ongoing formative evaluation. As our review of the dominant conceptual models used to guide the development of online learning (e.g., Instructional Systems of Design model), enterprise software (e.g., Rational Unified Process), and new products (e.g., New Products Management) concludes, iterative review and feedback are essential to successful development of new technologies. Although the importance of formative evaluation has been embraced throughout industry, several forces conspire to make such formative evaluation efforts particularly challenging in the context of grant-funded academic research in mental health.

The conceptual model that has been used to guide much of research on technologies to support evidence-based mental health practices may constitute one such factor. This model is variously referred to as the Stage Model of Psychotherapy Development (Onken, Blaine, & Battjes, 1997), the Psychotherapy Technology Model (Morgenstern & McKay, 2007), or the Systematic Replication Model (Hayes, 2002). This model dictates that researchers move sequentially from stage 1, which demonstrates the efficacy of an intervention in a randomized controlled trial, to stage 2, which tests the effectiveness of the intervention in the real world, and finally to stage 3, which focuses on the large-scale dissemination of the intervention to improve public health.

The overall process is unidirectional and assumes that the theoretically derived "active ingredients" of the intervention captured in the protocol of a stage 1 efficacy trial must be retained as the intervention moves through the remaining stages. According to this model, technology interventions that have been demonstrated as efficacious in a stage 1 study must not be substantially revised (formatively evaluated) in subsequent stages, as that would compromise the empirical support for the intervention. Similarly, the model dictates that technologies evaluated in stage 3 as vehicles for implementing a set of evidence-based practices (e.g., an online clinical training course or an online guided self-help program for clients) must support both the content and process of the original intervention. Substantial departure from the original, efficacious model is strongly discouraged at the implementation phase as it is thought to undermine the evidence base. As a result, researchers have relatively little opportunity to engage in the very ongoing, substantive, formative evaluation activities that have been identified as such a critical element of successful technology development efforts.

A related challenge to the widespread and systematic use of formative evaluation focuses on the allegiance of the investigators who conduct much of this research. Allegiance in the context of treatment outcome research is a belief in the superiority of a treatment (Leykin & DeRubeis, 2009). Several meta-analyses of treatments for a variety of different disorders have found a statistically significant association between the direction and magnitude of treatment outcome findings and the measured allegiance of the investigators (e.g., Luborsky et al., 1999, 2002). Given that researchers often devote years, if not their entire careers, to developing and validating a particular treatment, the finding that they may not be the most objective evaluators of their own interventions is hardly surprising. An unfortunate consequence of this allegiance may be that researchers are reluctant to solicit and attend to formative evaluation data that may substantially alter the fundamental structure, content, or process of the intervention they are evaluating.

A third challenge facing those evaluating technologies to support evidence-based behavioral health practices is the importance of attending to the context in which the technology is being implemented. Ideally, our evaluation efforts should attend to many levels of context ranging from the level of the individual client–therapist relationship, to the level of the organizational, social, and environmental context in which these relationships occur. On the micro level, this context includes the background, experience, and technical skills of both therapist and client, as well as the technology resources available to them (computer hardware, high-speed

Internet connection, handheld devices, software applications). On the macro level, the evaluation context includes leadership support for implementing the new technology, the personnel and financial resources that are available, and the degree to which the social or organizational care setting supports the implementation effort. The PARIHS and PRECEDE-PROCEED implementation frameworks can provide helpful guidance about how to practically conduct evaluations that are sensitive to these important contextual factors.

In conclusion, this volume describes myriad technology applications that strive to improve the lives of individuals with mental health issues. Ultimately, the true measure of their worth is the degree to which these interventions succeed in helping the individuals who use them to achieve behavioral goals, reduce psychiatric symptoms, and improve quality of life. Our hope is that this chapter has provided the reader with a more informed basis from which to evaluate the scientific literature in this area, has helped the reader to appreciate the importance of formatively evaluating new behavioral health technologies during all phases of the design and development process, and to appreciate the importance of understanding the context in which a new technology is being implemented.

References

APA Presidential Task Force on Evidence-Based Practice (2006). Evidence-based practice in psychology. *American Psychologist, 61*, 271–285.

Babor, T. F., Higgins-Biddle, J. C., Saunders, J. B., & Monteiro, M .G. (2001). *The alcohol use disorders identification test: Guidelines for use in primary care* (2nd ed.). Geneva, Switzerland: World Health Organization.

Beck, A. T., Steer, R. A., & Brown, G. K. (1996) *Manual for the Beck Depression Inventory-II*. San Antonio, TX: Psychological Corporation.

Bhola, H. S. (1990). Evaluating *"Literacy for Development" projects, programs and campaigns: Evaluation planning, design and implementation, and utilization of evaluation results*. Hamburg, Germany: UNESCO Institute for Education; DSE (German Foundation for International Development).

Bordens, K. S., & Abbott, B. B. (2008). *Research design and methods: A process approach* (7th ed.). Boston: McGraw-Hill.

Brenneman, J. (1989). When you can't use a crowd: Single subject testing. *Performance and Instruction, 28*, 22–25.

Campbell, D. T., & Stanley, J. C. (1968). *Experimental and quasi-experimental designs for research*. Chicago: Rand McNally & Company.

Cooper, R. G. (1993). *Winning at new products: Accelerating the process from idea to launch*. Reading, MA: Addison-Wesley.

Crawford, C. M., & DiBenedeto, C. A. (2003). *New products management* (7th ed.). New York: McGraw-Hill Higher Education.

Dick, W., & Carey, L. (1996*). The systematic design of instruction* (4th ed.). New York: Harper Collins.

Flagg, B. N. (Ed). (1990). *Formative evaluation for educational technologies.* Hillsdale, NJ: Erlbaum.

Gagne, R. M., Briggs, L. J., & Wagner, W. W. (1992). *Principles of instructional design* (4th ed.). New York: Harcourt Brace Janovich College Publishers

Glasgow, R. E., Green, L. W., Klesges, L. M., Abrams, D. B., Fisher, E. B., Goldstein, M. G., et al. (2006). External validity: We need to do more. *Annals of Behavioral Medicine, 31*(2), 105–108.

Green, L. W., & Glasgow, R. E. (2006). Evaluating the relevance, generalization and applicability of research: Issues in external validity. *Evaluation & the Health Professions, 29*(1), 126–153.

Green, L. W., & Kreuter, M. W. (2005) *Health promotion planning: An educational and environbehavioral approach* (4th ed.). Mountain View, CA: Mayfield.

Hayes, S. C. (2002). Getting to dissemination. *Clinical Psychology: Science and Practice, 9*(4), 410–415.

Humphreys, K., Weingardt, K. R., & Harris, A. H. S. (2007). Influence of subject eligibility criteria on compliance with national institutes of health guidelines for inclusion of women, monitories and children in treatment research. *Alcoholism: Clinical and Experimental Research, 31*(6), 988–994.

Iteration. (2009). In Merriam-Webster Online Dictionary. Retrieved July 19, 2009, from http://www.merriam-webster.com/dictionary/interation

Kahn, K. B. (2004). *The PDMA handbook of new product development* (2nd ed.). New York: John Wiley & Sons.

Kahneman, D., & Tversky, A. (1983). Choices, values and frames. *American Psychologist, 39,* 341–350.

Kroll, P., & Kruchten, P. (2003). *The Rational Unified Process made easy: A practitioner's guide to the RUP.* Boston, MA: Pearson Education.

Lachin, J. M., Matts, J. P., & Wei, L. J. (988). Randomization in clinical trials: Conclusions and recommendations. *Controlled Clinical Trials, 9*(4), 365–74.

Lambert, M. J., Whipple, J. L., Hawkins, E. J., Vermeersch, D. A., Nielsen, S. L., & Smart, D. W. (2003). Is it time for clinicians to routinely track patient outcomes? A meta-analysis. *Clinical Psychology: Science and Practice, 10*(3), 288–301.

Lewin, K. (1951). *Field theory in social science: Selected theoretical papers by Kurt Lewin.* D. Cartwright (Ed.). New York: Harper & Row.

Liebert, R. M., & Liebert, L. L. (1995). *Science and behavior: An introduction to methods of psychological research.* Englewood Cliffs, NJ: Prentice Hall.

Luborsky, L., Diguer, L., Seligman, D. A., Rosenthal, R., Krause, E. D., Johnson, S., et al. (1999). The researchers' own therapy allegiances: A "wild card" in comparisons of treatment efficacy. *Clinical Psychology: Science and Practice, 6,* 95–106.

Luborsky, L., Rosenthal, R., Diguer, L., Andrusyna, T. P., Berman, J. S., Levitt, J. T., et al. (2002). The dodo bird verdict is alive and well—mostly. *Clinical Psychology: Science and Practice, 9,* 2–12.

Marks, I. M., Cavanagh, K., & Gega, L. (2007). *Hands-on help: Computer-aided psychotherapy.* New York: Psychology Press.

Morgenstern, J., & McKay, J. R. (2007). Rethinking the paradigms that inform behavioral treatment research for substance use disorders. *Addiction, 102*(9), 1377–1389.

Morrison, G. R., Ross, S. M., & Kemp, J. E. (2003). *Designing effective instruction* (4th ed.). New York: John Wiley & Sons.

Onken, L. S., Blaine, J. D., & Battjes, R. N. (1997). Behavioral therapy research: A conceptualization of a process. In S. W. Henggeler & A. B Santos (Eds.), *Innovative approaches for difficult-to-treat populations* (pp. 477–485). Washington, DC: American Psychiatric Press.

Patton, M.Q. (1979). Evaluation of program implementation. *Eval Stud Rev Annu, 4*, 318–345.

Rosnow, R. L., & Rosenthal, R. (2008). *Beginning behavioral research: A conceptual primer* (6th ed.). Englewood Cliffs, NJ: Pearson/Prentice Hall.

Rossi, P., & Freeman, H. (1993) *Evaluation: A systematic approach*. Newbury Park, CA: Sage Publications.

Rycroft-Malone, J. (2004). The PARIHS framework—A framework for guiding the implementation of evidence-based practice. *Journal of Nursing Care Quality, 19*(4), 297–304.

Rycroft-Malone, J., Kitson, A., Harvey, G., McCormick, B., Seers, K., Titchen, A., et al. (2002). Ingredients for change: Revisiting a conceptual framework. *Quality and Safety in Health Care, 11*, 174–180.

Shadish, W., Cook, T., and Campbell, D. (2002). *Experimental and quasi-experimental designs for generalized causal inference*. Boston: Houghton Mifflin.

Spring, B. (2007). Evidence-based practice in clinical psychology: What it is, why it matters, what you need to know. *Journal of Clinical Psychology, 63*, 611–631.

Stetler, C. B., Legro, M. W., Wallace, C. M., Bowman, C., Guihan, M., Hagedorn, H., et al. (2006). The role of formative evaluation in implementation research and the QUERI experience. *Journal of General Internal Medicine, Suppl 2*, S1–S8.

Thiagarjan, S., Semmel, D., & Semmel, N. (1974). *Instructional development for training teachers of exceptional children: A sourcebook*. Bloomington, IN: Center for Innovation in Teaching the Handicapped.

Weathers, F. W., Litz, B. T., Herman, J. A., Huska, J. A., & Keane, T. M. (1993). *The PTSD Checklist (PCL): Reliability, validity and diagnostic utility*. Paper presented at the 9th annual conference of the International Society for Traumatic Stress Studies, San Antonio, TX.

Weingardt, K. R. (2004). The role of instructional design and technology in the dissemination of empirically supported, manual-based therapies. *Clinical Psychology: Science & Practice, 11*(3), 313–331.

Weingardt, K. R., Cucciare, M. A., Bellotti, C., & Lai, W. P. (In press). A randomized trial comparing two models of Web-based training in cognitive behavioral therapy for substance abuse counselors. *Journal of Substance Abuse Treatment*.

Weingardt, K. R., & Villafranca, S. W. (2005). Translating research into practice: The role of Web-based education. *Journal of Technology in Human Services, 23*(3/4), 259–273.

Weingardt, K. R., Villafranca, S. W., & Levin, C. (2006). Technology-based training in cognitive-behavioral therapy for substance abuse counselors. *Substance Abuse, 27*(3), 19–25.

chapter thirteen

Ethics in technology and mental health

Elizabeth Reynolds Welfel and Kathleen (Ky) T. Heinlen

The era in which Hans Eysenck (1952) concluded that psychotherapy was an insignificant influence on client functioning is long past. The last fifty years of rigorous research into psychotherapy efficacy and effectiveness have unequivocally demonstrated that therapy can be a powerful intervention for meaningful and long-lasting change in both the symptoms of psychological distress and general life satisfaction (for example, Lambert & Ogles, 2004; Seligman, 1995). Unfortunately, the evidence also unmistakably shows that psychotherapy produces significant deterioration effects in as many as 5 to 10% of those who seek it (Lambert & Ogles).

The therapist factors that influence whether a particular client is helped or harmed are numerous and include the accuracy of the professional's assessment and treatment of the problem, the sensitivity with which the professional manages the therapeutic alliance and the ruptures that occur within it, the degree to which the professional operates within his or her boundaries of competence and ethical standards for practice, and the professional's use or rejection of extreme and untested therapeutic modalities. Some deterioration effects are the result of egregious ethical violations such as sexual misconduct or the use of outrageous interventions such as rebirthing therapy, but others have been caused by well-intentioned professionals eager to benefit their clients—so eager, in fact, that their choices of assessments and interventions were not subject to sufficient reflection, ongoing evaluation of client progress, or attention to the evidence (or lack thereof) related to their use.

Well-intentioned professionals who continually seek better ways to relieve the suffering of their clients are the driving force behind the creation of many effective approaches to psychotherapy; their motivation to be effective clinicians is admirable, but at the same time, their very drive to find innovative methods to reach the most people and help them as quickly as possible has its inherent dangers. When that altruistic motivation is not grounded in a scientific frame of mind that attends to research evidence and the theoretical underpinnings of the innovation, it can backfire. And when the motivation to explore innovative care is not fundamentally altruistic but is driven instead by self-interest, the risk of harm is multiplied.

The development of online behavioral health services has been generated primarily by two forces—public interest in online health information (Madden & Fox, 2006) and professionals seeking to offer services to underserved populations, to provide more consistent aftercare to patients no longer receiving intensive services, and to provide some service to those unable or unwilling to attend traditional service (Maheu & Gordon, 2000). These are admirable goals worthy of the full attention of the mental health professions. And as much of this book demonstrates, preliminary evidence of the benefits of online behavioral health interventions is encouraging. They do indeed hold the promise to reach underserved populations in timely, efficient, and creative ways. However, because our history is fraught with well-intentioned interventions that were unintentionally harmful, we need to pause and reflect on the ethical dimensions of online behavioral health and identify strategies and guidelines to reduce the risk of harm.

If these modes of care are to realize their potential, online professionals also need to provide clearer evidence of their merits and limitations. They need to move beyond the rational analysis of why these modes of care ought to be beneficial and engage in rigorous research that objectively evaluates whether those potential benefits are being realized. They also need to "unbundle" online behavioral health and see the variety in the phenomena—evidence of the effectiveness of videoconferencing or online monitoring of face-to-face clients during or after therapy or hospitalization may not automatically translate into evidence of effectiveness of therapy or assessment via e-mail or chat.

The current chapter highlights the ethical issues in the various forms of online behavioral health, identifies their levels of risk, and reviews the guidelines the professions have offered to reduce those risks. Specifically, the chapter addresses ethical issues inherent in the electronic medium itself (such as the possibility of technology failure), and goes on to discuss ethical dimensions in online interventions that supplement face-to-face services. The chapter ends with an analysis of the ethical dimensions of online consultation and supervision activities for psychotherapists. Our aim is to encourage innovation based on the highest scientific standards of research and practice, not to dissuade professionals from exploring the potential benefits of this medium. The chapter refers to the codes of ethics of the American Medical Association (2001), American Psychological Association (2002), and American Counseling Association (2005), the nonbinding ethical standards that have been adopted by the International Society for Mental

Health Online (ISMHO, 2000), and the Ohio Psychological Association (2008), in addition to the writings and research of ethics scholars.

Ethical implications inherent in the technology

Consider the following case:

> Janine, a psychotherapist and former Army nurse, runs a biweekly support group for some of the combat veterans she works with in private practice. Between sessions, Janine offers e-mail exchanges with interested members to help them cope with issues that arise between face-to-face meetings. Most members send at least occasional e-mails to her. Evan, a veteran of the Afghanistan war who lost part of his right foot in combat, is one of the most frequent e-mail users. He has struggled with post-traumatic stress and marital conflict, but has coped well enough to stay sober for 2 years. One day Evan sends Janine an e-mail in which he explains that he is extremely distraught because his best friend has been seriously injured in a helicopter accident on his California army base. Janine does not receive that e-mail or the others Evan sends over the next 24 hours until nearly three days later. The cause of the delay was a server error of which Janine was unaware. Evan gets so frustrated that he quits the group, provokes a major disagreement with his wife, and begins drinking again to deal with the stress. His wife leaves him. When Janine calls Evan to find out why he failed to come to group, Evan hangs up the phone. Two days later he also trashes the letter she has mailed him, explaining the problem. She never hears from Evan again.

This case illustrates the ethical complications that arise from the use of the Internet as a mode for therapeutic contact. One of the commonly noted issues is the real possibility that technology will fail at crucial points—hard drives crash, e-mails get delayed, or human error intervenes when e-mails are inadvertently deleted or sent to the wrong address. What is particularly disconcerting is that these technology glitches are not always readily apparent to the user—the sent e-mail appears to have been delivered when that did not happen, as the case of Evan so vividly demonstrates, or the e-mail gets sent to a group rather than to an individual. Technology problems are not limited to the Internet; nearly everyone with a facsimile machine has received a fax intended for someone else. Anecdotal reports abound of therapists receiving faxed insurance documents with private health information about people who are not their clients. In addition, many students and professionals have been participating in distance learning or videoconferencing only to have the connections lost and the class time wasted. The *Telepsychology Guidelines* of the Ohio Psychological Association (OPA, 2008) address this problem with the following language:

3.0 INFORMED CONSENT AND DISCLOSURE

Psychologists inform clients about potential risks of disruption in the use of telepsychology, clearly state their policies as to when they will respond to routine electronic messages, and in what circumstances they will use alternative communications for emergency situations. Given the 24-hour, 7-day-a-week availability of an online environment, as well as the inclination of increased disclosure online, clinical clients may be more likely to disclose suicidal intentions and assume the psychologist will respond quickly (supplements APA Ethics Code Sec. 4.05).

Other standards that specifically advise professionals to inform clients about the possibility of technology failure are the *Code of Ethics* of the American Counseling Association (ACA, 2005) and *Standards for the Ethical Practice of Internet Counseling* of the National Board of Certified Counselors (NBCC, 2005). These codes use language similar to the wording in the OPA *Telepsychology Guidelines*.

Little research exists regarding the frequency of such technology problems in online behavioral health services, but in several studies of the content of e-therapy Web sites, researchers found low levels of compliance with these recommendations and standards (Heinlen, Welfel, Richmond, & O'Donnell, 2003; Heinlen, Welfel, Richmond & Rak, 2003a; Maheu & Gordon, 2000; Recupero & Rainey, 2006). In Heinlen, Welfel, Richmond, and Rak (2003), only 3% of their sample of 87 e-therapy sites operated by mental health professionals explicitly mentioned the possibility of technology failure. In a related study of 44 Web sites run by psychologists, Heinlen, Welfel, Richmond, and O'Donnell (2003) found that only 27% mentioned the risks of e-therapy, of which technology failure is but one.

Chapter thirteen: Ethics in technology and mental health 271

Research by Maheu and Gordon (2000) also suggests low compliance since less than half (48%) of their sample of 56 online practitioners indicated that they obtained informed consent prior to providing e-therapy services to clients. Physicians who use e-mail with patients appear to demonstrate the same disappointing levels of compliance with professional guidelines for informed consent for physician–patient e-mail communications (Brooks & Menachemi, 2006; Gaster et al., 2003).

Risks of unauthorized access

The possibility of unauthorized access, either from hackers or from individuals who share the client's or clinician's computer, is another ethical risk in using technology. The probability of unauthorized access from individuals in cyberspace is relatively low (see staysafeonline.org) especially when those using the Internet employ encryption software and passwords. Internet users should also understand that Internet service providers (ISP) may gain access to e-mail messages (Center for Democracy and Technology, 2004). However, little evidence suggests that the procedures of encryption, password protection, or the disclosure of risks of unauthorized access are occurring. Only 27% of the psychology sites in Heinlen, Welfel, Richmond, and O'Donnell (2003) stated that they used encryption or password protection, a figure consistent with their broader sample of mental health professionals (Heinlen, Welfel, Richmond, & Rak, 2003) in which 22% employed this security device for content of e-mail and chat transmissions. Only 16% of this sample disclosed the risks of an ISP threatening confidentiality. Interestingly, in both samples virtually all sites used encryption and related security measures for the transmission of credit card information. The ethical guidelines and recommendations of ethics scholars do not make such a distinction between financial and personal data, and they advise that encryption and passwords be used for all communications whenever possible. The American Medical Association (AMA) *Guidelines for Physician-Patient Electronic Communication* (2004) also direct that if patients refuse encryption, the risks of that choice should be disclosed prior to online communication.

Because the computers that many people use are shared with family, with coworkers, or with a larger group (such as other students at a university computer lab or other patrons of a public library), computer-based communications require special care if privacy is to be maintained. For example, a workplace computer is never a safe entry point since by law, all information on a work computer belongs to the employer. Moreover, many employers regularly monitor communications from office computers to ensure that employees are doing their jobs. (Several training programs for computer specialists offer courses on "ethical hacking" for the very purpose of monitoring employee e-mails.) Consumers entertaining the option of e-mail with professionals, even about mundane matters such as rescheduling appointments, should understand that an office computer is inappropriate if they wish to maintain the confidentiality of their behavioral health services.

Home-based computers offer more protection, since entry to any person's e-mail is usually password protected. However, not all data fall under this protection. When a computer has multiple users, the opening screen usually allows any user to click on any other user's window. With a second click of the mouse, that person can typically view the sites the other person has bookmarked or the sites recently visited. Only a password-protected screen saver prevents such access. Many people use passwords that people who know them can easily deduce. Heinlen, Welfel, Richmond, and O'Donnell (2003) found that only 16% of psychologists with e-therapy sites mentioned the risks to privacy from family or coworker access to the client's computer. No data have been collected regarding privacy protection issues when the therapist's computer is shared by others. The OPA guidelines (2008) address this issue:

> 5. ACCESS TO AND STORAGE OF COMMUNICATIONS
>
> Psychologists inform clients about who else may have access to communications with the psychologist, how communications can be directed to a specific psychologist, and if and how psychologists store information. (p. 6)

Risks of misunderstanding

Online behavioral health services vary considerably in the degree to which they mirror traditional practice. Online services in which participants can see and hear each other closely approximate traditional face-to-face care, while services delivered online via asynchronous e-mail exchanges vary more dramatically. Professionals utilizing technology as an adjunct to traditional services may find it necessary to clarify communication conducted in this medium to avoid the loss of contextual cues commonly present in previous communication. Those with lower levels of computer literacy or less experience in using the computer are likely to be especially vulnerable to misunderstanding a message sent to them or to miscommunication of their own ideas. Individuals who are ignorant of the meaning of *emoticons*, the symbols for emotions used in text communications, are especially vulnerable to misinterpreting a message. In fact, professional guidelines direct professionals to disclose this particular risk to consumers (ACA *Code of*

Ethics, 2005, A.12.g, A.12.h; ISMHO, 2000, Principle 1.a) and to inform them that such misunderstanding is even possible in videoconferencing due to limited bandwidth. Again, research reveals that only 14% of psychologists' Web sites mentioned this difficulty (Heinlen, Welfel, Richmond, & O'Donnell, 2003). Whether the communication difficulties in text-based communication are heightened when cultural differences are involved is an open question that no research has yet explored, unfortunately.

Data from those who have experienced e-therapy are mixed. Two studies of client attitudes toward this medium suggests some clients see the limits of text-based communications as frustrations in the development of the therapeutic alliance and as a reason that they prefer face-to-face interactions (Liebert, Archer, Munson, & York, 2006; Rochlen, Land, & Wong, 2004), but other research reports that online clients actually experienced a stronger therapeutic alliance than face-to-face clients, with no impact of text-based communication found (Cook & Doyle, 2002). In a recent study comparing psychological and psychiatric services delivered either online or face-to-face there were no significant differences in either the perceptions of the alliance or the overall satisfaction with services (Morgan, Patrick, & Magaletta, 2008).

A few other limitations related to the use of asynchronous e-mail itself present potential ethical issues. Because the Internet is a global medium, the online professional is likely to reside in a different time zone than the client, a fact not always apparent when communications are text based and possible at any hour. In addition, the issue of turnaround time, that is, the length of time between a client e-mail and a provider response, emerges. At a minimum, clients ought to be informed about the influence of these matters prior to initiation of such a service, though current compliance of e-therapy sites with this recommendation ranges from 43% to 50% of sites (Heinlen, Welfel, Richmond, & O'Donnell, 2003; Maheu & Gordon, 2000). The importance of such disclosure and consent is highlighted when clients are experiencing a crisis. It is also important to note that legal regulations may affect electronic communication. For example, Koocher and Morray (2000) point out that several jurisdictions are considering or have enacted legislation to limit online service to state residents to professionals licensed in that jurisdiction.

An important risk of misunderstanding derives from cultural differences, especially in light of the global nature of the Internet. In providing face-to-face interventions, one of the components for ethical practice is attention to multicultural considerations in diagnosis and treatment, as the APA *Ethical Principles* (2002), the ACA *Code of Ethics* (2005), and virtually all other standards for mental health service clearly indicate. It is not clear to what extent multicultural differences between the client and provider impact online communications as there has been virtually no research completed in this area. Two facts are clear though: first, the influence of client diversity on the structure, content, or delivery of online behavioral health services has been largely overlooked in most scholarly writing and empirical investigations on the topic; second, access to computer-based services has been partially a function of socioeconomic status. In her survey of 48 actual e-therapy users, the largest study of consumers published to date, Young (2005) found that the modal user was a middle-aged, European-American male with a college education. Only four people in her sample identified themselves as African American or Asian American. No one identified as Latino. Similarly, those with mobility and visual limits have also experienced restricted access to the Internet. In addressing this issue, the OPA *Telepsychology Guidelines* (2008) recommend "compliance with Section 508 of the Rehabilitation Act to make technology accessible for people with disabilities" (Section 2, p. 5). Even when access is available for clients who may not own a computer, language barriers (including unfamiliarity with emoticons and jargon used in electronic communications) may increase the risk of misunderstanding in this format. These risks are highlighted when the professional–client interaction is limited to online modes, but it exists even when online contact is a supplement to face-to-face interaction.

What is clear from the standards and research evidence from e-therapy, is that online behavioral health services demand a high level of technological sophistication and sensitivity to the needs and perspective of the service recipient. As the technology matures, the risks of simple problems such as computer crashes, nonfunctioning hardware, or limited bandwidth may diminish, but the responsibility of professionals to have sufficient knowledge to use the technology appropriately and to disclose to clients/patients its limits and risks will not disappear. Regardless of the method of electronic service delivery, there are risks for the client as well as the professional. Since technology has now become an inherent part of our culture and the manner in which we conduct business in behavioral health care, it is important to review all policies regarding electronic communication provided to clients in conjunction with traditional modes of service delivery. The avoidance of technology will probably not be an option for professionals in the future. Consequently, understanding the ethical implications inherent in its use is an integral part of our ethical responsibility to demonstrate best practices.

Undetermined standard of care

Consider the following case:

> Jonah, a 31-year-old resident of rural Nevada, was diagnosed with paranoid schizophrenia a decade

ago. He has been stable for several years now, thanks to regular visits to his therapist, good family support, and appropriate medication. Jonah had only one face-to-face appointment with his therapist 11 months ago. Since then his biweekly sessions have occurred via video conferencing because the therapist lives in Reno, nearly 240 miles away. Jonah goes to the community center in his small town where the videoconferencing equipment is set up and his therapist uses parallel equipment in her office. Using a fire-walled T1 line, they discuss Jonah's progress and concerns. Jonah was skeptical about the procedure when it began, but now says that he actually prefers not to have his therapist in the same room while he is talking about himself. He says he feels less anxious and thinks better. The therapist also noted a period of adjustment for Jonah and other video-clients and expressed some frustration with the complications of technology, but also believes that videoconferencing is the only feasible way that rural clients like Jonah can receive regular professional service.

To date, no well-accepted theoretical model for conducting this service has been developed, though Suler (2000) and Maheu (2003) have presented models that show promise. Consequently, videoconferencing as well as text-based online behavioral health services clearly fall within the definition of "an emerging area" as described by APA *Ethical Principles* (2002). Its language is as follows:

2.01 BOUNDARIES OF COMPETENCE

(c) In those emerging areas in which generally recognized standards for preparatory training do not yet exist, psychologists nevertheless take reasonable steps to ensure the competence of their work and to protect clients/patients, students, supervisees, research participants, organizational clients, and others from harm.

10.01 INFORMED CONSENT TO THERAPY

(b) When obtaining informed consent for treatment for which generally recognized techniques and procedures have not been established, psychologists inform their clients/patients of the developing nature of the treatment, the potential risks involved, alternative treatments that may be available, and the voluntary nature of their participation.

What are the reasonable steps professionals should take to ensure competence in online services in the absence of clearly defined standards of care? Generally, competence includes three components—formal training, supervised experience, and an attitude of diligence (Welfel, 2010). Formal training in online health means a structured program of study through systematic instruction, readings, and related activities. Malone, Miller, and Walz (2007) as well as Tyler and Sabella (2004) provide two training models for providing behavioral health care utilizing technology. Once a professional obtains a sufficient knowledge base, he or she needs to gain experience under the supervision of a qualified professional. Finally, the therapist must maintain an attitude of diligence to update competence as new data emerges and to address areas where improvement may be needed. The responsibility for diligence emerges as an especially important factor in an emerging area of practice since current knowledge and research is limited; practices that appear promising at one point may prove ineffective or even harmful when subjected to more rigorous study.

Therapists must also be cautious about assuming parallels to traditional service that have not yet been demonstrated. The higher the degree of variance from face-to-face care, the greater the caution needed about applying methods and findings from psychotherapy to online care. Thus, they must take what Sonne (1994) called a "risk preventive" stance to care. In prior research (Heinlen, Welfel, Richmond, & O'Donnell, 2003; Heinlen, Welfel, Richmond, & Rak, 2003; Shaw & Shaw, 2006), online therapists attempted to demonstrate their competence by identifying their areas of competence in traditional practice and using those as the areas for practice in online service. Almost no sites explained the training or credentials they obtained in online behavioral health specifically, although many claimed that they adhered to ISMHO principles.

As Section 10.01 of the APA code specifies, whenever professionals are implementing an innovative service, they must take special care with informed consent to ensure that clients understand that the treatment they are voluntarily entering into is not well established and may have unknown risks. While some treatments are well established in traditional face-to-face settings, the efficacy of using these same treatments with technology has not been clearly demonstrated and should be considered experimental. Therefore, practitioners offering services through technology have a responsibility to advise clients that services provided through technology are new and experimental. Treatments using electronic communications as adjuncts to face-to-face services have at least preliminary evidence of benefit, and therefore, they can be described to clients during informed consent with greater optimism about the outcome, though still innovative. Clients should also be informed of the other established forms of care available to them. The OPA *Guidelines* (2008) suggest psychologists disclose levels of experience and training in telepsychology as well as identify telepsychology as an innovative practice (Section 3, p. 5). ACA standards (2005) evoke similar language for computer-based communications. Again, compliance with this standard is disappointing in text-based forms of service, with no description of the potential risks for e-therapy mentioned in 73% of the e-therapy sites Heinlen, Welfel, Richmond, and O'Donnell (2003) investigated. This omission is especially problematic in light of their finding that 82% of their sample described the potential benefits of online service. Consumers who view these sites are seeing a skewed presentation of the merits and risks of online service, an action inconsistent with the intent of the codes and guidelines.

The ISMHO standard (2000) for competence draws a parallel between competence for in-person service and online service:

> **2.A BOUNDARIES OF COMPETENCE**
>
> The counselor should remain within his or her boundaries of competence and not attempt to address a problem online if he or she would not attempt to address the same problem in person.

The notion that anyone whom a professional can competently treat in the office can be competently treated online is a notion that has been challenged in the professional literature, for example, Ragusea and VandeCreek (2003). Can a psychologist skilled in the treatment of borderline personality disorder in traditional modes be considered competent to offer care, even aftercare, for those individuals online? In light of the absence of an established standard of online care for persistent or severe psychopathology or traumatic symptoms, any approach other than a risk-preventive approach to care seems foolhardy and potentially irresponsible, especially if the experimental nature of the medium and its other risks are not fully communicated to the client.

Evidence-based practice

An emphasis on providing clients with empirically supported behavioral health interventions has paralleled the increase in utilizing technology.

The APA defines *evidence-based practice* (EBP) as "the integration of the best available research with clinical expertise in the context of patient characteristics, culture, and preferences" (APA, 2005, p. 1). They further stipulate that applying empirically supported principles enhances public health. This is consistent with a traditional scientist-practitioner model and helps improve patient outcomes by applying relevant research to clinical practice. It stands to reason that standards of care and criteria for professional competence derived from research on traditional psychotherapy have the greatest applicability to computer-based services with visual and voice components.

The therapeutic alliance has been identified as one of the key factors accounting for success in therapy (Castonguay & Beutler, 2005; Lambert & Ogles, 2004). The communication between the client and the provider is central to developing a therapeutic alliance. The most obvious difference in delivering services electronically is the absence of nonverbal cues when utilizing text-based communication and the difficulty in reading cues via videoconferencing. This difference cannot be considered inconsequential, as the significance of nonverbal cues in interpersonal communication is well documented. In a large-scale study of 80 professionals delivering educational counseling electronically, participants reported the counseling more accessible than face-to-face services and more fruitful as well (Schultze, 2006). While there is some evidence that the therapeutic relationship established on the Internet is slightly stronger (Cook & Doyle, 2002), this represents an area where further research is required.

In an effort to provide services that enhance treatment outcomes, clinicians frequently respond to what clients report as therapeutic. Consider the following case:

> Tasha, a psychotherapist with a thriving private practice, uses a Web site to market her practice. On that Web site she includes extensive information about her contact training and credentials, areas of competence, fees and insurance matters, HIPAA statements, and the logistics of scheduling appointments. She also provides an e-mail address on the business cards that she gives to clients at their initial appointment. Recently several clients have initiated between-session e-mail contact with her, the content of which ranges from scheduling changes to disclosures about matters discussed in the last session. One client, diagnosed with moderate depression subsequent to her diagnosis as HIV positive, sends e-mails nearly every day. At first Tasha discouraged this contact because she was worried

about the extra workload, but when this client revealed that she found the e-mails very helpful in the management of her depressive symptoms, Tasha relented. However, Tasha did caution the client that she should never use e-mail to disclose a crisis and never send e-mails from work. Because a few other clients have also commented that between-session e-mails are helpful in maintaining therapeutic gains, Tasha now views this interaction as helpful, even though she cannot bill for it and it takes extra time each day to read the e-mails. She has begun to encourage selected clients to e-mail her between sessions and post-termination.

As this case illustrates, incorporating technology into the practice of behavioral health care has the potential to improve treatment satisfaction and perceived helpfulness.

As Cucciare & Weingardt (2007) point out, one of the advantages in delivering services via technology is that it is cost-effective and provides an extended reach for the individuals who may not otherwise have access to services. One of the significant findings in this literature is that individuals appear to feel more comfortable with disclosing more information over the computer than they have been in traditional face-to-face settings. In a study utilizing a telephone call center in conjunction with an online counseling service, adolescents reported feeling an increased sense of privacy as well as feeling more safe and "less emotionally exposed" when utilizing technology-based services (King et al., 2006). While there continues to be evidence that using technology in conjunction with traditionally based services represents the best practice, a few studies suggest this combination may not always be necessary. In a study of individuals with obesity, where outcomes were compared between those who received Internet only services and those who received Internet plus in-person treatment, the authors found no differences between the two methods of service delivery. (Micco, Gold, Buzzell, Leonard, Pintauro, & Harvey-Berino, 2007). This study highlights the value of research that examines the effectiveness of each component of online behavioral health.

Between-session homework completed online or electronically and sent to the therapist is another application of technology as an adjunct to service. This serves to reinforce work done between sessions as well as help the patient progress in treatment. Providers have an ethical responsibility in these situations to provide assistance to clients who may have adverse reactions to such work. This responsibility is similar to any other nontechnology-based intervention a provider may recommend the patient complete independently. Some providers have found that using e-mail on an as-needed, case-by-case basis is effective in addressing these concerns, is therapeutic, and helps build the alliance and improve outcomes for the patient. Caution should be used in employing e-mail and making sure that the patient understands the availability of the professional and the limits of such availability.

Videoconferencing is a common mode of online behavioral health care. This mode has been well received thus far, as it provides greater access to care and increased user satisfaction. Based on an extensive review of the literature, Hilty, Liu, Marks, and Callahan (2003) make several recommendations for improving the effectiveness of services utilizing this technology. First, they recommend having adequate technology to support the provision of services, a feature that is especially important in delivering services via videoconferencing. Adequate technology also includes implementing appropriate encryption software when supplementing videoconferencing with e-mail or text messages. Hilty et al. (2003) also emphasize the role of treatment coordination in enhancing outcomes. This recommendation seems to suggest again that having some face-to-face contact may enhance outcomes when providing technologically based services. Psychiatrists have taken the lead in using videoconferencing and related technologies that closely parallel the face-to-face meeting, using it to facilitate contact with rural patients and patients who experience difficulty in attending face-to-face appointments, as the case of Jonah illustrates (Elford et al., 2000; Jerome & Taylor, 2000). In fact, the American Psychiatric Association published guidelines to offer members assistance in finding resources for videoconferencing (APA, 1998). The American Telehealth Association (2009) has also published a draft version of guidelines for videoconferencing.

Increasing numbers of psychologists, psychiatrists, and other mental health professionals appear to find Web-based resources helpful for practice management, research, professional development, and communication with colleagues (Maheu, Whitten, & Allen, 2001; McMinn, Buchanan, Ellens, & Ryan, 1999; Salib & Murphy, 2003; Welfel & Bunce, 2003). For example, Salib and Murphy (2003) report that 80% of their sample of independent practitioners used electronic billing, 64% sought professional information on the Internet, and 57% communicated with other professionals via e-mail. Psychologists generally view these activities as ethical and helpful to professional practice (McMinn et al., 1999). Consumers also value Web-based information about psychotherapy and medical practices and tend to seek more online information than professionals generally provide (Moyer, Stern, Dobias, Cox, & Katz, 2002; Palmiter & Renjilian, 2003).

Less information is available about the extent of electronic communication with clients when it supplements face-to-face care. Salib and

Murphy (2003), under the auspices of the APA Practice Directorate, surveyed psychologists regarding their attitudes toward the use of technology in practice. Not surprisingly, they found an increase in the use of technology in practice management, online contact with other professionals, and e-mail contact with clients (both as an adjunct to therapy and as e-therapy) when compared to previous studies (McMinn et al., 1999; VandenBos & Williams. 2000) Those who responded to the online version of the survey demonstrated substantially higher use of technology in practice than those who responded via postal service. Forty-two percent of professionals who responded to the online survey indicated they received and responded to client e-mails, but only 15% of those who responded to the paper-and-pencil version of the questionnaire endorsed that action. In a mailed survey of APA members in clinical practice, Welfel and Bunce (2003) reported that 40% of their sample experienced at least one e-mail contact with a client, with a modal number of 2 contacts per client. One percent of respondents used between-session online chat. The contents of these contacts were divided among therapeutic issues (28%), billing and scheduling (25%), and interestingly, initiating, terminating, or restarting therapy (19%). A small percentage of clients used e-mail to discuss issues related to the therapeutic alliance (6%; Welfel & Bunce). Respondents characterized the between-session therapeutic content as either information about between-session developments in clients' lives or as topics that clients felt too embarrassed to discuss face to face. Participants in this study expressed varying reactions to client e-mail contact. Some encouraged future contact (16%), others discouraged it (13%), but most (70%) left the decision about future e-mail contact to the judgment of the client.

Physicians also engage in extensive e-mail contact for research and contact with professional colleagues (Brooks & Menachemi, 2006). Some have e-mail communication with patients, though the rate of patient e-mail has varied widely in research, ranging from 16.6 % to 72% of those surveyed depending on location and type of practice (Brooks & Menachemi; Gaster et al., 2003; Hussain, Agyeman, & Das Carlo, 2004). Physicians are also divided in their reactions to this mode of contact, with much of their caution derived from worries about patient privacy and workload (Brooks & Menachemi; Houston, Sands, Nash, & Ford, 2003).

The potential ethical problems in online behavioral health lessen when e-mail and online chat are used as supplements to traditional service and when the mode of communication closely approximates the face-to-face experience. Client identity is more easily verified, data to reliably assess client problems can be gathered more readily, and confusion in communications can be more quickly identified and addressed. In other words, the scope of information available to both parties is significantly greater. However, the possibility of technology failure and increased risks to privacy from unauthorized users remains, along with the possibility that the client will use text-based communications to disclose a crisis or dangerousness to self or others. Thus, the standard of care mandates that clinicians employ encryption and related privacy protections and that they implement careful and comprehensive informed consent whenever online behavioral health methods are used.

Also, since consumers are interested in online contact with health care professionals and actively seek it out (Madden & Fox, 2006), clinicians should anticipate that some clients will initiate e-mail contact with them even if they omit their e-mail addresses from documents they provide clients, and even if they do not offer a clinical Web page. E-mail addresses are difficult to hide. When professionals develop Web pages to market their services, as Tasha had done in the case described above, the likelihood that clients will initiate online contact with them probably increases dramatically. Thus, clinicians need to develop policies and procedures for responding to client e-mail and communicate them to clients as part of the informed consent process in psychotherapy. The APA *Ethical Principles* (2002) specifies in Section 4.02(c), "Psychologists who offer services, products, or information via electronic transmission inform clients/patients of the risks to privacy and limits of confidentiality." The ACA *Code of Ethics* (2005) Section A.12.h uses similar language. Welfel and Bunce (2003), the only known study that examined the ethical practices of professionals when using e-mail as a supplement to face-to-face care, showed disappointing levels of compliance with informed consent procedures. Only 36% of professionals interacting with clients via e-mail expanded their informed consent procedures to specifically discuss risks and limits to this mode of contact. Only 11% took the precaution of encrypting e-mails to clients or asking clients to do the same. Because minors receiving psychotherapy are especially likely to be comfortable with electronic communication, professionals need to obtain consent from parents for this supplement to treatment for their children. Welfel and Bunce reported that only 23% of the psychologists who had exchanged e-mails with minor clients had obtained explicit parental consent for this activity. Less than 1% of the sample also checked with their liability insurers about coverage for e-mail contact with clients. In short, this preliminary evidence suggests that when therapists interact with clients via e-mail, they do so without full examination of the ethical implications of their actions. This state of events can be remedied rather simply with better communication with clients about privacy issues, compromises to confidentiality, possible technology problems, and boundaries for using e-mail, including avoiding the use of e-mail to discuss a crisis or an impulse to harm self or others.

Technology as an educational and practice management resource

Increasing numbers of psychologists, psychiatrists, and other mental health professionals appear to find Web-based resources helpful for practice management, research, professional development, and communication with colleagues (Maheu, Allen, & Whitten, 2001; McMinn et al., 1999; Salib & Murphy, 2003; Welfel & Bunce, 2003). For example, Salib and Murphy reported that 80% of their sample of independent practitioners used electronic billing, 64% sought professional information on the Internet, and 57% communicated with other professionals via e-mail. Psychologists generally view these activities as ethical and helpful to professional practice (McMinn et al.). Consumers also value Web-based information about psychotherapy and medical practices, and tend to seek more online information than professionals generally provide (Moyer et al., 2002; Palmiter & Renjilian, 2003).

Psychologists and psychiatrists are also referring clients to online health and mental health Web sites to help them better understand their problems and offer additional strategies for change. More than one third (35%) of the Welfel and Bunce (2003) sample engaged in this practice. Because of the enormous potential of the Web for educating consumers about health and illness, this strategy has the potential to make therapy more effective and efficient. However, the variability in the quality and accuracy of content on Web sites means that professionals have an obligation to help their clients and patients find the credible sites and not to make a global recommendation to do an online search for a topic. For example, clients with eating disorders can find both therapeutic sites to promote positive coping with their disorder and sites that describe strategies for subverting care and becoming more determined anorectics. Therapists should also routinely query clients about their use of the Web to gather information and support so that they can correct any erroneous conclusions and attempt to intervene when clients are using Web sites and chat rooms to resist therapeutic change.

There are potential risks in having patients read online content provided on a clinician's Web site. Some clients may believe that with the information they have, they are equipped to handle the situation without professional intervention, and while there may be instances where this is the case for many clients, the assistance of a professional is indicated. Another risk or difficulty with patients who obtain information on the Internet is that they may determine that they don't need treatment at all and fail to get help when needed. Fox (2010) reports that 22% of Americans with Web access searched the Web for mental health information. When their loved ones were experiencing a health crisis, the percent of people looking to the Web for information, support, and referral resources rose to nearly 36% (Madden & Fox, 2006). And one in four Americans has searched for information about prescription drugs online (Fox, 2004). When clients are experiencing embarrassing problems, such as sexual fetishes or even obsessive-compulsive behaviors, they may be especially likely to seek out online sources of information and support in order to avoid direct disclosure of their problems.

Clinicians should also be alert to the complications that may arise from social networking sites. If they have a page on one of these sites, they need to be careful about the content they include and the access they allow. Similarly, mental health professionals ought to be aware of the involvement of their clients on social networking sites and discuss with them, as appropriate, what disclosures about involvement in therapy on those sites may mean for their personal privacy. Media reports of prospective employers accessing a candidate's page and declining to hire based on the content of the page have become frequent in recent days.

Online consultation and supervision

Psychotherapists have enthusiastically endorsed the Internet as a means of professional-to-professional communication and exchange of knowledge. We conduct literature searches, share research findings, participate in online continuing education, and join professional listservs that discuss issues of particular interest to us (Collins, 2007: McMinn et al., 1999; Murphy, 2003). All of these activities serve to increase the likelihood of competent service to clients and none has been viewed as controversial or ethically problematic. Because of both legal and ethical concerns, some professionals are unsure about the ethics of online supervision to novice therapists or consultation about specific clients. Is this caution justified? Can the ethical and legal issues be dealt with so that online supervision and consultation are utilized more fully? The caution stems from many of the same concerns about client privacy and confidentiality and accuracy of information upon which to make decisions that present in online assessment and treatment. To fully respond to the ethical dimension of online supervision and consultation, the two activities must be analyzed separately.

Online supervision assumes that the supervisee is not fully capable of independent practice either because he or she is not yet license eligible or because he or she is developing an entirely new area of competency. The latter may occur if a child therapist elects to undergo training in geriatric care, a specialization so far removed from the original area of competence that training and supervision are required. When a professional acts as a clinical supervisor, he or she is primarily responsible for client welfare, and that responsibility for client welfare takes precedence over the obligation to the trainee. In an emergency or a situation in which the trainee has provided potentially harmful care, the supervisor is obligated to intervene and thus needs to be able to access client records and even see the client if necessary (Welfel, 2006). When supervision takes place online and at a distance, the capacity of the supervisor to fulfill these duties is limited at best. Thus, the standard of care for online supervision must encompass a workable plan for the supervisor to reach the client if needed. This access could be accomplished through co-supervision with another credentialed professional at the supervisee's location, but some arrangements must be made to meet this duty.

The second major ethical obligation of clinical supervisors is to ensure that they are getting accurate and complete information from a supervisee about what is happening in therapy and what the client appears to be experiencing. The best way to achieve this level of understanding is through direct observation of the supervisee–client sessions, either at the time they occur or soon thereafter through a recording of the session. Supervisees, because they are novices, are sometimes unaware of important dynamics in the session, and even if not contentious about communicating with supervisors, they may inadvertently omit crucial client information. In the absence of an exact record of the session, supervisors have no means to gather this information. At other times, supervisees are too embarrassed to disclose their mistakes, or too worried about a poor performance evaluation to be honest, and in some situations, they are engaging in actions that they recognize as unethical or harmful to the client, but are too self-interested or compromised to resist. Online supervision cannot protect client welfare in any of these situations if no means exists for the supervisor to determine what is actually transpiring in session. Videoconferencing or real-time audio transmissions of sessions most closely parallel the access of a face-to-face supervisor, and thus are the least ethically problematic modes of online supervision. In fact, because videoconferencing or real-time audio transmission of sessions allows the supervisee access to professionals with special expertise in their fields, in such a circumstance, online supervision can be ethically and clinically preferable to face-to-face supervision with a less experienced or less knowledgeable professional. Rural professionals have noted that such access is especially valuable to them (Kanz, 2001; Wood, Miller & Hargrove, 2005).

Of course, online supervision is vulnerable to the same compromises to confidentiality and technology failure as online assessment and treatment. Consequently, even videoconferencing can be ethically problematic if it is done without proper client and supervisee consent, and without encryption, password protection, and the other steps already discussed at length to prevent unauthorized access to client information. The informed consent process must specifically disclose the nature of the online supervision, its risks and benefits, and the extent of client information that will be made available to the supervisor. Clients should be told that supervisors will have access to their records and identifying information.

Consultation, in contrast, by its very nature refers to communication between two credentialed professionals so that the professional being consulted does not take on the burden of client welfare to the same degree as the supervisor. The threats to client confidentiality that derive from the Internet are no different, nor are the recommendations for use of encryption and password protection, but the need to disclose identifying client information in most circumstances is removed. Thus, explicit client consent to online consultation is not as essential as long as the online consultation is truly free of data that could reveal the client's identity. A full disclosure of the consultation process to the client would certainly not be prohibited under any circumstances, and it may well be therapeutically beneficial to make such a disclosure even when not mandated.

The codes of ethics apply to online supervision and consultation activities to the same degree that they apply to online assessment and therapy. Promotion of the best interests of the client as the highest priority is still the highest professional value. When deciding whether to engage in online supervision and consultation, professionals should use that value as their central criterion, just as they would in determining whether any other online activity is ethically justified.

When professionals are using professional listservs to discuss practice and professional issues with colleagues, they must also use care to guard private client information (Collins, 2007). When questions about cases are posted on a listserv, it is important for the professional to disguise identifying information, assess whether the audience to whom client information is being divulged is truly qualified to assist, and know whether entries on a listserv those who receive this email are able to forward it to non-listserv members. If the message with client data can be forwarded to those who are not members of the listserv, the risks of online consultation increase.

Conclusion

In addition to the professional standards, guidelines, and emerging body of literature, professionals should rely on what Haas and Malouf (2005) refer to as the "clean well-lit room standard." In plain terms, this standard means that any activity that a professional would feel uncomfortable presenting to his or her peers in a public forum is one that is likely to be ethically problematic. In contrast, if a therapist can picture such a presentation without any fear of disapproval from respected colleagues, that action is likely to be consistent with professional ethics. Therapists considering the implementation of online therapeutic service or supervision would be well advised to rely on this standard as the first step in evaluating the wisdom of that action. It is quite possible to design many online services that promote client welfare, but that can only be done with careful reflection, deliberate attention to standards, and recognition of the potential limits and benefits of such innovative treatment.

References

American Counseling Association. (2005). *Code of ethics and standards of practice*. Retrieved October 1, 2008, from http://www.counseling.org/Resources/CodeOfEthics/TP/Home/CT2.aspx

American Medical Association (2001). *Code of Ethics*. Retrieved July 6, 2009 from http://www.ama-assn.org/ama/pub/physician-resources/medical-ethics/code-medical-ethics.shtml

American Medical Association. (2004). *Guidelines for physician-patient electronic communication*. Retrieved October 1, 2008, from http://www.ama-assn.org/ama/pub/category/2386.html

American Psychiatric Association. (1998). Resource document on telepsychiatry via videoconferencing. Retrieved October 6, 2008, from http://www.psych.org/Departments/HSF/UnderservedClearinghouse/Linkeddocuments/telepsychiatry.aspx

American Psychological Association. (1997). *APA statement on services by telephone, teleconferencing and Internet*. Retrieved October 8, 2008, from http://www.apa.org/ethics/stmnt01.html

American Psychological Association. (2002). *Ethical principles and code of conduct*. Retrieved October 8, 2008, from http://www.apa.org/ethics/code2002.html

American Psychological Association. (2005). *Policy statement on evidence-based practice*. Retrieved July 6, 2009 from http://www2.apa.org/practice/ebpstatement.pdf

American Telemedicine Association. (2009). Practice guidelines for videoconferencing-based telemental health. Retrieved July 15, 2009 from: http://www.americantelemed.org/files/public/standards/MH%20Guidelines_Public%20Comment%20_2_.pdf

Brooks, R. G., & Menachemi, N. (2006). Physicians' use of email with patients: Factors influencing electronic communication and adherence to best practices. *Journal of Medical Internet Research, 8*. Retrieved August 15, 2006, from http://www.jmir.org/2006/1/e2/

Castonguay, L. G., Larry E., & Beutler, L .G. (2005) *Principles of therapeutic change that work*. New York: Oxford University Press.

Center for Democracy and Technology. (2004). Email privacy protection called into question by federal appeals court decision. Retrieved August 23, 2006, from http://www.cdt.org/publications/pp_10.13.shtml

Collins, L. H. (2007). Practicing safer listserv use: Ethical use of an invaluable resource. *Professional Psychology: Research and Practice, 38*(6), 690–698.

Cucciare, M. A., & Weingardt, R. (2007). Integrating information technology into the evidence-based practice of psychology. *Clinical Psychologist, 11*(2), 61–70.

Doyle, C., & Cook, J. E. (2002). Working alliance in online therapy as compared to face-to-face therapy: Preliminary results. *CyberPsychology & Behavior, 5,* 95–105.

Elford, D. R., White, H., Bowering, R., Ghandi, A., Maddigan, B., St. John, K., House, M., Harnett, J., West, R., & Battcock, A. (2000). A randomized, controlled trial of child psychiatric assessments conducted using a videoconferencing system. *Journal of Telemedicine and Telecare, 6,* 73–82.

Eysenck, H. J. (1952). The effects of psychotherapy: An evaluation. *Journal of Consulting Psychology, 16,* 319–324.

Fox, S. (2004). Prescription drugs online. Pew Internet & American Life Project. Retrieved October 6, 2008, from http://www.pewinternet.org/pdfs/PIP_Prescription_Drugs_Online.pdf

Fox, S. (2006). *Online health search 2006.* Pew Internet & American Life Project. Retrieved October 1, 2008, from http://www.pewinternet.org/pdfs/PIP_Online_Health_2006.pdf

Gaster, B., Knight, C. L., DeWitt, D. E., Sheffield, J. V., Assefi, N. P., & Buchwald, D. (2003). Physicians' use of and attitudes toward electronic mail for patient communication. *Journal of General Internal Medicine, 18,* 385–389.

Haas, L. J., & Malouf, J. L. (2005). *Keeping up the good work: A practitioner's guide to mental health ethics* (2nd ed.). Sarasota, FL: Professional Resource Exchange.

Heinlen, K. T., Welfel, E. R., Richmond, E. N., & O'Donnell, M. S. (2003). The nature, scope and ethics of psychologists' e-therapy Websites: What consumers find when surfing the Web. *Psychotherapy: Theory, Research, Practice, Training, 40,* 112–124.

Heinlen, K. T., Welfel, E. R., Richmond, E. N., & Rak, C. F. (2003). The scope of WebCounseling: A survey of services and compliance with NBCC's guidelines for Internet counseling. *Journal of Counseling and Development, 81,* 61–69.

Hilty, D. M., Weiling, L., Marks, S., & Callahan, E. J. (2003). The effectiveness of telepsychiatry: A review. CPA Bulletin de líAPC, October, 10–17.

Houston, T. K., Sands, D. Z., Nash, B. R., & Ford, D. E. (2003). Experiences of physicians who frequently use e-mail with patients. *Health Communication, 15*(4), 515–525.

Hussain, N., Agyeman, A., & Das Carlo, M. (2004). Access, attitudes, and concerns of physicians and patients toward e-mail use in health-related communication. *Texas Medicine, 100,* 50–57.

International Society for Mental Health Online (2000). *Suggested principles for the online provision of mental health services.* Retrieved October 6, 2008, from http://www.ismho.org/builder/?p=page&id=214

Jerome, L. W., & Taylor, C. (2000). Cyberspace: Creating a therapeutic environment for telehealth applications. *Professional Psychology: Research and Practice, 31*(5), 478–483.

Kanz, J. E. (2001). Clinical-supervison.com: Issues in the provision of online supervision. *Professional Psychology: Research and Practice: 32,* 415–420.

King, R., Bambling, M., Lloyd, C., Gomurra, R., Smith, S., Reid, W., et al. (2006). Online counselling: The motives and experiences of young people who choose the Internet instead of face to face or telephone counselling. *Counselling & Psychotherapy Research, 6*(3), 169–174.

Koocher, G. P., & Morray, E. (2000). Regulation of telepsychology: A survey of state attorneys general. *Professional Psychology: Research and Practice, 31,* 503–508.

Lambert, M. J., & Ogles, B. M. (2004). The efficacy and effectiveness of psychotherapy. In M. J. Lambert (Ed). *Bergin and Garfield's handbook of psychotherapy and behavior change* (5th ed., pp. 141–193). New York: Wiley.

Liebert, T., Archer, J., Jr., Munson, J., & York, G. (2006). An exploratory study of client perceptions of Internet counseling and the therapeutic alliance. *Journal of Mental Health Counseling, 28,* 69–83.

Madden, M., & Fox, S. (2006). Finding answers online in sickness and in health. Pew Internet and American Life Project. Retrieved October 6, 2008, from http://www.pewinternet.org/pdfs/PIP_Health_Decisions_2006.pdf

Maheu, M. (2003). The online clinical practice management model. *Psychotherapy: Theory, research, practice, training, 40,* 20–32.

Maheu, M. M., Allen, A., & Whitten, P. (2001). *E-health, telehealth, and telemedicine: A guide to start-up and success.* San Francisco: Jossey-Bass.

Maheu, M. M., & Gordon, B. L. (2000). Counseling and therapy on the Internet. *Professional Psychology: Research and Practice, 31,* 484–489.

Mallen, M. J., Vogel, D. L., & Rochlen, A. B. (2005). The practical aspects of online counseling. *The Counseling Psychologist, 33,* 776–818.

Malone, J. F., Miller, R. M., & Walz, G. R. (2007). *Distance counseling: Expanding the counselor's reach and impact.* Ann Arbor, MI: Counseling Outfitters.

McMinn, M. R., Buchanan, T., Ellens, B. M., & Ryan, M. K. (1999). Technology, professional practice and ethics: Survey findings and implications. *Professional Psychology: Research and Practice, 30,* 165–172.

Micco, N., Gold, B., Buzzell, P., Leonard, H., Pintauro, S., & Harvey-Berino, J. (2007). Minimal in-person support as an adjunct to Internet obesity treatment. *Annals of Behavioral Medicine, 33*(1), 49–56.

Miller, T. W. (2006). Telehealth issues in consulting psychology practice. *Consulting Psychology Journal: Practice and Research, 58,* 82–90.

Morgan, R. D., Patrick, A. R., & Magaletta, P. R. (2008). Does the use of telemental health alter the treatment experience? Inmates' perceptions of telemental health versus face-to-face treatment modalities. *Journal of Consulting and Clinical Psychology, 76*(1), 158–162.

Moyer, C. A., Stern, D. T., Dobias, K. S., Cox, D. T., & Katz, S. J. (2002). Bridging the electronic divide: Patent and provider perspectives on email communication in primary care. *American Journal of Managed Care, 8,* 427–433.

Murphy, M. J. (2003). Computer technology for office-based psychological practice: Application and factors affecting adoption. *Psychotherapy: Theory, Research, Practice, Training, 40,* 10–19.

National Board for Certified Counselors (2005). *Standards for the ethical practice of web counseling.* Retrieved October 6, 2008, from http://www.nbcc.org/webethics2

Newman, M. G. (2004). Technology in psychotherapy: An introduction. *Journal of Clinical Psychology, 60,* 141–145.

Ohio Psychological Association. (2008). Telepsychology Guidelines. Retrieved July 14, 2009 from: http://www.ohpsych.org/resources/1/files/Comm%20Tech%20Committee/TelepsychologyGuidelinesApproved041208.pdf

Palmiter, D., & Renjilian, D. (2003). Clinical Web pages: Do they meet expectations? *Professional Psychology: Research and Practice, 34*, 164–169.

Ragusea, A. S., & VandeCreek, L. (2003). Suggestions for the ethical practice of online psychotherapy. *Psychotherapy: Theory, Research, Practice, Training. 40*, 94–102.

Recupero, P. R., & Rainey, S. E. (2006). Characteristics of e-therapy Websites. *Journal of Clinical Psychiatry, 67*, 1435–1440.

Reed, G. M., McLaughlin, C. J., & Milholland, K. (2000). Ten interdisciplinary principles for professional practice in telehealth: Implications for psychology. *Professional Psychology: Research and Practice, 31*, 170–178.

Richard, J., Werth, J. L., & Rogers, J. R. (2000). Rational and assisted suicidal communication on the Internet: A case example and discussion of ethical and practice issues. *Ethics and Behavior, 10*, 215–238.

Rochlen, A. B., Land, L. N., & Wong, Y. J. (2004). Male restrictive emotionality and evaluations of online versus face-to-face counseling. *Psychology of Men and Masculinity, 5*, 190–200.

Salib, J. C., & Murphy, M. J. (2003). Factors associated with technology in private practice settings. *Independent Practitioner, 23*, 72–76.

Schultze, N. (2006). Success factors in Internet-based psychological counseling. *CyberPsychology & Behavior, 9*(5), 623–626.

Seligman, M. E. P. (1995). The effectiveness of psychotherapy: The *Consumer Reports* study. *American Psychologist, 51*, 1072–1079.

Shaw, H. E., & Shaw, S. F. (2006). Critical ethical issues in online counseling: Assessing current practices with an Ethical Intent Checklist. *Journal of Counseling and Development, 84*, 41–53.

Sonne, J. L. (1994). Multiple relationships: Does the new ethics code answer the right questions? *Professional Psychology: Research and Practice, 25(4)*, 336–343.

Suler, J. R. (2000). Psychotherapy in cyberspace: A 5-dimension model of online and computer-mediated psychotherapy *CyberPsychology and Behavior, 3*, 151–160.

Tyler, J. M., & Sabella, R. A. (2004). *Using technology to improve counseling practice.* Alexandria, VA: American Counseling Association.

VandenBos, G. R., & Williams, S. (2000). The Internet versus the telephone: What is telehealth anyway? *Professional Psychology: Research and Practice, 31*(5), 490–492. Welfel, E. R. (2006). *Ethics in counseling & psychotherapy: Standards, research and emerging issues.* Belmont, CA: Cengage.

Welfel, E. R., & Bunce, R. (2003, August). *How psychotherapists use electronic communication with current clients.* Paper presented at the annual meeting of the American Psychological Association, Toronto.

Wood, J. A. V., Miller, T. W., & Hargrove, D. S. (2005). Clinical supervision in rural settings: A telehealth model. *Professional Psychology: Research and Practice, 36*, 173–179.

Young, K. S. (2005). An empirical examination of client attitudes towards online counseling. *CyberPsychology & Behavior, 8*, 172–177.

Index

A

Acceptability of Internet, 170–171
Accessibility of technology
 cognitive skills for, 76–79, 118–119, 128
 disabilities and, 129, 274
 for obesity interventions, 170–171
Access to care
 as advantage of technology (*see* Advantages of technological access to care)
 for anxiety disorders, 27–28
 disadvantages of technology and, 129, 178–179, 215, 228, 274
 low income and, 85
 in obesity Internet-based programs, 170–171
 for posttraumatic stress disorder, 48
 for smoking cessation, 109–110
Access to information, 271–272
Accident victims, 51
Acrophobia, 34
ADDIE (analysis, design, development, implementation, evaluation), 253–255
Adolescents
 advantages of technology for, 279
 body image and eating disorders of, 154, 156
 smoking cessation program for, 125
 weight loss and maintenance in, 185–190
Advantages of technological access to care. *See also* Advantages of technology
 for body image and eating disorders, 155, 156–157
 in diabetes management, 209
 for obesity, 170–171
 overview of, 227, 279
 for pain management, 135, 136
 for posttraumatic stress disorder, 52–54
 in rural areas, 109–110, 274–275, 280
 in smoking cessation, 109–110, 113–114, 129
Advantages of technology. *See also* Disadvantages of technology
 access to care (*see* Advantages of technological access to care)
 for anxiety disorders, 27–28, 29–30, 36–37
 for body image and eating disorders, 153, 155–157, 158, 160, 161, 162–163

291

cost effectiveness, 53, 136–137, 190, 226–227, 279
for diabetes management, 199, 203–204, 207, 209
disclosure facilitated, 57, 227, 279, 281, 284
enhancement of evidence-based practice, 28–30, 54–55, 277–283
ethical issues and, 268
language availability, 126
for mood disorders, 3, 5–6, 7, 8–9, 11, 16, 18
for obesity, 170–171, 179–180, 190
for pain management, 135, 136, 137, 142
for posttraumatic stress disorder, 52–55
psychoeducational programs, 55
for schizophrenia, 74, 85
for smoking cessation, 109, 113–115, 116, 120, 126
for substance use disorders, 92–93, 104, 109, 113–115, 116, 120, 126, 227
Agile software development, 255
Alcohol use. *See also* Substance use disorders
assessment of, 29, 95–96, 97, 234
implementation and, 237
Allegiance and bias, 262
Analysis, design, development, implementation, evaluation (ADDIE), 253–255
Anonymity, 53–54
Anorexia nervosa, 152, 158, 283
Anxiety disorders. *See also* Posttraumatic stress disorder (PTSD)
assessment of, 4, 28–29
chronic pain and, 134
clinical skills needed with technology, 35–36
clinician involvement in, 56
disadvantages and ethical dilemmas of technology, 36–37
efficacy literature review, 30–35, 51, 226
enhancements of technology, 28–30
implementation for, 37–38, 226
overview of, 27, 38
self-help or minimal-contact devices, 27–28, 31–32
Assessment
of anxiety disorders, 4, 28–29
for body image and eating disorders, 158, 160–161

in diabetes management, 201, 205, 206, 207
individualizing treatments and, 58, 96
of mood disorders, 4–5, 59
for pain management, 138–141, 142
reliability of, 4, 28–29
of substance use disorders, 29, 95–96, 97, 234
Assessment and Personalized Feedback for Problem Drinking, 96
Attention and pain experiences, 143–144
Attitude of diligence, 276
Attrition, 56, 114–115, 227
Audio transmissions, 285. *See also specific technology*
Automated calling systems, 207, 208–209, 215
Availability heuristic, 249

B

Barriers
depression and, 10
health insurance as, 21–22
to implementation, 236–238
overcoming with technology, 52–54
posttraumatic stress disorder and, 46–47, 48
provider-level, 204–205
Beating the Blues software, 8, 10
Become an Ex, 123, 125–126
BED (binge eating disorder), 152, 156
Behavioral logs, 138
Behavioral Self-Control Program for Windows (BSCPWIN), 97
Behavioral therapy, 134. *See also* Cognitive-behavior therapy (CBT)
Bias and research design, 249, 250, 262
Bibliotherapy, 113, 127
Binge eating
in bulimia nervosa, 152
Internet-delivered treatment for, 189
risk factors for, 151
self-help CD-ROMs for, 156
technology as adjunct for, 158–159
Binge eating disorder (BED), 152, 156
Biofeedback, 137–138
Bipolar disorder. *See also* Mood disorders
overview of, 3
self-help for, 13, 14
self-monitoring of, 20–21

Index

Blood pressure in diabetes management, 211, 212
Body image and eating disorders. *See also* Obesity Internet-based programs
  computer-assisted psychotherapy (CAT), 153, 154, 155–156
  Internet-delivered programs, 154–155, 156–157, 159–161, 189, 283
  issues in computer-delivered interventions, 160–163
  overview of, 151–153, 163–164
  psychoeducation and prevention, 151, 153–155, 283
  support groups, 159–160
  technology as adjunct to therapy, 157–159
Boundaries of competence, 275, 277
Brain stimulation, minimally invasive, 144–146
BSCPWIN (Behavioral Self-Control Program for Windows), 97
Bulimia nervosa
  chat-room technology for, 159–160
  overview of, 152
  self-help CD-ROMs for, 156
  self-monitoring for, 158
  telecommunications for, 157–158, 159
Bulletin boards
  for obesity, 172, 173, 174, 175, 176, 184
  for pain management, 135
  for smoking cessation, 119
  for substance use disorders, 96
Burn patients, 143–144
Bus bombing trauma, 52
BusWorld, 52

C

Calling systems, automated, 207, 208–209, 215
CAT. *See* Computer-assisted psychotherapy (CAT) and care
Causal inferences, 249
CBT. *See* Cognitive-behavior therapy
CBT4CBT program, 98–99
CCBT (computer-assisted cognitive-behavior therapy), 6–9, 239–241
CD-ROM technology. *See also* Computer-assisted psychotherapy (CAT); Computer programs
  for body image and eating disorders, 154, 156
  for self-management of diabetes, 207, 208
CDs for pain management, 141–142
Cellular technologies. *See also specific types of technology*
  for body image and eating disorders, 158
  for depression, 18
  for diabetes management, 206–208
  for obesity, 192
  for pain management, 137, 138
  for smoking cessation, 115, 128
Center for Epidemiologic Studies Depression Scale (CES-D), 4–5
CES-D (Center for Epidemiologic Studies Depression Scale), 4–5
Chat-room technology
  for body image and eating disorders, 156, 159–160
  ethical issues and, 271, 281, 283
  for obesity, 173, 177, 181, 185
  privacy concerns and, 161
  therapeutic relationships and, 162
Children, 169–170, 185–190
Cholesterol in diabetes management, 205, 206, 211, 212
Chronic Care Model, 203
Chronic illness management. *See also* Posttraumatic stress disorder (PTSD)
  diabetes, 200, 201, 203, 215–216
  pain, 133, 134–135
  schizophrenia, 70–71
ChronoRecord, 20–21
Claustrophobia, 33
Clean well-lit room standard, 286–287
Clinical decision support systems, 204–205
Clinician(s)
  burdens to, 163, 281
  incentives for, 236, 241
  involvement in care (*see* Clinician involvement)
  reminders to, 234
  skills needed with technology, 35–36
  support systems for, 100, 204–205
  therapist-assisted treatments and (*see* Therapist-assisted treatments)
  training of (*see* Clinician training)
Clinician involvement. *See also* Therapeutic relationships
  anxiety disorders, 56

body image and eating disorders, 156–157, 162, 163
burden to therapists and, 163, 281
clinical skills needed with technology, 35–36
diabetes management, 203
dropout rates and, 227
mood disorders, 6, 8, 9, 10–12, 21
outcomes and, 267–268, 278
posttraumatic stress disorder, 50, 54–57, 59
repetitious tasks and, 28, 54
smoking cessation, 112
substance use disorders, 98, 112
Clinician training
boundaries of competence and, 275
for implementation of technology, 233–234
in online health, 276
in smoking cessation, 109
technology failure during, 270
for weight control programs, 190
ClinQMS (Web-based Clinical Quality Management System), 140–141
Cognitive-behavior therapy (CBT)
for anxiety disorders, 27, 29–30, 31–32, 226
for body image and eating disorders, 154–155, 156, 157, 158
clinical skills needed with technology, 35–36
clinician training on, 233
computer-assisted, 6–9, 36–37, 239–241
mobile phones with, 18, 36–37
for obesity, 189
for pain management, 134
for posttraumatic stress disorder, 47, 48–50, 59
psychoeducational programs in, 240
research design and, 259
for social phobia, 226
for substance use disorders, 94, 98–99
telephone-administered, 17–18
therapeutic alliance absence and, 55–56
Cognitive skills for technology use, 76–79, 118–119, 128
Colby and Colby program (Overcoming Depression), 6–7, 13–14
Combat-related posttraumatic stress disorder, *see* Military personnel with posttraumatic stress disorder

Committed Quitters Stop Smoking Plan, 94
Community Reinforcement Approach (CRA), 98
Compensatory behaviors, 155, 156
Computer-assisted cognitive-behavior therapy (CCBT), 6–9, 239–241
Computer-Assisted Cognitive Imagery System, 142–143
Computer-assisted psychotherapy (CAT) and care
for anxiety disorders, 31–32, 33–34
for body image and eating disorders, 153, 154, 155–156
cognitive-behavior therapy, 6–9, 239–241
for diabetes management, 204, 207
disadvantages and ethical dilemmas of, 36–37
flexibility of, 57–58
implementation of, 37–38
for mood disorders
clinical use of, 9–12, 21–22
ethics of, 12–13
programs for, 5–9, 10
for obesity, 191
for pain management, 137–138, 142–143
for posttraumatic stress disorder, 48–50
for substance use disorders, 97, 98
Computerized Lifestyle Assessment, 95–96
Computer programs. *See also* CD-ROM technology; Computer-assisted psychotherapy (CAT)
development of, 37–38, 255–256
for diabetes management, 206
for mood disorders, 5–9, 10
for social phobia, 226
for substance use disorders, 99–103
Confidentiality. *See also* Privacy concerns
breaching, 59, 161
ethical issues and, 284
Web hosting packages and, 101, 102
Confounding, 249
Consultation, 159, 286
Context of implementation, 229–230, 258–261, 262–263
Continuing education credit, 135, 233
Control. *See* Self-management
Control Your Depression, 125
Coordinated care management, 199
Costs
as advantage of technology, 53, 136–137, 190, 226–227, 279
of diabetes, 200

Index

as disadvantage of technology, 190, 212, 215
of implementation, 36
of obesity, 169
CQ Plan, 117
CRA (Community Reinforcement Approach), 98
Crises and supervision, 284–285
Culture, 229–230, 273–274
Customized treatment. *See* Flexibility of Internet-delivered programs
Customized Web design tools, 101–102

D

DCU (Drinker's Check-up) program, 97
Decision support systems, 204–205
DElivery of Self-TRaining and Education for Stressful Situations (DE-STRESS) program, 59–61
Dependent variables, 250
Depression. *See also* Mood disorders
assessment of, 4–5, 59
barriers and, 10
body image and eating disorders and, 151, 160–161
cellular technologies for, 18
computer-assisted psychotherapy for, 7–12
e-mail contact for, 16–17, 278–279
ethical issues and, 270
natural language applications for, 6–7
support groups for, 13, 14, 15
Depression and Bipolar Support Alliance, 13, 14
Depression and Related Affective Disorders Association (DRADA), 13, 14
Design, experimental, 248–252, 259
Design issues
for instructional systems, 248, 253–254
seeking assistance for, 102–103
usability studies on, 77–80, 118–119
DE-STRESS (DElivery of Self-TRaining and Education for Stressful Situations) program, 59–61
Developmental testing, 254
Diabetes management
burden of diabetes, 200
complexity and challenges of, 200–202
emergence of technology in, 203–204
future research, 214–216
overview of, 199–200, 216
patient-centered technology
automated calling systems, 207, 208–209, 215
handheld, portable, mobile devices, 206–208
Internet-delivered, 205, 207, 209–214, 215
provider-centered technology, 204–205
self-management (*see* Self-management of diabetes)
type 1 diabetes, 210
type 2 diabetes, 200, 205, 210, 212
Diagnosis and assessment. *See* Assessment
Diary methods, 19–20
Dieting, 151, 153–154
Digital pens, 138
Diligence, attitude of, 276
Disability
access to technology and, 129, 274
tobacco use and, 110
usability studies, 77–80
Disadvantages of technology. *See also* Advantages of technology
access to care, 129, 178–179, 215, 228, 274
for anxiety disorders, 36–37
attrition, 56, 114–115, 227
body image and eating disorders, 158, 160, 162
burden to therapists, 163, 281
cognitive skills needed, 76–79, 118–119, 128
common factors eliminated, 228
costs, 190, 212, 215
diabetes management, 212, 215–216
ethical issues not addressed, 277
failure of connections, 269–271
iatrogenic effects, 127–128
Internet speed, 178–179
for mood disorders, 5–6, 11–12, 15, 16–17, 18
need for clinicians reduced, 228
nonverbal cues absent, 278
for obesity, 171, 177, 178–179, 181, 190, 192
for posttraumatic stress disorder, 55–59
risks of misunderstanding, 272–274
smoking cessation, 126–128
substance use disorders, 104, 126–128
sustainability, 215–216
treatment modality preferences, 191, 237

treatment not sought after Internet program, 283
Disclosure
  as advantage of technology, 57, 227, 279, 281, 284
  informed consent guidelines and, 270–271
Discussion boards
  for obesity, 172, 173, 174, 175, 176, 184
  for pain management, 135
  for smoking cessation, 119
  for substance use disorders, 96
Disease registries, 204–205
Diversity, 273–274
Domain names, 100
DRADA (Depression and Related Affective Disorders Association), 13, 14
Drinker's Check-up (DCU) program, 97
Drinking Less program, 96
Driving phobia, 33, 51
Dropout rates, 56, 114–115, 227
Drug use, 98–99
DVDs for relaxation, 141–142

E

Eating disorder(s). See Body image and eating disorders; specific eating disorders
Eating disorder not otherwise specified (EDNOS), 152
Ecological momentary assessment, 158
EDNOS (eating disorder not otherwise specified), 152
Education. See Clinician training; Psychoeducational programs
EHR systems. See Electronic health record systems
Electronic health record (EHR) systems
  diabetes management and, 203, 205
  implementation and, 235
  integrated personal health record (PHR) and, 214
  Internet-delivered treatment and, 212
  overview of, 216
E-mail contact
  for anxiety disorders, 32
  for body image and eating disorders, 157–159
  for depression, 16–17, 278–279
  disclosure and, 281
  emotional bracketing in, 58

ethical issues and, 270–272, 273, 274, 277, 278–279, 280, 281, 282
for obesity
  weight loss, 172, 174, 175, 176, 178, 179–180, 191
  weight maintenance, 181, 182, 184
for posttraumatic stress disorder, 55, 59–61
privacy concerns with, 36–37
for smoking cessation, 95, 114, 115, 119, 123–124, 125
therapeutic relationships and, 273
therapist availability and burden with, 163
EMDR (eye movement desensitization and reprocessing), 47
Emergencies and supervision, 284–285
Emoticons, 272
Emotional bracketing, 58, 162, 272
Emotional detachment, 54
Empathy, 228
Empowerment, 73–74
Encryption, 271, 280, 281–282, 285
Enhancement of evidence-based practice, 28–30, 54–55, 277–283
Epidemiology, 45–46, 231, 232
ERP (exposure and response prevention), 32
Ethical issues. See also Privacy concerns
  for anxiety disorders, 36–37
  case studies, 269, 274–275, 278–279
  codes of ethics on, 268–269, 270, 272, 275–276, 277, 282, 286
  in consultation and supervision, 284–286
  in education and practice management, 283–284
  evidence-based practice and, 277–283
  experimental design and, 251
  inherent in technology, 269–274
  for mood disorders, 12–13
  overview of, 267–269, 286–287
  in posttraumatic stress disorder treatment, 58–59
  in smoking cessation, 127–128
  for substance use disorders, 103–105, 127–128
  of telehealth, 19, 276–277
  undetermined standard of care, 274–277
Evaluation
  in context, 258–261, 262–263
  of evidence-based practice, 111

Index 297

formative, 247–248, 252–258, 259, 261–262
  in implementation process, 37, 230, 241, 253–255, 260–261
  overview of, 247–248, 261–263
  research 101, 248–252
Evidence-based practice
  definition of, 258–259, 278
  enhancement of, 28–30, 54–55, 277–283
  for implementation, 233–236
  for smoking cessation, 110–112
  Web site on, 92
Evidence in implementation, 229
Experimental design, 248–252
Exposure and response prevention (ERP), 32
Exposure therapy
  for anxiety disorders, 30, 32, 33, 34, 51
  clinician involvement and, 56
  for posttraumatic stress disorder, 33, 47, 49, 50–52
  virtual reality
    for anxiety disorders, 30, 32, 33, 34, 51
    clinical skills needed with technology, 35
    costs of, 36
    for posttraumatic stress disorder, 33, 50–52
External validity, 251, 261
Eye movement desensitization and reprocessing (EMDR), 47

F

Face-to-face interaction. *See* Clinician involvement; Therapeutic relationships
Facilitation in implementation, 230–231
Family psychoeducation (FPE), 70, 72–73, 80–81
Fast repetitive TMS (frTMS), 145
FE (formative evaluation), 247–248, 252–258, 259, 261–262
Fear of driving or flying, 33–34, 51. *See also* Phobias
Feasibility determination, 239–240
Feedback
  for alcohol use, 96
  in evaluation, 247–248, 254
  in implementation, 230, 234
  for obesity
    children and adolescents, 188, 190

weight loss, 172, 173, 174, 175, 176, 178, 181, 190
weight maintenance, 182, 183, 184, 185
  on self-monitoring, 181, 185
Field trials, 254
5-A model, 201, 202, 213–214
Flexibility of Internet-delivered programs
  for alcohol use, 96
  for diabetes management, 210
  for obesity, 178, 191
  for posttraumatic stress disorder, 57–58
  for smoking cessation, 115–116, 117–118
Flight phobia, 33–34
FMA (Food, Mood, and Attitude), 154
Food, Mood, and Attitude (FMA), 154
Foot disease screening, 205, 206
Formative evaluation (FE), 247–248, 252–258, 259, 261–262
Forums
  chat rooms (*see* Chat-room technology)
  discussion boards (*see* Discussion boards)
  listservs for professionals, 286
  for smoking cessation, 119, 123–124
FPE (family psychoeducation), 70, 72–73, 80–81
French programs for smoking cessation, 122, 123, 124, 126
FrTMS (fast repetitive TMS), 145

G

GAD (generalized anxiety disorder), 33, 34
Generalized anxiety disorder (GAD), 33, 34
German programs, 126
Go Nosmoke program, 93–94
Good Days Ahead program
  clinician involvement, 10–11
  effectiveness of, 9–10
  overview of, 7–8
Grief reactions, 55

H

Hamilton Depression Rating Scale (HDRS), 4
Handheld devices. *See* Palmtop devices
HbA1c, 205, 206, 210, 211, 212, 213, 214
HDRS (Hamilton Depression Rating Scale), 4
Headaches, 134, 137

Health insurance
  anxiety disorders and, 27–28
  as barrier to technology, 21–22
  billing policies and, 163
  depression and, 10, 17–18
  for e-mail contact, 279
  managed care, 180
  obesity Internet-based programs, 170
Health Insurance Portability and Accountability Act (HIPPA), 103–104
HIPAA (Health Insurance Portability and Accountability Act), 103–104
HIPTeens, 186
Home-based health monitors, 138
Homework
  for anxiety disorders, 29–30
  ethical issues and, 279–280
  for posttraumatic stress disorder, 60
  for substance use disorders, 98
HTML (Hypertext Markup Language), 101
Humanizing interventions, 115–116. *See also* Clinician involvement
Human-technology team, 12. *See also* Clinician involvement
Hybrid programs, 191. *See also* Computer-assisted psychotherapy (CAT); Therapist-assisted treatments
Hypertext Markup Language (HTML), 101
Hypnosis, 134

I

Iatrogenic effects, 127–128
IBGMS (Internet-Based Blood Glucose Monitoring System), 212
IDEATel (Informatics for Diabetes Education and Telemedicine) project, 211–212
Illicit drug use, 98–99. *See also* Substance use disorders
Imagery, 134
Implementation
  advantages of, 226–227 (*see also* Advantages of technology)
  for anxiety disorders, 37–38, 226
  barriers to, 236–238
  context of, 229–230, 258–261, 262–263
  costs of, 36
  evaluation in, 37, 230, 241, 253–255, 260–261
  evidence-based strategies for, 233–236
  example initiative using, 238–241
  overview of, 225–226, 241–242
  PARIHS, 229–231, 238, 260
  PRECEDE-PROCEED, 231–234, 235, 238–241, 260–261
Incentives for clinicians, 236, 241
Independent variables, 250
Individualized treatment. *See* Flexibility of Internet-delivered programs
Informatics for Diabetes Education and Telemedicine (IDEATel) project, 211–212
Information technology (IT). *See specific types of technology*
Informed consent guidelines
  competence in Internet treatment and, 275–276
  disclosure and, 270–271
  e-mail contact and, 282
  privacy issues and, 281–282
  supervision online and, 285–286
Instructional Systems Design (ISD) model, 248, 253–254
Insurance. *See* Health insurance; Liability insurance
Integrated personal health record (PHR), 214. *See also* Electronic health record (EHR) systems
Interactive voice response (IVR) technology
  for depression, 4, 17–18
  for pain management, 136
  for tobacco use, 94
Interactivity, 118, 178, 185
Internal validity, 249–250, 251, 261
Internet-Based Blood Glucose Monitoring System (IBGMS), 212
Internet-delivered programs. *See also* Forums
  academic skills for, 129
  accessibility of, 170–171
  for adolescents, 279
  advantages of, 113, 116, 170–171, 179–180, 190 (*see also* Advantages of technology)
  for anxiety disorders, 31, 33
  for body image and eating disorders, 154–155, 156–157, 159–161, 189, 283
  for diabetes management, 205, 207, 209–214, 215
  disclosure and, 57, 284
  ethical issues in, 36–37, 268–269, 271, 273, 275–276, 283

Index

flexibility of, 57–58, 96, 115–116, 117–118, 178, 191, 210
implementation of, 37–38
information versus intervention, 121, 128
interactivity of, 118, 178, 185
for mood disorders, 5–6, 16–17, 18–19
multimedia features of, 178–179
for obesity (*see* Obesity Internet-based programs)
for pain management, 135
for posttraumatic stress disorder, 33, 48–50, 57
quality control of information in, 58
research about, interpreted, 249–250
for schizophrenia (*see* Internet-delivered treatment for schizophrenia)
for smoking cessation, 93–95, 99, 113–116, 117–118, 119–120
for substance use disorders, 93–95, 96, 99, 113–116, 117–118, 119–120
support groups, 159–160
therapeutic relationships and, 18–19, 20–21
therapist-assisted treatments, 48–50
Internet-delivered treatment for schizophrenia
clinical trial of, 81–85
components of Web-site, 80–81
design for accessibility needs, 76–80
therapeutic foundations of, 72–76
Intranet systems, 140–141
iPods, 135, 141–142. *See also* Palmtop devices
ISD (Instructional Systems Design) model, 248, 253–254
IT (information technology). *See specific types of technology*
Italian programs, 126
Iterative software development, 255–256
IVR technology. *See* Interactive voice response technology

J

J'Arrete (I Quit), 123, 125

K

Курению – Нет (No to Smoking), 123, 126

L

Leadership for implementation, 230
Legal regulations, 273
Liability insurance, 282
Listservs for professionals, 286
Logs for pain management, 138
Low-income populations
access to care for, 85
diabetes and, 202, 209
Internet use by, 128, 170
obesity care for, 170
tobacco use by, 110

M

Madrid + Salud Programa Dejar de Fumar, 123
Major depressive disorder (MDD), 3–4. *See also* Depression; Mood disorders
Managed care, 180
Marketing implementation, 235–236
Mass media, 235–236
MDD (major depressive disorder), 3–4. *See also* Depression; Mood disorders
Meal replacements, 174, 177
Medication-assisted treatment
computer-assisted therapy compared to, 12
for smoking cessation, 112, 116, 123–124, 125–126 (*see also* Nicotine replacement therapies)
for substance use disorders, 91, 112, 116, 123–124, 125–126
Message boards
for obesity, 172, 173, 174, 175, 176, 184
for pain management, 135
for smoking cessation, 119
for substance use disorders, 96, 119
Military personnel with posttraumatic stress disorder (PTSD)
advantages of technology for, 52–53
ethical issues and, 269
internet-assisted therapy, 48, 49
meaning making and, 56
prevalence rates, 46
virtual reality exposure therapy for, 33, 50, 51
Minimal-contact therapy, 27–28, 33–34
Minimally invasive brain stimulation, 144–146
Misunderstandings, 272–274

Mobile phones
  body image and eating disorders, 158
  depression, 18
  diabetes management, 206–208
  obesity, 192
  pain management, 137, 138
  smoking cessation, 115, 128
Monitors of health, 138, 212. *See also* Self-monitoring
Mood disorders
  body image and eating disorders and, 151, 160–161
  chronic pain and, 134
  computer-assisted psychotherapy for
    clinical use of, 9–12, 21–22
    ethics of, 12–13
    programs for, 5–9, 10
  diagnosis and assessment, 4–5, 59
  ethical issues for, 12–13, 270
  minimally invasive brain stimulation for, 146
  online psychotherapy for, 5–6, 16–17, 18–19
  overview of, 3–4, 21–22
  self-help programs for, 5, 13–15, 17–18
  self-monitoring technology for, 19–21
  smoking cessation and, 114, 123–124, 125
  telepsychiatry for, 15–19, 278–279
MoodGym program, 13–14
Motivation, 120–121
Motivational Enhancement System, 98
Motor vehicle accident victims, 51
MP3s for relaxation, 141–142. *See also* Palmtop devices
Multifamily groups, 72–73, 82–84
Multimedia features, 7–12, 153, 178–179
Multiple time-series designs, 259
MycareTeam Web site, 209, 210

N

NAMI (National Alliance on Mental Illness), 13, 14
National Alliance on Mental Illness (NAMI), 13, 14
Natural language applications, 6–7
Needs, self-perceived, 75
Never Smoke Again, 123
New Products Management (NPM) framework, 248, 253, 257–258
Nicotine replacement therapies
  with Internet programs, 115, 116, 119–120
  quit rates for, 112
  sustained compliance with, 94
  technology-assisted program compared to, 114
9/11 attack, 50–51
Nonequivalent control group design, 251
NPM (New Products Management) framework, 248, 253, 257–258

O

Obesity Internet-based programs. *See also* Body image and eating disorders
  accessibility and acceptability of, 170–171
  advantages of, 170–171, 179–180, 190
  for children and adolescents, 185–190
  computer-assisted programs and, 191
  future directions, 190–192
  login frequency importance, 177–179
  in managed care and workplace settings, 180
  overview of, 169–170, 192
  weight loss, 171–179, 180–181, 185–189, 191, 279
  weight maintenance, 181–185, 188–190
Obsessive-compulsive disorder (OCD), 32, 34–35
OCD (obsessive-compulsive disorder), 32, 34–35
1-2-3 Smokefree, 115
Online psychotherapy. *See* Internet-delivered programs
OPTS (Outpatient Psychosocial Triage System), 141
Outcomes
  clinician involvement and, 267–268, 278
  of obesity Internet-based programs, 171
  in pain management, 138–141
Outpatient Psychosocial Triage System (OPTS), 141
Overcoming Depression, 6–7, 13–14
Overweight, 156. *See also* Obesity Internet-based programs

P

Pain management
  applied physiological interventions, 137–138

Index

chronic pain aspects, 134–135
computerized hypnosis, 142–143
Internet technology as resource for, 135
minimally invasive brain stimulation, 144–146
overview of, 133, 146
relaxation messages, 141–142
screening and outcomes assessment, 138–141
symptom tracking and behavioral logs, 138
telemedicine, 135–137
virtual reality techniques, 143–144
Palmtop devices
for anxiety disorders, 30, 31–32, 226
for bipolar disorder, 20
for body image and eating disorders, 157–159
for diabetes management, 206–208
iPods, 135, 141–142
MP3s, 141–142
for pain management, 135, 138, 141–142
personal digital assistants (PDAs), 18, 138, 191–192, 206
Panic disorders, 31–32
PARIHS (Promoting Action on Research Implementation in Health Services), 229–231, 238, 260
Password protection, 271, 272, 285
PDAs (personal digital assistants), 18, 138, 191–192, 206. *See also* Palmtop devices
Peer-to-peer support groups. *See also* Self-help
for depression, 15
need for, 72
for obesity, 185
for schizophrenia, 73–75
for smoking cessation, 119
Personal digital assistants (PDAs), 18, 138, 191–192, 206. *See also* Palmtop devices
Personal health record (PHR), 214. *See also* Electronic health record (EHR) systems
Personalized treatment. *See* Flexibility of Internet-delivered programs
Phobias, 32, 33–34, 51, 226
Phones. *See* Mobile phones; Telephone-administered programs

PHR (personal health record), 214. *See also* Electronic health record (EHR) systems
Physical activity for obesity, 178, 179–180
Pilot testing, 254
Podcasts for pain management, 135
Policy, Regulatory, and Organization Constructs in Educational and Environbehavioral Development (PROCEED), 231–234, 235, 238–241, 260–261
Portuguese programs, 126
Posttraumatic stress disorder (PTSD). *See also* Anxiety disorders
advantages of technology for, 52–55
barriers to care, 46–47, 48
challenges to technology for, 55–59
cognitive-behavior therapy for, 47, 48–50, 59
computer-assisted psychotherapy for, 48–50
DE-STRESS program for, 59–61
epidemiology of, 45–46
ethical issues and, 269
Internet-assisted treatment for, 33, 48–50, 57
overview of, 45, 46, 61–62
stigma of, 46–47, 53
virtual reality exposure therapy for, 33, 50–52
Potentially traumatizing events (PTEs), 45–46
Practice management, 283–284
PRECEDE-PROCEED (Predisposing, Reinforcing, and Enabling Constructs in Educational Diagnosis and Evaluation-Policy, Regulatory, and Organization Constructs in Educational and Environbehavioral Development), 231–234, 235, 238–241, 260–261
Predisposing, Reinforcing, and Enabling Constructs in Educational Diagnosis and Evaluation-Policy, Regulatory, and Organization Constructs in Educational and Environbehavioral Development (PRECEDE-PROCEED), 231–234, 235, 238–241, 260–261
Pre/post design, 251

Prevention
    of body image and eating disorders, 151, 153–155, 283
    of obesity for children and adolescents, 188–190
    of smoking relapse, 120, 123–124, 126
Privacy concerns
    with chat rooms, 161
    with e-mail contact, 36–37
    encryption and, 281–282
    ethical issues and, 269–270, 271, 284
    of online psychotherapy, 36–37
    posttraumatic stress disorder treatment and, 58–59
    professional listservs and, 286
    social networking sites and, 284
    of substance use disorders treatment, 103–104
    Web hosting packages and, 101, 102
Probability, 249
Problem solving, 75
PROCEED (Policy, Regulatory, and Organization Constructs in Educational and Environbehavioral Development) framework, 231–234, 235, 238–241, 260–261
Professional listservs, 286
Progress trackers. *See also* Record keeping; Self-monitoring
    in diabetes management, 210
    for obesity, 172, 173, 174, 175, 176, 178, 188
    for pain management, 138
    for smoking cessation, 120, 125
Promoting Action on Research Implementation in Health Services (PARIHS), 229–231, 238, 260
Psychoeducational programs
    for adolescents, 154
    for body image and eating disorders, 153–155, 283
    clinical decision support, 205
    cognitive-behavior therapy based, 240
    on depression, 13–14
    for diabetes management, 201, 202, 205, 207, 211
    ethical issues in, 283–284
    for families, 70, 72–73, 80–81
    for obesity, 175, 185–186
    for pain management, 135
    for posttraumatic stress disorder, 49, 50, 55, 60
    for schizophrenia, 75–76, 80–81
    for smoking cessation, 114, 120
    for substance use disorders, 94, 114, 120
Psychopharmacology. *See* Medication-assisted treatment
PTEs (potentially traumatizing events), 45–46
PTSD. *See* Posttraumatic stress disorder
Public interest in smoking cessation, 113–114
Public speaking anxiety, 32
Purging, 151, 158–159. *See also* Body image and eating disorders

Q

Qualitative versus quantitative evaluation, 253
Quasi-experimental designs, 251
Quit 4 Life, 123
QUITLINK, 118
QuitNet, 119, 123
QuitNet program, 94
Quit Smoking, 124
Quitsmoking, 122
QuitSmokingNetwork program, 94

R

RAD (rapid application development) software, 37–38
Randomization, 250
Randomized controlled trials (RCT), 250–251
Rapid application development (RAD) software, 37–38
Rational United Process (RUP), 248, 253, 255–256
RCT (randomized controlled trials), 250–251
Real-time audio transmissions, 285
Record keeping. *See also* Progress trackers; Self-monitoring
    for anxiety disorders, 30
    for body image and eating disorders, 158
    in diabetes management, 206, 210
    for mood disorders, 19, 20–21
    for obesity, 192
    for pain management, 138

Index

Recovery, 71, 91, 120
Registries, disease, 204–205
Relapse prevention, 120, 123–124, 126
Relaxation messages, 141–142
Reliability
 of assessment, 4, 28–29
 definition of, 251–252
 of self-monitoring, 19
Reminders
 to clinicians, 234
 in diabetes management, 205–206, 212
 for mood disorders, 16
 overview of, 234–235
Repetitive TMS (RTMS), 145
Research 101, 248–252. *See also* Evaluation
Resources. *See* Web sites and resources
Risk factors
 for body image and eating disorders, 151, 153, 160
 for diabetes, 202
 for misunderstandings, 272–274
RTMS (repetitive TMS), 145
RUP (Rational United Process), 248, 253, 255–256
Rural areas
 access to care in, 109–110, 274–275, 280
 barriers to care in, 52
 Internet use in, 170
 obesity care in, 170
 serious mental illness in, 71–72
 supervision and, 285
Russian programs, 122, 126

S

Safety issues, 19, 128, 160–161. *See also* Ethical issues
Sample size, 250
San Francisco Stop Smoking Site, 124
Schizoaffective disorder, 81–85
Schizophrenia
 background on, 69–71
 overview of, 69, 85–86
 videoconferencing for, 274–275
 weaknesses of current system for, 71–72
 Web-based intervention
  clinical trial of, 81–85
  components of Web-site, 80–81
  design for accessibility needs, 76–80
  therapeutic foundations of, 72–76
Schizophrenia On-Line Access to Resources (SOAR), 77–81

Screening. *See also* Assessment
 for alcohol use, 96
 for body image and eating disorders, 160–161
 in diabetes management, 201, 205, 206
 pain management and, 138–141
Screening and Brief Intervention Program, 96
Second-hand smoke, 110
Security issues. *See* Privacy concerns; Safety issues
Self-disclosure
 as advantage of technology, 57, 227, 279, 281, 284
 informed consent guidelines and, 270–271
Self-efficacy, 73–74
Self-esteem, 151, 153–154
Self-help
 for anxiety disorders, 27–28, 31–32
 for body image and eating disorders, 154–155, 156
 for mood disorders, 5, 13–15, 17–18
 for schizophrenia, 73–74
 for substance use disorders, 91
Self-management
 for chronic illness, 70–71
 of diabetes (*see* Self-management of diabetes)
 for obesity, 184
 schizophrenia and, 73–74
 in smoking cessation, 113, 118, 120–121, 125
Self-management of diabetes
 automated calling systems, 207, 208–209, 215
 CD-ROM technology, 207, 208
 complexity and challenges of, 201–202
 handheld, portable, mobile devices, 206–208
 Internet-delivered, 207, 209–214, 215
 overview of, 199
Self-monitoring. *See also* Progress trackers; Record keeping
 of alcohol use, 97
 for anxiety disorders, 30
 for body image and eating disorders, 158
 cellular technologies for obesity, 192
 as confounding research variable, 249–250
 in diabetes management, 212, 213

feedback on, 181, 185
of mood disorders, 19–21
for obesity
  children and adolescents, 187, 188
  weight loss, 172, 173, 175, 176, 179, 180
  weight maintenance, 182, 183, 184, 185
for posttraumatic stress disorder, 49, 60–61
Sensory stimuli in virtual reality therapy, 52, 54
September 11 attack, 50–51
Serious mental illness (SMI), 69–72, 76–80. *See also* Schizophrenia
Set Your Body Free, 157
Shame, 53, 155–156, 162–163
Short message service. *See* SMS (short message service)
Single-subject design, 259
Smartphones, 18, 206
SMI (serious mental illness), 69–72, 76–80. *See also* Schizophrenia
Smoking cessation. *See also* Substance use disorders
  clinician incentives to promote, 236
  considerations for, 126–128
  dangers of tobacco use, 110
  e-mail contact for, 95, 114, 115, 119, 123–124, 125
  evidence-based treatments for, 110–112
  future directions, 128–129
  improvements in programs for, 116–119
  Internet-based programs, 93–95, 99, 113–116, 117–118
  Internet intervention as resource for clinician, 119–121
  overview of, 109–110
  text messaging for, 115, 128, 157–159
  tobacco use prevalence, 110–111
  Web sites for, 121–126
SMS (short message service)
  for body image and eating disorders, 157–159
  encryption and, 280
  for pain management, 138
  for smoking cessation, 115, 128, 157–159
  therapist availability and burden with, 163
SnowWorld, 143–144
SOAR (Schizophrenia On-Line Access to Resources), 77–81

Social networking sites, 284
Social phobia, 32, 226
Social support. *See also* Chat-room technology; Message boards; Support groups
  health status and, 74–75
  for obesity, 179, 181, 185, 189, 190
  for smoking cessation, 119
  social networking sites, 284
Software. *See also* CD-ROM technology; Computer-assisted psychotherapy (CAT)
  development of, 37–38, 255–256
  for diabetes management, 206
  for mood disorders, 5–9, 10
  for social phobia, 226
  for substance use disorders, 99–103
Spanish programs
  for diabetes management, 209
  for smoking cessation, 114, 122, 123, 124, 125, 126
Spider phobia, 34
Stage-Gate Process, 257
Stage Model of Psychotherapy Development, 261–262
Standards of care, 274–277, 286–287
Stopsmoking, 121, 125
Stop Smoking Center, 124
Stopsmoking program, 94
Stop-Tabac program, 95
Stop Tobac Coach, 124, 126
Stress-vulnerability-coping-competence model, 72–73
StudentBodies program, 154–155, 189
StudentsBodies2-BED, 156
Substance use disorders
  alcohol use, 95–97, 99, 234
  assessment of, 29, 95–96, 97, 234
  background on, 92–93
  ethical issues and, 103–105
  illicit drug use, 98–99
  implementation issues with, 227, 233, 234, 237, 239–241
  Internet Web site and software development, 99–103
  overview of, 91, 105–106
  technology-assisted programs for, 93–99
  tobacco use, 93–95, 99 (*See also* Smoking cessation)
Suicidal ideation

Index

body image and eating disorders and, 160–161
computer-assisted psychotherapy and, 11, 12
ethical issues and, 270
Summative evaluations, 247, 252
Supervision, 284–286
Support groups. *See also* Self-help; Social support
for body image and eating disorders, 159–160
for depression, 13, 14, 15
emotional bracketing in, 162
need for, 72
for obesity, 185
for pain management, 135
for schizophrenia, 73–75
for smoking cessation, 119
Support systems for clinicians, 100, 204–205
Sustainability, 215–216
Symptom tracking. *See* Progress trackers
Systematic Replication Model, 261–262

T

Tailored treatment. *See* Flexibility of Internet-delivered programs
Technology support, 100
Ted program, 94
Teleconferencing, 16
Telehealth. *See also specific types of technology*
for anxiety disorders, 32
for body image and eating disorders, 157–159
ethical issues and, 19, 276–277
goals for, 76
for mood disorders, 15–19, 278–279
for obesity, 182, 184
overview of, 69
for pain management, 135–137
for posttraumatic stress disorder, 59–61
treatment adherence and, 56
Telephone(s). *See* Mobile phones; Telephone-administered programs
Telephone-administered programs
for adolescents, 279
automated calling systems, 207, 208–209, 215
for body image and eating disorders, 161

for depression, 17–18
for obesity, 181, 184
for pain management, 135–136
treatment adherence and, 56
troubleshooting, 162
Terrorist bus bombing trauma, 52
Text messaging
for body image and eating disorders, 157–159
encryption and, 280
for pain management, 138
for smoking cessation, 115, 128, 157–159
therapist availability and burden with, 163
Therapeutic alliance. *See* Therapeutic relationships
Therapeutic Interactive Voice Response (TIVR), 136. *See also* Interactive voice response (IVR) technology
Therapeutic relationships. *See also* Clinician involvement
body image and eating disorders and, 162–163
disclosure and, 281
e-mail contact and, 273
empathy and, 228
for obesity, 179–180
online therapy and, 18–19, 20–21
outcomes and, 267–268, 278
posttraumatic stress disorder and, 55–57
self-management support and, 71
technology overuse cautions and, 104
Therapist(s). *See* Clinician(s)
Therapist-assisted treatments. *See also* Clinician involvement
for anxiety disorders, 33
internet-assisted therapy, 48–50
for obesity, 173, 184, 187, 188, 191
in pain management, 141–142
for posttraumatic stress disorder, 48–50
for substance use disorders, 98
Time-series designs, 251, 259
TIVR (Therapeutic Interactive Voice Response), 136. *See also* Interactive voice response (IVR) technology
TMS (transcranial magnetic stimulation), 145
Tobacco use. *See* Smoking cessation
Trackers. *See* Progress trackers

Training for patients. *See* Internet-delivered programs; Psychoeducational programs
Training of clinicians. *See* Clinician training
Transcranial magnetic stimulation (TMS), 145
Trauma. *See also* Posttraumatic stress disorder (PTSD)
 cognitive-behavior therapy for, 48
 meaning making and, 56
 potentially traumatizing events and, 45–46
 and shame, 53
 terrorist bus bombing, 52
 virtual reality exposure therapy for, 50–51
Triage system, 141
Trichotolomania, 32
Trouble on the Tightrope: In Search of Skateboard Sam, 154
Type 1 diabetes, 210
Type 2 diabetes, 200, 205, 210, 212

U

Usability studies, 77–80, 118–119

V

Validity, 249–250, 251–252, 261
Value versus breadth of Web sites, 117
Vehicle accident victims, 51
Veterans. *See* Military personnel with posttraumatic stress disorder
Vicarious exposure and response prevention (ERP), 32
Vida Sin Tabaco, 124
Video-based programs
 for pain management, 135
 for smoking cessation, 115
 videoconferencing
  for anxiety disorders, 30
  for depression, 16
  diabetes management, 211
  ethical issues and, 272–273, 280
  for schizophrenia, 274–275
  for supervision, 285
Virtual reality, 143–144
Virtual reality exposure therapy (VRET)
 for anxiety disorders, 30, 32, 33, 34, 51
 clinical skills needed with technology, 35
 costs of, 36
 for posttraumatic stress disorder (PTSD), 33, 50–52, 54
VRET. *See* Virtual reality exposure therapy

W

Walkers in Darkness, 13, 14–15
War veterans. *See* Military personnel with posttraumatic stress disorder
Waterfall model, 255
WATI (Web-Assisted Tobacco Interventions), 126
Web-Assisted Tobacco Interventions (WATI), 126
Web-based Clinical Quality Management System (ClinQMS), 140–141
Web-based programs. *See* Internet-delivered programs; Web sites and resources
Webcasts for pain management, 135
Web sites and resources
 alcohol use, 97
 diabetes management, 206, 209, 210
 domain name registration, 100
 electronic medical records, 235
 evidence-based practice, 92
 Health Insurance Portability and Accountability Act, 104
 Health on the Net Foundation, 127
 mood disorders, 4–5, 9, 13–14, 16, 20–21
 pain management, 135, 138, 143–144
 self-monitoring, 20–21
 smoking cessation, 94–95, 110, 112, 121–126
 substance use disorders, 94–95, 97, 99–103, 110, 112, 121–126
 therapeutic reminders, 16
 value versus breadth of, 117
 Web authoring applications and tools, 102
Web hosting companies, 100
Weight loss and management. *See* Body image and eating disorders; Obesity Internet-based programs
Weight Loss Maintenance Trial, 181
What you see is what you get (WYSIWYG) editors, 101

Wireless-enabled pain drawings, 138
Workbooks, 28
Working alliance, 18–19. *See also* Therapeutic relationships
Workplace settings, 180
World Trade Center attack trauma, 50–51
WYSIWYG (what you see is what you get) editors, 101